DIGITAL SYNTHESIZERS AND TRANSMITTERS FOR SOFTWARE RADIO

Digital Synthesizers and Transmitters for Software Radio

by

JOUKO VANKKA

Helsinki University of Technology,
Finland

A C.I.P. Catalogue record for this book is available from the Library of Congress.

ISBN-10 1-4020-3194-7 (HB) Springer Dordrecht, Berlin, Heidelberg, New York
ISBN-10 1-4020-3195-5 (e-book) Springer Dordrecht, Berlin, Heidelberg, New York
ISBN-13 978-1-4020-3194-6 (HB) Springer Dordrecht, Berlin, Heidelberg, New York
ISBN-13 978-1-4020-3195-3 (e-book) Springer Dordrecht, Berlin, Heidelberg, New York

---

Published by Springer,
P.O. Box 17, 3300 AA Dordrecht, The Netherlands.

Printed on acid-free paper

Printed in the Netherlands.

Contents

Acknowledgements

A significant part of this work was conducted during project-work funded by the Technology Development Center (Tekes) and the Academy of Finland. Personal grants were received from the Nokia Foundation, Jenny and Antti Wihuri Foundation, and the Electronic Engineering Foundation. I would like to acknowledge my sincere gratitude to Jaakko Ketola, Marko Kosunen, Jonne Lindeberg, Johan Sommarek, Ilari Teikari and Olli Väänänen for generously providing assistance during the development of the material presented in this book.

Acknowledgements

A significant part of this work was [illegible] supported by the Technology [illegible] and the [illegible] of [illegible]. [illegible] grants were [illegible] the [illegible] Foundation [illegible] and [illegible] [illegible] to acknowledge my [illegible] gratitude to Jackie, Kevin, [illegible] and [illegible] for [illegible] and [illegible] during the development of the [illegible] research [illegible].

Preface

The approach adopted in this book will, it is hoped, provide an understanding of key areas in the field of digital synthesizers and transmitters. It is easy to include different digital techniques in the digital synthesizers and transmitters by using digital signal processing methods, because the signal is in digital form. By programming the digital synthesizers and transmitters, adaptive channel bandwidths, modulation formats, frequency hopping and data rates are easily achieved. Techniques such as digital predistortion for power amplifier linearization, digital compensation methods for analog I/Q modulator nonlinearities and digital power control and ramping are presented in this book. The flexibility of the digital synthesizers and transmitters makes them ideal as signal generators for software radio. Software radios represent a major change in the design paradigm for radios in which a large portion of the functionality is implemented through programmable signal processing devices, giving the radio the ability to change its operating parameters to accommodate new features and capabilities. A software radio approach reduces the content of radio frequency (RF) and other analog components of traditional radios and emphasizes digital signal processing to enhance overall transmitter flexibility. Software radios are emerging in commercial and military infrastructure. This growth is motivated by the numerous advantages of software radios, such as the following:

1. Ease of design—Traditional radio design requires years of experience and great care on the part of the designer to understand how the various system components work in conjunction with one another. The time required to develop a marketable product is a key consideration in modern engineering design, and software radio implementations reduce the design cycles for new products, freeing the engineer from much of the iteration associated with

analog hardware design. It is possible to design many different radio products using a common RF front-end with the desired frequency and bandwidth in conjunction with a variety of signal processing software.

2. Ease of manufacture—No two analog components have precisely identical performance; this necessitates rigorous quality control and testing of radios during the manufacturing process. However, given the same input, two digital processors running the same software will produce identical outputs. The move to digital hardware thus reduces the costs associated with manufacturing and testing the radios.

3. Multimode operation—the explosive growth of wireless has led to a proliferation of transmission standards; in many cases, it is desirable that a radio operates according to more than one standard.

4. Use of advanced signal processing techniques—the availability of high speed signal processing on board allows the implementation of new transmitter structures and signal processing techniques. Techniques such as digital predistortion for power amplifier linearization, digital compensation methods for analog I/Q modulator errors and digital power control and ramping, previously deemed too complex, are now finding their way into commercial systems as the performance of digital signal processors continues to increase.

5. Flexibility to incorporate additional functionality—Software radios may be modified in the field to correct unforeseen problems or upgrade the radio.

Figure 1 shows a block diagram of the conventional digital modulator. It consists of the following blocks: clipping circuit (Chapter 17 and Chapter 18), pulse shaping filters (Chapter 11), interpolation filters (Chapter 11), resamplers (Chapter 12), quadrature direct digital synthesizer (Chapters 4, 7, 8 and 9), inverse sinc filter (Chapter 13) and D/A converter (Chapter 10). The

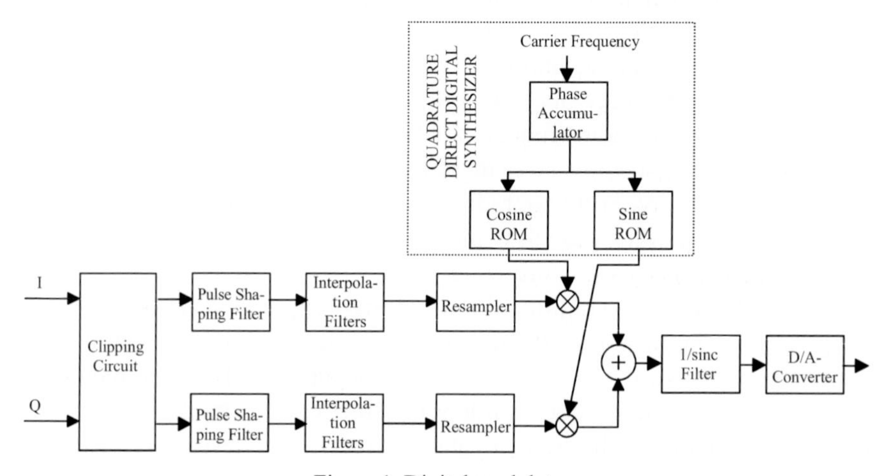

Figure 1. Digital modulator.

alternative method of translating the baseband-centered spectrum to a programmable carrier center frequency is to use the CORDIC rotator (Chapter 6) instead of the quadrature direct digital synthesizer, two mixers and an adder. Three design examples of the digital modulator are presented (Chapters 14, 15 and 16).

Chapter 1 provides a basic introduction to transmitter architectures. The classic transmitter architecture is based upon linear power amplifiers and power combiners. Most recently, transmitters have been based upon a variety of different architectures including Envelope Elimination and Restoration (EER), polar loop, LInear amplification with Nonlinear Components (LINC), Combined Analogue Locked Loop Universal Modulator (CALLUM), LInear amplification employing Sampling Techniques (LIST) and transmitters based on bandpass sigma delta modulators.

Power amplifier linearization techniques are used both to improve linearity and to allow more efficient, but less linear, methods of operation. The three principal types of linearization are feedback, feedforward and predistortion. The combination of digital signal processing (DSP) and microprocessor control allows a widespread use of complicated feedback and predistortion techniques to improve power amplifier efficiency and linearity, as shown in Chapter 2.

In Chapter 3, methods and algorithms to compensate analog modulator errors are reviewed, while in Chapter 4, a description of the conventional direct digital synthesizer (DDS) is given. It is easy to include different modulation capabilities in DDSs by using digital signal processing methods, because the signal is in digital form. By programming the DDSs, adaptive channel bandwidths, modulation formats, frequency hopping and data rates are easily achieved. The digital circuits used to implement signal-processing functions do not suffer the effects of thermal drift, aging and component variations associated with their analog counterparts. The flexibility of the DDSs makes them ideal as signal generator for software radios. Recursive sinusoidal oscillators are presented in Chapter 5.

In Chapter 6, it is seen that circular rotation can be implemented efficiently using the CORDIC algorithm, an iterative algorithm for computing many elementary functions. The CORDIC algorithm is studied in detail. The finite word length effects in the CORDIC algorithm are investigated. Redundant implementations of the CORDIC rotator are overviewed and the hybrid CORDIC algorithms are reviewed.

The DDS is shown to produce spurs (spurious harmonics), as well as the desired output frequency, in Chapter 7. Different noise and spur sources are studied in detail. In Chapter 8, a study is made of how additional digital techniques (for example, dithering, error feedback methods) may be incorporated in the DDS in order to reduce the presence of spurious signals at the DDS output. The spur reduction techniques used in the sine output direct digital synthesizers are reviewed.

In Chapter 9, an investigation into the blocks of the DDS, namely a phase accumulator, a phase to amplitude converter (conventionally a sine ROM) and a filter, is carried out. Different techniques used to accelerate the operation speed of the phase accumulator are considered. Different sine memory compression and algorithmic techniques and their trade-offs are investigated.

D/A converters, along with the power amplifier, are the most critical components in software radio transmitters. Unfortunately, the development of D/A converters does not keep up with the capabilities of digital signal processing utilizing faster technologies. The different techniques used to enhance D/A converter static and dynamic performance are reviewed in Chapter 10.

The pulse shaping and interpolation filters are the topic of Chapter 11. Different methods of designing the pulse shaping filters are reviewed. The multirate signal processing is particularly important in software radio transmitters, where sample rates are low initially and must be increased for efficient subsequent processing.

The multi-standard modulator has to be able to accept data with different symbol rates. This fact leads to the need for a re-sampler that performs a conversion between variable sampling frequencies. There are several methods of realizing the re-sampler with an arbitrary sampling rate conversion. In Chapter 12, the design of the polynomial-based interpolation filter using the Lagrange method is presented. Some other polynomial-based methods are also discussed.

Three different designs to compensate the sinc(x) frequency response distortion resulting from D/A converters by using digital FIR filters are represented in Chapter 13. The filters are designed to compensate the signal's second image distortion.

The design and implementation of a DDS with the tunable (real or complex) 1-bit $\Delta\Sigma$ D/A converter are described in Chapter 14. Since the 1-bit $\Delta\Sigma$ D/A converter has only one bit, the glitch problems and resulting spurious noise resulting from the use of the multi-bit D/A converter are avoided.

In traditional transmit solutions, a two-stage upconversion is performed in which a complex baseband signal is digitally modulated to the first IF (intermediate frequency) and then mixed to the second IF in the analog domain. The first analog IF mixer stage of the transmitter can be replaced with this digital quadrature modulator, as shown in Chapter 15.

In Chapter 16, the digital IF modulator is designed using specifications related to GSM, EDGE and WCDMA standards. By programming a GSM/EDGE/WCDMA modulator, different carrier spacings, modulation schemes, power ramping, frequency hopping and symbol rates can be achieved. By combining the outputs of multiple modulators, multicarrier signals can be formed or the modulator chips can be used for steering a phased array antenna. The formation of multi-carrier signals in the modulator increases the base station capacity

In a WCDMA system, the downlink signal typically has a high Peak to Average Ratio (PAR). In order to achieve a good efficiency in the power amplifier, the PAR must be reduced, i.e. the signal must be clipped. In Chapter 17, the effects of several different clipping methods on Error Vector Magnitude (EVM), Peak Code Domain Error (PCDE) and Adjacent Channel Leakage power Ratio (ACLR) are derived through simulations. A very straightforward algorithm for implementing a peak windowing clipping method is also presented.

In conventional base station solutions, the carriers transmitted are combined after the power amplifiers. An alternative to this is to combine the carriers in the digital domain. The major drawback of combining digital carriers is a strongly varying envelope of the composite signal. The high PAR sets strict requirements for the linearity of the power amplifier. High linearity requirements for the power amplifier lead to low power efficiency and therefore to high power consumption. In Chapter 18, the possibility of reducing the PAR by clipping is investigated in two cases, GSM and EDGE.

List of Abbreviations

ACI	Adjacent channel interference
ACLR	Adjacent channel leakage power ratio
ACP	First adjacent channel power
ADC	Analog-digital-converter
ALT1	Second adjacent channel power
ALT2	Third adjacent channel power
AM-AM	Amplitude-dependent amplitude distortion
AM-PM	Amplitude-dependent phase distortion
ASIC	Application specific integrated circuit
BiCMOS	Bipolar complementary metal-oxide-semiconductor
BPF	Bandpass filter
CALLUM	Combined analogue locked loop universal modulator
CATV	Cable Television
CDMA	Code division multiple access
CF	Crest factor
CFBM	Cartesian feedback module
CIA	Carry increment adder
CIC	Cascaded-integrator-comb
CICC	Custom integrated circuits conference
CLK	Clock
CMOS	Complementary metal-oxide-semiconductor

CORDIC	Co-ordinate digital computer
CP	Carry Propagation
CS	Carry Save
CSD	Canonic signed digit
CSFR	Constant scale factor redundant
D/A	Digital to analog
DAC	Digital to analog converter
DAMPS	Digital-advanced mobile phone service
dB	Decibel
dBc	Decibels below carrier
dBFS	Decibels below full-scale
DCORDIC	Differential CORDIC
DCT	Discrete cosine transform
DDFS	Direct digital frequency synthesizer
DDS	Direct digital synthesizer
DECT	Digital enhanced cordless telecommunications
DEMUX	Demultiplexer
DFF	Delay-flip-flop
DFT	Discrete Fourier transform
DNL	Differential non-linearity
DPLL	Digital phase locked loop
DRC	Design rule check
DSP	Digital signal processing
EDGE	Enhanced data rates for global evolution
EER	Envelope elimination and restoration
EF	Error feedback
ETSI	European telecommunications standards institute
EVM	Error vector magnitude
FET	Field-effect transistors
FFT	Fast Fourier transform
FIR	Finite impulse response
FPGA	Field programmable gate array
GCD	Greatest common divisor
GMSK	Gaussian minimum shift keying

GPRS	General packet radio service
GSM	Groupe spécial mobile
HPF	High-pass filter
HSCSD	High-speed circuit switched data
IC	Integrated circuit
IDFT	Inverse discrete Fourier transform
IEE	Institution of electrical engineers
IEEE	Institute of electrical and electronics engineers
IEICE	Institute of electronics, information and communication engineers
IF	Intermediate frequency
IIR	Infinite impulse response
IMD	Inter-modulation distortion
INL	Integral non-linearity
ISI	Inter-symbol interference
ISM	Industrial, scientific and medicine
ISSCC	International solid-state circuits conference
LE	Logic element
L-FF	Logic-flip-flop
LINC	Linear amplification with nonlinear components
LIST	Linear amplification employing sampling techniques
LMS	Least-mean-square
LMS	Least mean squares algorithm
LO	Local oscillator
LPF	Low-pass filter
LSB	Least significant bit
LTI	Linear time invariant
LUT	Look-up table
LVS	Layout versus schematic
MAE	Maximum amplitude error
MSB	Most significant bit
MSD	Most significant digits
MUX	Multiplexer

NCO	Numerically controlled oscillator
NEG	Negator
NRZ	Non-return-to-zero
NTF	Noise transfer
OSC	Oscillator
P/I	Pipelining/interleaving
PA	Power amplifier
PAR	Peak to average ratio
PCDE	Peak code domain error
PFD	Phase/frequency detector
PLD	Programmable logic device
PLL	Phase-locked loop
PPM	Part per million
PSK	Phase shift keying
QAM	Quadrature amplitude modulation
QDDS	Quadrature direct digital synthesizer
QM	Quadrature modulator
QMC	Quadrature modulator compensator
QPSK	Quadrature phase-shift keying
R/P	Rectangular-to-polar
RF	Radio frequency
RLS	Recursive least squares algorithm
RLS	Recursive least squares
RMS	Root-mean-square
RNS	Residue number system
ROM	Read-only memory
RTL	Register transfer level
RZ	Return-to-zero
RZ2	Double RZ
RZ2c	Double complementary
SFDR	Spurious free dynamic range
SIR	Signal-to-interference ratio
SMS	Short message services
SNDR	Signal to noise and distortion ratio

SNR	Signal-to-noise ratio
TDD	Time division duplex
TDD-WCDMA	Time division duplex WCDMA
TDMA	Time division multiple access
TEKES	Technology development center
VCO	Voltage controlled oscillator
VHDL	Very high speed integrated circuit HDL
VHF	Very high frequency
VLSI	Very large scale integration
VMCD	Voltage Mode Class-D
WCDMA	Wideband code division multiple access
XOR	Exclusive or
$\Delta\Sigma$	Delta sigma

Chapter 1

1. TRANSMITTERS

This chapter provides a basic introduction to transmitter architectures. The classic transmitter architecture is based upon linear power amplifiers and power combiners. Most recently, transmitters have been based upon a variety of different architectures including Envelope Elimination and Restoration (EER), polar loop, LInear amplification with Nonlinear Components (LINC), Combined Analogue Locked Loop Universal Modulator (CALLUM), LInear amplification employing Sampling Techniques (LIST) and transmitters based on bandpass sigma delta modulators.

1.1 Direct Conversion Transmitters

The principle of the direct conversion transmitter is presented in Figure 1-1. In direct conversion transmitters, the band limited baseband signals are converted directly up to the radio frequency with in-phase and quadrature carriers. The band-pass filter after the signal summation is used to suppress the

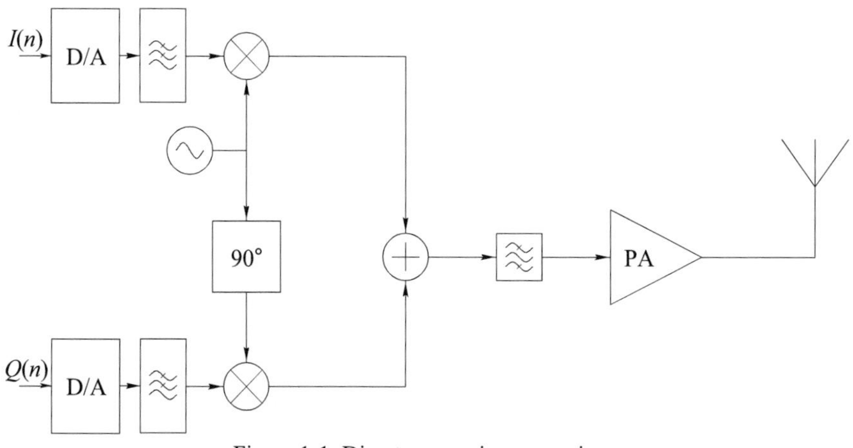

Figure 1-1. Direct conversion transmitter.

out of band signals generated by the harmonic distortion of the carrier. The direct conversion transmitter is theoretically simple (no IF components) and potentially suitable for high integration level solutions. The drawbacks are: an I,Q mixer is needed at RF frequency, there is LO-leakage at RF frequency (filtering impossible) and VCO pulling.

The image rejection is given by

$$R = 10\log_{10}\left(\frac{1 + 2\Delta G\cos(\Delta\theta) + \Delta G^2}{1 - 2\Delta G\cos(\Delta\theta) + \Delta G^2}\right), \tag{1.1}$$

where ΔG is the gain mismatch and $\Delta\theta$ is the phase mismatch. For instance, with a 5 degree phase mismatch and 0.1 dB amplitude mismatch, the maximum achievable single sideband suppression is only 27.2 dB, as shown in Figure 1-2.

The strong signal at the output of the power amplifier may couple to the local oscillator (LO), which is usually a voltage controlled oscillator, causing the phenomenon known as injection pulling [Raz98]. This means that the frequency of the local oscillator is pulled away from the desired value. The severity of the injection pulling is proportional to the difference between the frequency of the local oscillator and the frequencies at the output of the PA. By taking advantage of this, the problem of injection pulling can be alleviated by using an offset LO direct-conversion structure. In this structure, the carrier signal is formed by mixing two lower frequency signals. An additional band-pass filter is needed to filter away the undesired carrier at frequency. Another solution is to generate the LO signal from a lower frequency VCO by the frequency multiplication or from higher frequency VCO by frequency division. The VCO frequency is harmonically dependent on the

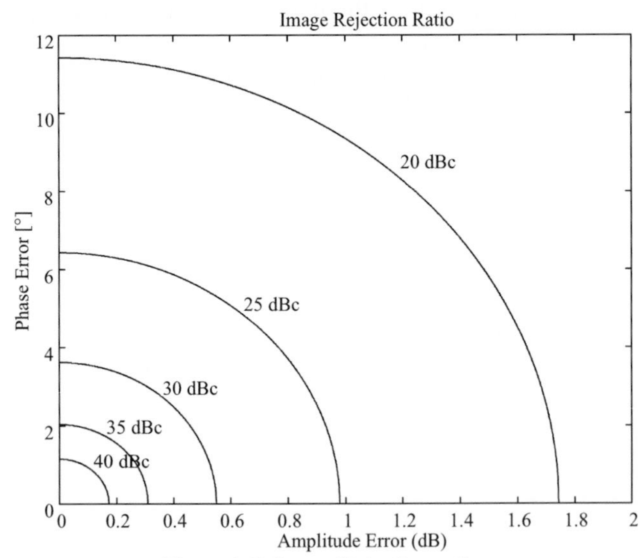

Figure 1-2. Image-Rejection ratio.

LO signal; the pulling rejection is not, therefore, as advantageous as in the offset VCO. Reported direct conversion transmitters using CMOS are, for example, [Ors99], [Lee01], [Ger01], and [Liu00]. The benefits and drawbacks of using direct conversion architecture in a transmitter are heavily dependent on the particular case, i.e. on application area, modulation method and technologies.

1.2 Dual-Conversion Transmitter

The injection pulling can also be avoided by using a dual conversion transmitter presented in Figure 1-3. In this structure, the baseband data is first upconverted to the intermediate frequency and then to the desired radio frequency. The dual conversion transmitter has advantages. First, the quadrature modulation is performed at the fixed lower frequency leading to the better matching between I and Q. Second, the additional attenuation of the adjacent channel spurs and noise may be achieved by using a band-pass filter at the IF. The hardware can be partly shared with the receiver (same oscillator frequencies). The drawbacks are complexity (more components), lower integration level, impedance matching required for external components and more power consumption. The stopband attenuation of the image reject filter at the RF frequency has hard requirements due to high frequency and the high attenuation factor because the signal component at the image frequency has the same power as the desired sideband. The first analog IF mixer stage of the transmitter in Figure 1-3 can be replaced with a digital quadrature modulator as shown in Figure 15-1.

1.3 Transmitters Based on VCO modulation

The constant envelope modulator can simply be implemented by direct modulation of a voltage controlled oscillator (VCO) [Bax99]. Ideally the VCO output frequency can be expressed as

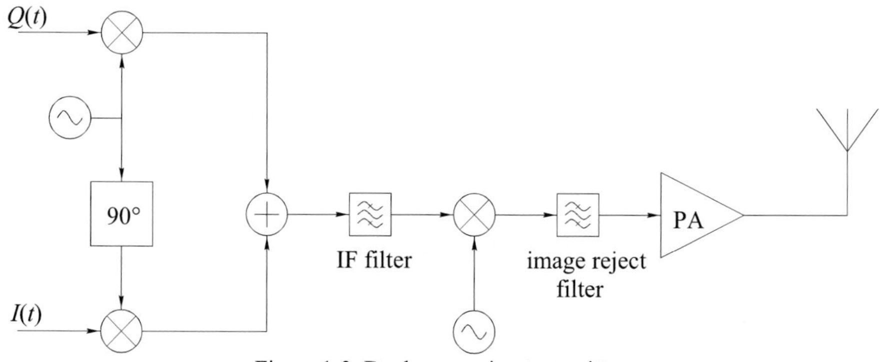

Figure 1-3. Dual conversion transmitter.

$$f_{out} = f_o + K_v V_{tune}, \quad (1.2)$$

where f_o is the base output frequency of the VCO, K_v is the VCO sensitivity in Hz/V and V_{tune} is the input voltage that tunes the VCO. In principle, this produces a FM signal proportional to the modulating signal. There are many disadvantages, however, to this approach:

* Frequency drift: change in the VCO frequency due to tuning voltage drift
* Frequency pushing: change in the VCO frequency due to change in the power supply voltage
* Load pulling: change in the VCO frequency due to change in the VCO load

The change in VCO frequency can be compensated so that the receiving radio end tells the error to the transmitting radio, which tunes the modulating signal (V_{tune}) in order to compensate changes in the tuning slope. In wireless communication systems using time division multiple access (TDMA), such as DECT, data is transmitted in bursts with inactive periods in-between. Figure 1-4 presents a DECT architecture that utilizes these inactive periods between bursts to force the VCO frequency to match the desired channel frequency by a closed PLL [Bax99]. During transmit bursts the PLL loop is open and the incoming data modulates the VCO. Since the transmit burst duration is short (< 500 μs) in DECT and the requirements on the frequency error are not very tight (<50 kHz) [Bax99], the frequency drift in the VCO during the burst can be made so low that it is tolerable. Frequency pushing caused by the switching and power ramping of the power amplifier (PA) is also a problem. Another more severe problem is frequency pulling caused by changes in the input impedance of the PA when it is switched or ramped. While these problems can be overcome in the DECT system, they render direct modulation unsuitable for standards that have strict frequency control

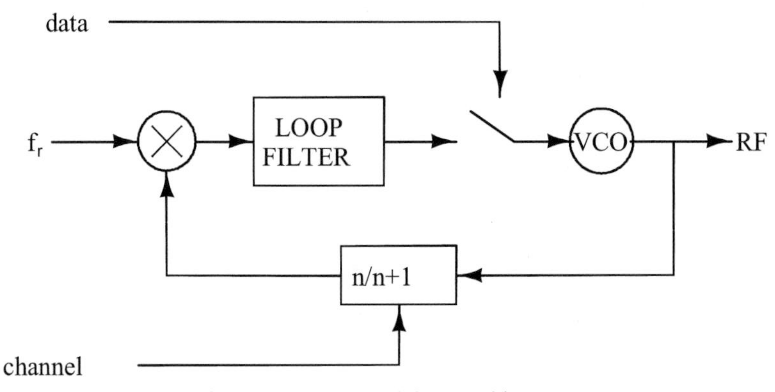

Figure 1-4. VCO modulator architecture.

specifications, such as GSM [Bax99].

In an indirect modulation scheme, the problems of VCO drift and instability are overcome by digitally modulating a synthesizer rather than directly modulating a VCO as in a simple direct modulator. In indirect modulation, the modulating signal is injected while the PLL is closed [Bax99], this makes it possible to constantly maintain accurate frequency control. An indirect modulator architecture is illustrated in Figure 1-5 [Ril94], [Per97], [Bax01], and [McM02]. The architecture comprises an FIR filter and a frequency synthesizer. The FIR filter filters the data bits. It consists of an oversampling counter, a ROM look-up-table and a small amount of random logic. The FIR filter taps stored in the ROM are quantised to single-bit. A reference frequency f_r is needed to phase-lock the VCO to a stable source. The delta-sigma modulator ($\Delta\Sigma$) and dual modulus divider comprise a fractional-N frequency synthesizer. The key feature of this synthesizer approach is that it uses a digital $\Delta\Sigma$ to generate a bit stream b(n), which embodies the higher resolution of the k-bit input within the long term average of b(n). By making the k-bit input to the $\Delta\Sigma$ a function of time, the instantaneous frequency can be directly manipulated.

The advantage of this technique is both that no mixers are needed to upconvert the modulating signal to the carrier frequency and that the RF signal is inherently band-limited to suppress noise. The disadvantage of this technique is that the modulation bandwidth must be less than the synthesizer bandwidth to avoid any loop suppression of the modulating signal. Since the synthesizer closed-loop bandwidth is usually narrow in order to suppress the quantization noise of the $\Delta\Sigma$ modulator, the maximum bandwidth is limited. This problem, however, can be tackled by equalizing the signal entering the

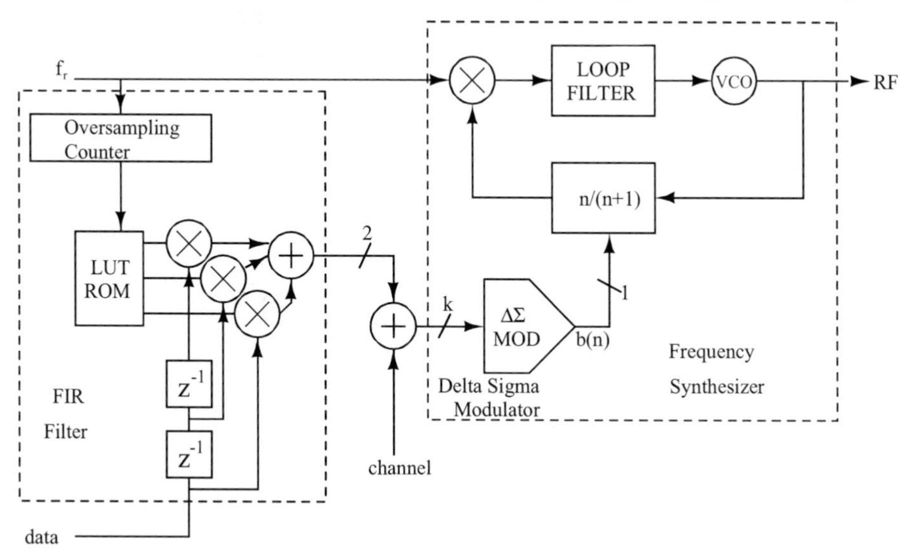

Figure 1-5. Indirect GMSK modulator with $\Delta\Sigma$–fractional-N-synthesizer.

synthesizer with an equalization filter that precompensates the suppression of high frequency components in the PLL [Per97].

Another approach to indirect modulation is to fix the divider modulus to a value corresponding to the desired channel frequency and vary the reference frequency f_r instead. The reference frequency is replaced by an analog modulator that produces the desired modulated signal at some intermediate frequency (IF). Then the synthesizer output becomes $f_{out}(t)=Nf_r(t)$, implying that this kind of upconversion scales up the frequency deviation of the modulating signal. Any multiplication of the reference frequency results in a degraded phase noise and spurs spectrum inside the loop bandwidth per the classical 20 $\log_{10}(N)$ rule.

1.4 Offset-PLL Architecture

The offset-PLL is suited for systems using constant envelope modulation, such as the GSM [Yam97], [Irv98]. The PLL operates as a narrowband filter centered around f_{RF}, suppressing the out-of-band noise generated by the reference, as shown in Figure 1-6. The TX-SAW filter can be replaced with this structure. The phase comparator generates an error signal by comparing the modulated reference IF signal and the feedback signal. The PFD output controls the frequency of the TX-VCO such that the VCO output frequency is modulated with the original GMSK data at the center frequency of the RF channel. The tradeoff is between the TX noise level and the phase error, which are related to the loop bandwidth. For example, if the loop bandwidth is designed narrow in order to increase the suppression of the TX noise, the range in which the input phase variation can be reproduced becomes correspondingly narrow, so the phase error gets bigger. However, widening the bandwidth may result in excessive wideband noise in the transmitted RF signal. This, in turn, necessitates additional filtering.

1.5 Envelope Elimination and Restoration (EER)

The envelope elimination and restoration (EER) technique combines a

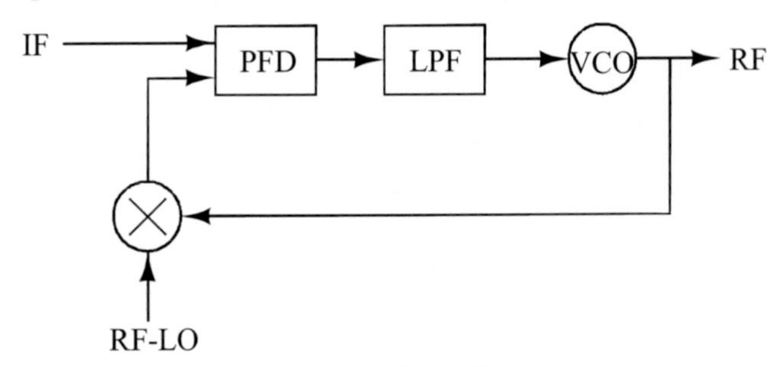

Figure 1-6. Offset-PLL.

highly efficient, but nonlinear RF PA, with a highly efficient envelope amplifier to implement a high-efficiency linear RF PA. The technique was first presented by Kahn [Kah52]. In its classic form, a limiter eliminates the envelope, allowing the constant-amplitude phase modulated carrier to be amplified efficiently by class-C, class-D, class-E, or class-F RF PAs [Su98] as shown in Figure 1-7. Amplitude modulation of the final RF PA restores the envelope to the phase-modulated carrier, creating an amplified replica of the input signal. The EER is based upon the principle that any narrow-band signal can be produced by simultaneous amplitude (envelope) ($A(n)$) and phase modulations ($P(n)$):

$$\begin{aligned} RF_{in} &= I(n)\cos(\omega_{in}\, n) + Q(n)\sin(\omega_{in}\, n) \\ &= A(n)\cos(\omega_{in}\, n - P(n)), \end{aligned} \tag{1.3}$$

$$\text{where} \quad A(n) = \sqrt{I(n)^2 + Q(n)^2}\,,\; P(n) = \arctan(Q(n)/I(n)),$$

where arctan is the four quadrant arctangent of the quadrature phase data ($Q(n)$) and in-phase ($I(n)$). The two most important factors affecting the linearity are the envelope bandwidth and alignment of the envelope and phase modulations.

The ERR is suitable for narrowband systems because of the bandwidth expansion that is associated with the polar representation of the signal in (1.3). The envelope and phase modulators need to amplify at least 2-3 times the RF bandwidth to meet ACLR and EVM requirements, as shown in Table 1-1. The synchronization requirements are shown in Table 1-2. The wider bandwidth requires a higher sampling frequency in DSP (Figure 1-8).

Theoretically, the EER can achieve 100 % efficiency. However, usually a linear predriver amplifier is needed before the power amplifier in order to increase the power of the constant envelope signal to a high enough level to

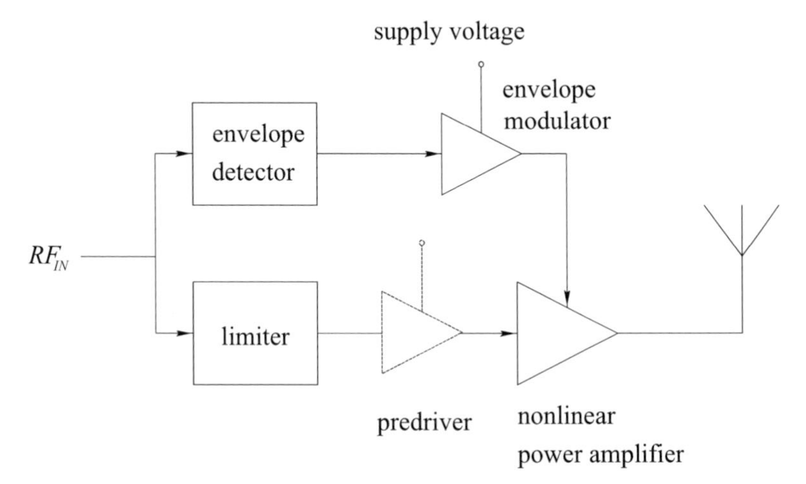

Figure 1-7. Envelope elimination and restoration block diagram.

keep the PA in saturation, which decreases the efficiency. The efficiency can be increased by modulating the drive level of the predriver. This introduces a small amount of amplitude modulation to the PA-input signal, but the spectral degradation due to this is not large enough to be a problem. Maximal efficiency is achieved when this modulation decreases the PA input signal amplitude to zero when the envelope is zero. This, however, causes the PA gain to decrease nonlinearly near these zero points so intermodulation products are introduced to the amplified spectrum. The problem can be alleviated by limiting the minimal drive level of the predriver to the minimal point where the PA is still in saturation. This decreases the efficiency slightly, but decreases/improves the distortion/linearity considerably [Raa99].

The delay mismatch between envelope and phase path in an EER transmitter should be considered carefully because it causes inter-modulation distortion (IMD), which not only degrades the modulation but also causes the spectral regrowth. In [Raa96] it has been shown that the resultant IMD for two-tone input can be approximated as

$$IMD = 2\pi B_{RF}^2 \tau_{EER}^2, \qquad (1.4)$$

where B_{RF} is the bandwidth of the RF signal and τ_{EER} is the corresponding path-delay difference. To reduce IMD, the envelope feedback provides delay equalization and amplitude linearization to increase the IMD suppression [Sta99]. The phase and envelope feedback provides delay equalization and phase/amplitude linearization as described in [Raa98].

The efficiency of an EER system depends primarily on the efficiencies of the envelope modulator and the nonlinear RF amplifier, assuming that the output power is sufficiently high to ensure that the power consumption of the signal processing devices is negligible. If this assumption is made, then the efficiency of the EER system can be regarded as simply the product of the

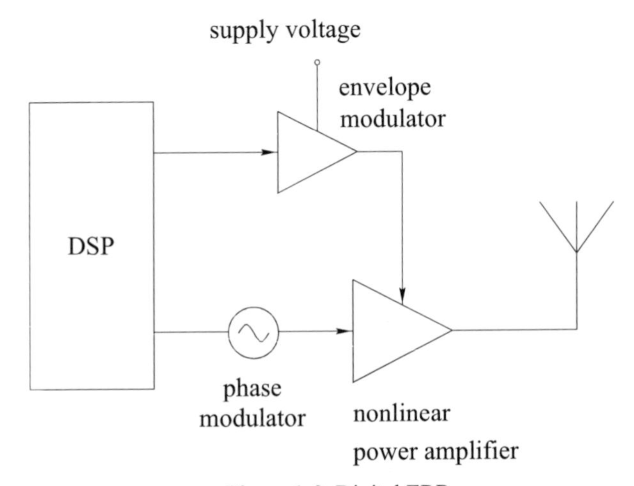

Figure 1-8. Digital ERR.

efficiencies of the envelope modulator (n_{AF}) and the nonlinear power amplifier (n_{RF})

$$n = n_{AF}\, n_{RF}\,. \qquad (1.5)$$

A typical system employing a class-C power amplifier (n_{RF} = 0.6) and a class-S (pulse-width modulation) audio amplifier (n_{AF} = 0.9) will yield an overall efficiency of around 54%. If the switching RF power amplifier is employed with a basic efficiency of, say, 80%, then the overall efficiency will increase to around 72%. The experimental results of the EER transmitter are presented in [Su98], [Raa98], [Sta99], and [Raa99].

In a modern implementation, both the envelope and phase-modulated carrier are generated by a DSP, as shown in Figure 1-8. The CORDIC algorithm can be used to generate the envelope and phase modulated carrier.

1.6 Polar-Loop Transmitter

This technique was proposed by Petrovic et al. [Pet79]. The principle is illustrated in Figure 1-9. It is closely related to the envelope elimination and restoration (EER) technique (see section 1.5) in that it completely avoids the nonlinear characteristic of the amplifier. The input signal is an intermediate frequency signal. This signal is split up into its polar components, amplitude and phase, and compared with their respective counterparts of the amplifier output signal. The resulting phase error signal controls a VCO that feeds the amplifier with a constant envelope but phase modulated signal. Equally, the amplitude error signal modulates the collector voltage of the power amplifier. Thus a phase-locked loop is used to track the phase and a classical feedback circuitry to track the amplitude. Note that, unlike a conventional transmitter, the channel frequency is set by the local oscillator in the feedback chain.

Although this feedback arrangement is applicable to any form of modulation it is most suitable for narrowband systems because of the bandwidth

Table 1-1. Needed bandwidth (assuming perfect synchronization) [Nag02]

NADC	90kHz
EDGE	600kHz
IS-95	3.75MHz
UMTS	15MHz

Table 1-2. Degree of synchronization (assuming ideal phase and envelope modulation) [Nag02]

NADC	±666ns
EDGE	±100ns
IS-95	±16ns
UMTS	±4ns

expansion that is associated with the polar representation of the signal. Nevertheless, it has shown promising results with spurious emission at about 60dB below the main signal for narrowband applications, typically a couple of kHz modulation bandwidth, with carrier frequencies ranging from 100MHz to 950MHz [Che68], [Pet84].

1.7 Linear amplification with Nonlinear Components (LINC)

In the mid 1930's, the so called outphasing technique was introduced to overcome increasing problems with the cost and power efficiency of high power AM-broadcast transmitters [Chi35]. When the technique was rediscovered in early 1970's by Cox [Cox75b], it became better known as LINC, an acronym for 'LInear amplification with Nonlinear Components'. Cox suggested a solution that was suitable for modulation schemes exhibiting both amplitude and phase variations. Like the envelope elimination and restoration technique described previously, the LINC scheme avoids the nonlinear characteristic of the power amplifier by feeding it with a constant envelope signal. But when it comes restoring the envelope, LINC is completely different. Two phasors with equal amplitudes are generated from the input signal in the signal component separator. These phasors are amplified separately in highly power efficient amplifiers and finally recombined to form an amplified replica of the input signal (see Figure 1-10). The signal in (1.3) can be split into two constant amplitude, but phase modulated signals:

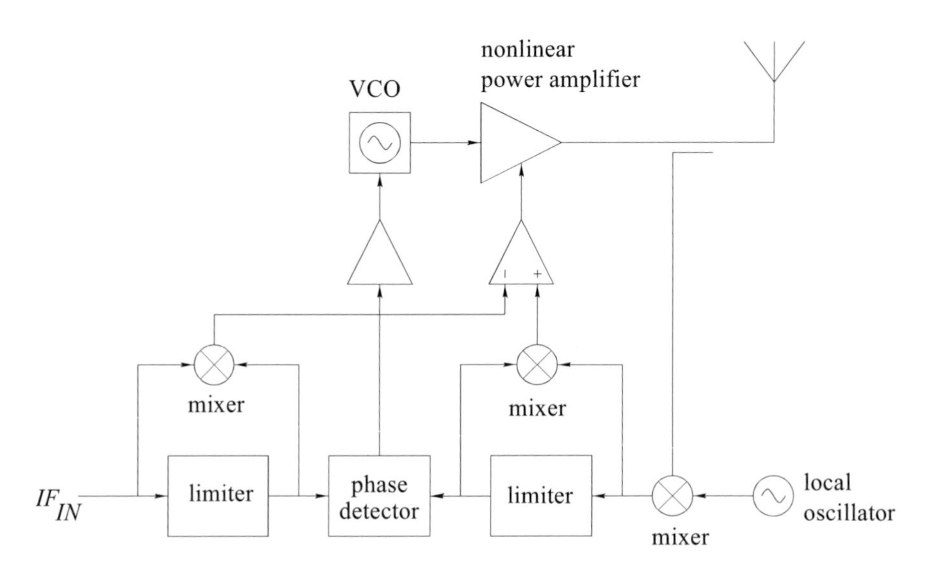

Figure 1-9. Polar-loop transmitter.

$$RF_{in} = A(n)\cos(\omega_{in}\, n - P(n)) = A_{\max}\cos(D(n))\cos(\omega_{in}\, n - P(n)) \\ = s_1(n) + s_2(n), \tag{1.6}$$

where

$$s_1(n) = A_{\max}\cos(\omega_{out}\, n - P(n) + D(n))/2, \tag{1.7}$$

and

$$s_2(n) = A_{\max}\cos(\omega_{out}\, n - P(n) - D(n))/2, \\ \text{where} \quad D(n) = \arccos(A(n)/A_{\max}). \tag{1.8}$$

These relations indicate that the generation of the constant amplitude signals is nontrivial. Earlier papers suggested a completely analogue solution, in which the signal component separator operated at some intermediate frequency or directly at the carrier frequency [Cox75b], and [Rus76]. The complexity of these systems prevented the technique from becoming widely accepted. Today, the evolution of DSP techniques has made it possible to implement the signal component separator completely in software using a standard DSP device [Het91]. With this scheme, all processing is executed at baseband, while one quadrature modulator for each amplifier arm follows the separator to translate the baseband signals to the desired carrier frequency, see Figure 1-11. However, the bandwidth of the phasors is substantially larger than that of the original input signal (due to nonlinear operations in (1.6)), so the DSP and D/A converters (four of them are needed for baseband operation) need to operate with sampling rates at least some 15–20 times the bandwidth of the input signal [Sun95b]. This has a significant impact on the power consumption of the DSP and D/A converters, as it is roughly proportional to the clock frequency. In Figure 1-11, DSP techniques are shown to generate the phasors at baseband as *I–Q* pairs, one quadrature

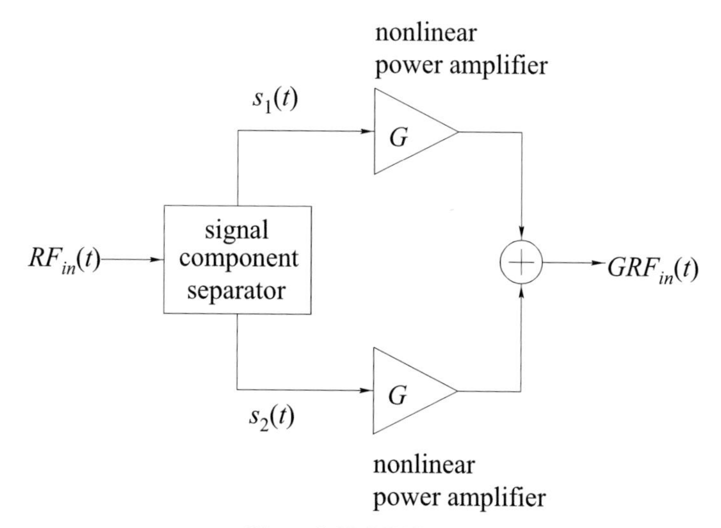

Figure 1-10. LINC transmitter.

modulator must be used in each branch to upconvert the signal to the desired carrier frequency. Quadrature modulators also suffer from gain and phase imbalance (see Figure 1-2), as well as dc offset (carrier leakage), that results in an unwanted residual spectrum at the transmitter output, and therefore degrades the system linearity [Sun00]. In [Shi00], it was attempted to develop a signal component separator architecture based on analog integrated circuit (IC) techniques to avoid the need for highly balanced quadrature modulators and high-speed D/A converters, as would be required in a DSP-based realization. The feedback loop in [Shi00] limits the signal bandwidth so the scheme is suited for single carrier modulation techniques with a limited amplitude variation range.

For the practical implementation of the LINC, the two amplifiers in the two channels must be very accurately matched regarding amplitude and phase (typically 0.1-dB amplitude matching and 0.5 phase matching) [Tom89], [Sun95a]. These specifications are extremely difficult to meet in an open-loop fashion. A "phase-only" correction was proposed in [Tom89]. In this algorithm, the phase difference between two amplifier branches is used as a guide for the correction. The phase imbalance is detected by multiplying the outputs of two amplifiers; hence, any imbalance after the power amplifiers is ignored. Besides, careful design is required to prevent the additional phase imbalance introduced by the measurement circuit. A simplex search algorithm was proposed in [Sun95a] to correct for both gain and phase errors. The correction of these errors relies on the measurement of the out-of-band emission, which requires a long data sequence for each iteration. This requirement sets a lower limit on the calibration time of approximate 1–2 s, which is a consideration in real-time applications. A direct search method was proposed in [Dar98] to correct the gain imbalance as well as the consequent phase imbalance due to AM–PM transition. This technique is

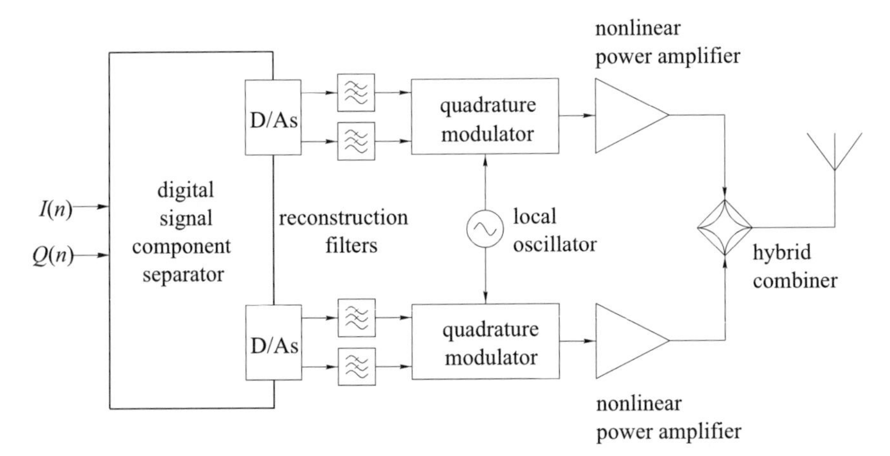

Figure 1-11. LINC transmitter with digital separator operating in baseband.

based on the evaluation of the in-band distortion by downconverting the LINC output and subtracting it from the input signal with an extra digital-to-analog (D/A) branch. The subtraction has to be quite accurate for the complete cancellation of the in-band signal. A DSP based calibration scheme is proposed in [Zha00], in which the evaluation of path imbalance (both gain and phase) is based on the measurement of a set of simple down-converted and low-pass filtered calibration signals. The application of this technique is limited, since the calibration is not transparent to data transmission. An alternative calibration scheme, which operates continuously in the background during regular data transmission, thus requiring no interruption of the transmitted signal for calibration, has been developed in [Zha01]. In the approach of [Zha01], the gain and phase imbalances are characterized by exchanging two LINC vector components and controlling a down-conversion loop.

Efficiency is probably the most difficult problem with LINC. Using a conventional hybrid combiner as shown in Figure 1-11 is a convenient solution, since it can provide high isolation and well defined impedances. But it also has a major disadvantage in that it is a power combiner that requires the input signals to be identical to avoid power losses. If the two input signals are uncorrelated, the loss will be 3dB, while, for the LINC transmitter, the loss is sometimes even worse, depending on the modulation scheme [Sun94]. Combining techniques that are more efficient exist, but they require the amplifiers to act as ideal voltage sources, since the load impedance for each amplifier varies with such signal combiners [Raa85].

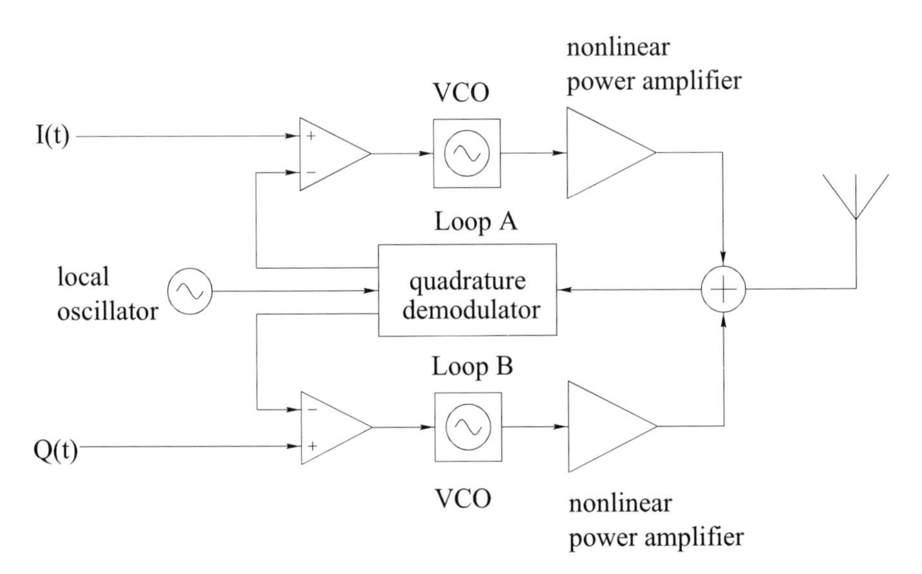

Figure 1-12. CALLUM block diagram.

1.8 Combined Analogue Locked Loop Universal Modulator (CALLUM)

The Combined Analogue Locked Loop Universal Modulator (CALLUM) proposed by Bateman [Bat92], [Bat98] is related to the LINC technique in that it combines two constant amplitude phasors to form the output signal. But, instead of having a signal component separator, the two constant amplitude phasors are generated by means of two feedback loops (see Figure 1-12). The baseband equivalent of the transmitter output signal is obtained with a quadrature demodulator and compared with the corresponding input signal. The resulting error signal controls one VCO in each loop that in turn drives a power amplifier. Note that the channel frequency is set by the local oscillator in the feedback path. The CALLUM consists of two phase-locked loops, loop A and loop B. Loop A will only maintain lock whilst the phase of the RF output signal is within ±90° of the sine of the local oscillator vector, while loop B will only maintain lock while the phase of the RF output signal is within ±90° of the cosine of the local oscillator vector. Consequently, there exists a stability region within which both loops can achieve and maintain lock; this is shown in Figure 1-13. Loop A is stable in the first and second quadrants, while loop B is stable in the first and fourth quadrants. For the technique to be truly useful, a method must be found to extend the stable region of operation to all quadrants. Work has been carried out to solve this problem by including additional signal processing within the basic CALLUM modulator [Cha95]. One option is to control the sign of the VCO

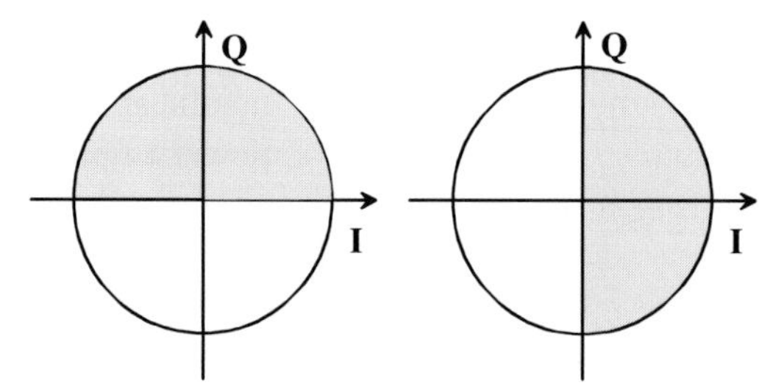

Stability region for loop A Stability region for loop B

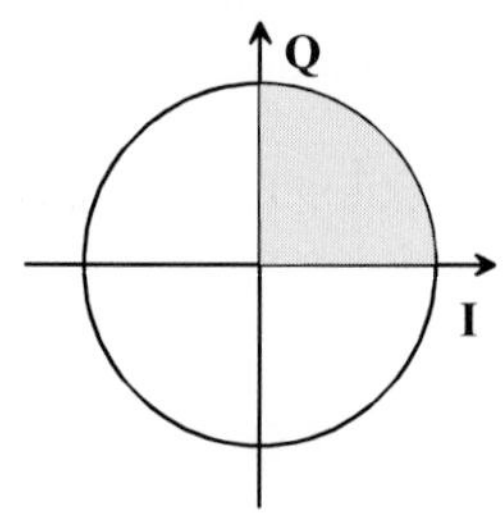

Stability region for loops A&B

Figure 1-13. Stability regions for the basic CALLUM modulator.

input signals in order to ensure that the loop is always operating in its stable region [Jen98], [Jen99].

Some attention has been directed towards characterizing the closed-loop behavior of CALLUM [Cha95]. It is reported that the amplitude and phase step response are very different. Both are functions of the input signal such that the phase time constant increases with decreasing input signal amplitude, whereas the amplitude time constant increases with the input signal amplitude.

Experimental systems have been reported so far in [Bat92], [Jen98]. Measured results gave –50dB ACI when operating at 160MHz carrier frequency and 2kHz modulation bandwidth [Bat92].

Another VCO-derived synthesis method, called the vector locked-loop (VLO) [Das92], operates in a similar manner, but the required signal processing utilizes polar (magnitude and phase) rather than Cartesian signals. The system consists of two cross-coupled phase-locked loops employing phase and magnitude detection so that both phase and magnitude may be employed as feedback signals. The main drawback of this technique, in comparison with the CALLUM technique described below, lies in the difficulty of realizing appropriately low-distortion magnitude and phase detectors with a suitable broadband response and operating at high carrier frequencies [Ken00].

1.9 LInear amplification employing Sampling Techniques (LIST)

Traditionally, pulse width modulation (PWM), delta modulation and delta sigma modulation techniques in RF linear amplification have been utilizable only at low frequencies. Linear amplification employing sampling techniques (LIST) attempts to bring the advantages of the delta modulation techniques to RF amplification in higher frequencies.

The basic structure of a LIST transmitter is shown in Figure 1-14. The I and Q signals are fed directly into a delta coder [Cox75a], in which the original information is converted to a data-stream of value $\pm K$

$$\begin{aligned} \Delta I(n) &= K\Delta[i(n)] \\ \Delta Q(n) &= K\Delta[q(n)], \end{aligned} \tag{1.9}$$

where $\Delta[]$ represents the delta coding. The bandpass filter at the output (Figure 1-14) is required to perform the reconstruction of the delta-coded signals; this is equivalent to using a low-pass filter at baseband. A low-pass filter is a suboptimal solution (with the optimal solution being an integrator); however, it is adequate in most cases [Ken00].

The delta-coded signals (including image products) are then quadrature upconverted and fed to the two nonlinear power amplifiers. The binary output ($\pm K$) from the delta coders results in a carrier phase shift of $\pm$ 180° during the upconversion process; this is more commonly known as Phase Reversal Keying (PRK). The resulting upconverted signals are given by

$$\begin{aligned} \Delta I_{RF}(t) &= G\,\Delta I(t)\cos(\omega_0 t) \\ \Delta Q_{RF}(t) &= G\,\Delta Q(t)\sin(\omega_0 t), \end{aligned} \tag{1.10}$$

where G is the gain of the amplifier. The power amplifiers would be constructed from, for example, class-E stages, with no loss of signal fidelity (since the signals are constant envelope at this point in the system) and with an excellent potential efficiency. Following the nonlinear amplification, the two paths are combined and fed to a bandpass filter. The combination process should, ideally, be lossless, although this is unlikely to be the case in practice. Any imbalance between the two paths should not result in degradation in linearity performance, but will result in an image signal. This image can be arranged to be in-band and hence the suppression required can be relatively modest. As result, a gain and phase error between the two paths of 0.3 dB and 3° may well be adequate for most applications [Ken00].

The bandpass filter (in Figure 1-14) must be sufficiently effective to remove the many out of band products generated by the LIST modulator, and hence is one of the main drawbacks of this technique. Usage of fast digital logic enables a high sampling frequency, so that unwanted products can be

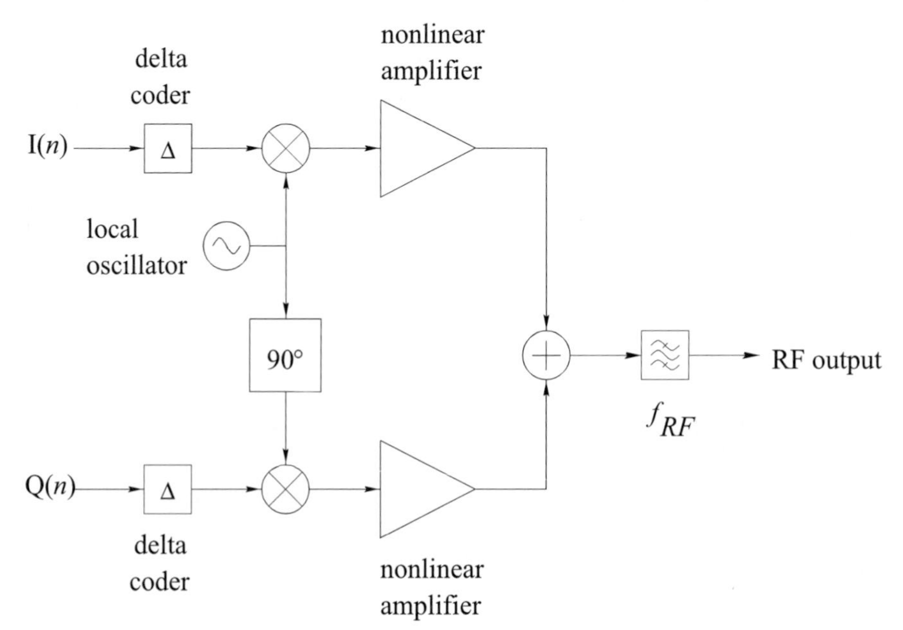

Figure 1-14. Basic architecture of a LIST transmitter.

pushed far from band, and bandpass filter specifications can be more relaxed.

1.10 Transmitters Based on Bandpass Delta Sigma Modulator

The conventional approach to employing switching mode amplifiers for signals with a time-varying envelope is the class-S amplifier, widely used for audio power amplification. The switching mode amplifier can be implemented as a voltage mode class-D (VMCD) amplifier, schematically shown in Figure 1-15. The signal is passed through a bandpass delta sigma modulator with 1-bit (two-level) output, and the resulting binary signal is fed into a VMCD amplifier. Finally, a bandpass filter is used at the amplifier output prior to the load. The output of the bandpass delta sigma modulator is a binary signal, in which the quantization noise associated with the digitization is spectrally shaped so that it lies largely outside of the band of interest. The output filter provides nonzero conductance essentially only in the band of interest, so there is no power dissipation associated with spectral components that do not reach the load. Amplifiers using the bandpass delta sigma algorithm to generate digital signal streams that encode analog communication signals of interest have also been investigated [Jay98], [Asb99], [Iwa00], and [Key01].

Since the pulse density modulated signal is linear within a narrow bandwidth and the class-D amplifier is a linear stage for digital signals, the filtered output should likewise be linear. With the combination of the bandpass delta sigma modulator and class-D amplifier, this simple amplifier topology has the potential to provide both high linearity and efficiency. A wide bandwidth pre-driver is necessary because the delta sigma modulated signal is broadband, and any significant lowpass and highpass filtering would corrupt the encoded signal. The class-D amplifier (Figure 1-15) has no rejection of errors introduced in the power stage due to power supply modulation, jitter,

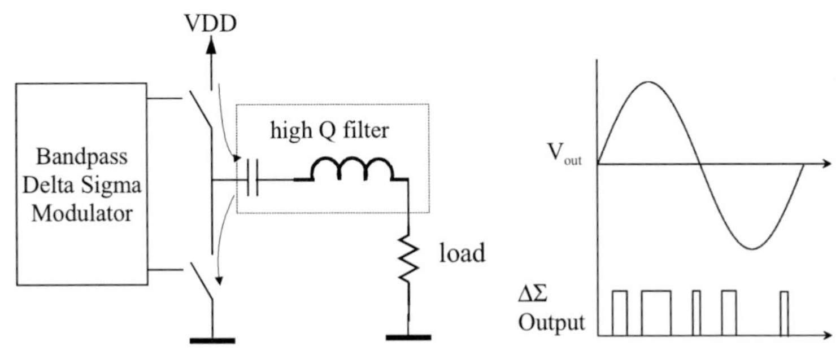

Figure 1-15. Schematic structure of voltage mode class-D amplifier, together with representative output and drive signals.

switching artifacts, etc.

The principal sources of power loss in the VMCD amplifier (Figure 1-15) are [Ham95], [Jay98]:

1) power lost in the transistors during the ON–OFF transients;
2) energy associated with discharging parasitic capacitance associated with the output terminals of the transistors;
3) power needed to operate the bandpass delta-sigma modulator, and pre-driver for the output amplifier;
4) power lost in the transistors and diodes associated with their on voltage;
5) power lost in the passive elements, such as the output filter.

The contributions listed as 1)–3) above increase as the rate of switching of the amplifier input increases, and thus require minimizing the sampling frequency of the bandpass delta-sigma modulator. If the entire system is accounted for, the power dissipated by the bandpass delta sigma modulator and the pre-driver has to be considered.

VMCD PAs with power outputs of 100 W to 1 kW are readily implemented at HF, but are seldom used above lower VHF because of losses associated with the drain capacitance [Raa93]. In order to reduce the capacitive loss, the voltage across the switch should be zero when it turns on or off. This is called zero-voltage-switching (ZVS) and can be achieved with current mode class-D (CMCD) amplifiers, where the transistor drain capacitance can become part of the output filter. Another way to reduce turn-on and turn-off loss is minimizing the series inductive loss by zero-current-switching (ZCS), where the current is always zero when the switch turns on or off. The ZCS is, however, less important than the ZVS at hundreds of megahertz switching frequencies [Kob01]. The VMCD power amplifier achieves the ZCS condition. The signal frequency component is only a small

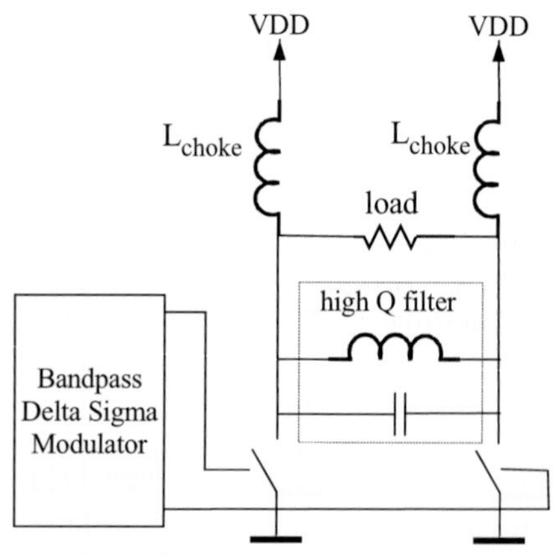

Figure 1-16. Schematic structure of current mode class-D amplifier.

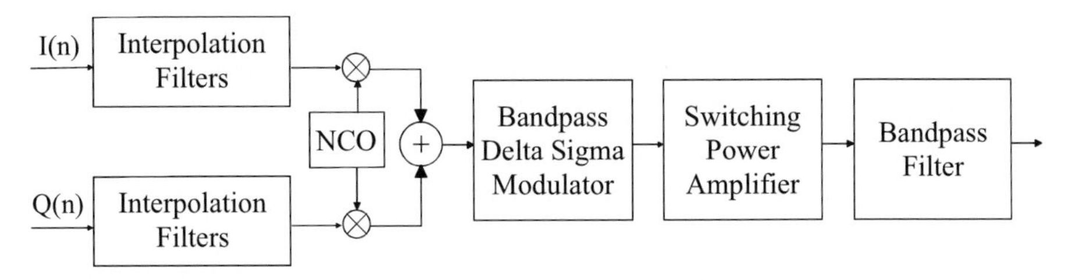

Figure 1-17. Transmitter based on bandpass delta sigma modulator.

portion of the overall ΔΣ modulated signal as shown in Figure 14-8. Therefore the switching frequency is not the same as signal frequency (output filter is tuned to signal frequency), so the drain-to-source currents and drain-to-source voltages have different frequencies; the voltage changes at switching frequency and the current at signal frequency at the VMCD amplifier or vice versa at the CMCD amplifier. Thus, it is impossible to achieve the ZVS or the ZCS when the class-D amplifiers are driven with ΔΣ modulated signals. Only the power at signal frequency is passed to the load, and the switching activity due to the outside band power will cause switching losses. For these reasons, when compared to the square wave with fixed switching frequency, the efficiencies of circuits driven with delta-sigma modulated signals drop dramatically. According to the simulations, the VMCD amplifiers give better efficiency than the CMCD amplifiers with ΣΔ-modulated signal [Som04]. It is possible to achieve with the VMCD amplifiers high efficiencies at audio frequencies [Var03].

Bandpass delta sigma modulators can be implemented with analog or digital inputs. Figure 1-17 shows the structure of a representative digital transmitter [Spl01].

REFERENCES

[Asb99] P. Asbeck, J. Mink, T. Itoh, and G. Haddad, "Device and Circuit Approaches for Next-Generation Wireless Communications," Microwave J., Vol. 42, No. 2, pp. 22–42, Feb. 1999.

[Bat92] A. Bateman, "The Combined Analogue Locked Loop Universal Modulator (CALLUM)," In Proceedings of the 42nd IEEE Vehicular Technology Conference, May 1992, pp. 759-763.

[Bat98] A. Bateman, and C. K. Yuen, "Method and Apparatus for Amplifying, Modulating and Demodulating," U. S. Patent 5,719,527, Feb. 17, 1998.

[Bax01] W. T. Bax, and M. A. Copeland, "A GMSK Modulator Using a ΣΔ Frequency Discriminator-based Synthesizer," IEEE J. Solid-State Circuits, Vol. 36, No. 8, pp. 1218-1227, Aug. 2001.

[Bax99] W. T. Bax, "Modulation and Frequency Synthesis for Wireless Digital Radio," PhD Thesis, Carleton University, October 1999.
[Cha95] K. Y. Chan, and A. Bateman, "Linear modulators Based on RF synthesis: Realization and Analysis," IEEE Transactions on Circuits and Systems-I: Fundamental Theory and Applications, Vol. 42, No. 6, pp. 321-333, June 1995.
[Che68] E. M. Cherry, and D. E. Hooper, "Amplifying Devices and Low-Pass Amplifier Design," John Wiley and Sons, Inc., New York, 1968.
[Chi35] H. Chireix, "High Power Outphasing Modulation," Proceedings IRE, Vol. 23, No. 11, Nov. 1935, pp. 1370-1392.
[Cox74] D. C. Cox, "Linear Amplification with Nonlinear Components," IEEE Trans. Commun., Vol. COM-22, pp. 1942–1945, Dec. 1974.
[Cox75a] D. C. Cox, "Linear Amplification by Sampling Techniques: A New Application for Delta Coders," IEEE Trans. on Comm., Vol. 23, No. 8, pp. 793-798, Aug. 1975.
[Cox75b] D. C. Cox, and R. P. Leck, "Component Signal Separation and Recombination for Linear Amplification with Nonlinear Components," IEEE Transactions on Communications, Vol. 23, pp. 1281-1287, Nov. 1975.
[Dar98] S. Ampem-Darko, and H. S. Al-Raweshidy, "Gain/Phase Imbalance Cancellation Technique in LINC Transmitters," Electron. Lett., Vol. 34, No. 22, pp. 2093–2094, Oct. 1998.
[Das92] M. K. Dasilva, "Vector Locked Loop," U. S. Patent 5,105,168, Apr. 14, 1992.
[Fen97] J. Fenk, "Highly Integrated RF-ICs for GSM and DECT," IEEE Radio Frequency Integrated Circuits (RFIC) Symposium, 1997, pp. 69-72.
[Ger01] D. Gerna, A. Giry, D. Manstretta, D. Belot, and D. Pache, "1W 900 MHz Direct Conversion CMOS Transmitter for Paging Applications," RFIC Symposium 2001 Digest of Technical Papers, pp. 191-194.
[Het91] S. A. Hetzel, A. Bateman, and J. P. McGeehan, "LINC Transmitter," Electronics Letters, Vol. 27, No. 10, pp. 844-846, May 1991.
[Irv98] G. Irvine, and et al., "An Up-Conversion Loop Transmitter IC for Digital Mobile Telephones," ISSCC Digest of Technical Papers, Feb. 1998, pp. 364-365.
[Iwa00] M. Iwamoto, A. Jayaraman, G. Hanington, P. F. Chen, A. Bellora, W. Thornton, L. E. Larson, and P. M. Asbeck, "Bandpass Delta-Sigma Class-S Amplifier," Electronics Letters, Vol. 36, 12, pp. 1010-1012, June 2000.
[Jay98] A. Jayaraman, P. F. Chen, G. Hanington, L. Larson, and P. Asbeck, "Linear High-Efficiency Microwave Power Amplifiers Using Bandpass Delta-Sigma Modulators," IEEE Microwave and Guided Wave Letters, Vol. 8, 3, pp. 121–123, March 1998.

[Jen98] D. J. Jennings, and J. P. McGeehan, "Hardware Implementation of Optimal CALLUM Transmitter," Electronics Letters, Vol. 34, No. 19, pp. 1816-1817, Sept. 1998.

[Jen99] D. J. Jennings, and J. P. McGeehan, "A High-Efficiency RF Transmitter Using VCO-Derived Synthesis: CALLUM," IEEE Transactions on Microwave Theory and Techniques, Vol. 47, No. 6, June 1999, pp. 715-721.

[Kah52] L. R. Kahn, "Single-Sideband Transmission by Envelope Elimination and Restoration," Proceedings IRE, Vol. 40, pp. 803-806, July 1952.

[Ken00] P. B. Kenington, "High-Linearity RF Amplifier Design," Norwood, MA: Artech House, 2000.

[Key01] J. Keyzer, J. Hinrichs, A. Metzger, M. Iwamoto, I. Galton, and P. Asbeck, "Digital Generation of RF Signals for Wireless Communications with Band-Pass Delta-Sigma Modulation," IEEE MTT-S Digest, pp. 2127-2130, 2001.

[Kob01] H. Kobayashi, J. M. Hinrichs, and P. M. Asbeck, "Current-Mode Class-D Power Amplifiers for High-Efficiency RF Applications," IEEE Transactions on Microwave Theory and Techniques, Vol. 49, No. 12, pp. 2480–2485, Dec. 2001.

[Lee01] K-Y. Lee, S-W. Lee, Y. Koo, H-K. Huh, H-Y. Nam, J-W. Lee, J. Park, K. Lee, D-K. Jeong, and W. Kim, "Full-CMOS 2.4 GHz Wideband CDMA Transmitter and Receiver with Direct Conversion Mixers and DC-Offset Cancellation," Symposium on VLSI Circuits 2001 Digest of Technical Papers, pp. 7-10.

[Liu00] T-P. Liu, and E. Westerwick, "5-GHz CMOS Radio Tranceiver Front-End Chipset," IEEE J. Solid-State Circuits, Vol. 35, pp. 1927-1933, Dec. 2000.

[McM02] D. R. McMahill, and C. G. Sodini, "A 2.5-Mb/s GFSK 5.0-Mb/s 4-FSK Automatically Calibrated ΣΔ Frequency Synthesizer," IEEE J. Solid-State Circuits, Vol. 37, No. 1, pp. 18-26, Jan. 2002.

[Nag02] P. J. Nagle, D. P. Burton, E. P. Heaney, and F. J. McGrath, "A Wideband Linear Amplitude Modulator for Polar Transmitters Based on the Concept of Interleaving Delta Modulation," ISSCC Digest of Technical Papers, Feb. 2002, pp. 296-297.

[Ors99] P. Orsatti, F. Piazza, and Q. Huang, "A 20-mA-Receive, 55-mA-Transmit, Single-Chip GSM Tranceiver in 0.25-um CMOS," IEEE J. Solid-State Circuits, Vol. 34, pp. 1869-1880, Dec. 1999.

[Per97] M. H. Perrott, T. L. Tewksbury, and C. G. Sodini, "A 27-mW CMOS fractional-N Synthesizer Using Digital Compensation for 2.5-Mb/s GFSK Modulation," IEEE J. Solid-State Circuits, Vol. 32, No. 12, pp. 2048-2060, Dec. 1997.

[Pet79] V. Petrovic, and W. Gosling, "Polar-loop Transmitter," Electronics Letters, Vol. 15, No. 10, pp. 286-288, May 1979.

[Pet84] V. Petrovic, and C. N. Smith, "Reduction of Intermodulation Distortion by Means of Modulation Feedback," In IEE Colloquium on Intermodulation-Causes, Effects and Mitigation, London, April 9, 1984, pp. 8/ 1-8.
[Raa85] F. H. Raab, "Efficiency of Outphasing RF Power-amplifier Systems," IEEE Transactions on Communications," Vol. 33, No. 10, pp. 1094-1099, Oct. 1985.
[Raa93] F. H. Raab, and D. J. Rupp, "HF Power Amplifier Operates in Both Class B and Class D," in Proc. RF Expo West, San Jose, CA, Mar. 17–19, 1993, pp. 114–124.
[Raa96] F. H. Raab, "Intermodulation Distortion in Kahn-technique Transmitters," IEEE Trans. On Microwave Theory and Techniques, Vol. 44, No. 12, December 1996, pp. 2273-2278.
[Raa98] F. H. Raab, and et al., "High Efficiency L-band Kahn-technique Transmitter," Proc of IEEE MTT-S, Baltimore, USA, Vol. 2, June 1998, pp. 585-588.
[Raa99] F. H. Raab, "Drive Modulation in Kahn-technique Transmitters," IEEE MTT-S Microwave Symposium Digest, 1999, Vol. 2, pp. 811-814.
[Raz98] B. Razavi, "RF Microelectronics," Prentice-Hall PTR, New Jersey, USA, 1998.
[Ril94] T. Riley, and M. Copeland, "A Simplified Continuous Phase Modulator Technique," IEEE Transactions on Circuits and Systems-II, Vol. 41, No. 5, May 1994, pp. 321-328.
[Rus76] A. J. Rustako, Jr., and Y. S. Yeh, "A Wide-Band Phase-Feedback Inverse-Sine Phase Modulator with Application Toward a LINC amplifier," IEEE Transactions on Communications, Vol. 24, pp. 1139-1143, Oct. 1976.
[Shi00] B. Shi, and L. van Sundström, "A 200-MHz IF BiCMOS Signal Component Separator for Linear LINC Transmitters," IEEE J. Solid-State Circuits, Vol. 35, pp. 987–993, July 2000.
[Som04] J. Sommarek, "Digital Modulators with On-Chip D/A Converters," Licentiate's Thesis, Helsinki University of Technology, 2004.
[Spl01] A. Splett, and et al., "Solutions for Highly Integrated Future Generation Software Radio Basestation Transceivers," in Proc. IEEE Custom Integrated Circuits Conf., 2001, pp. 511-518.
[Sta99] J. Staudinger, and et al., "800 MHz Power Amplifier Using Envelope Following Techniques," Proc. of the IEEE Radio and Wireless Conference (RAWCON) 1999, Denver, Colorado, Aug. 1999, pp. 301-304.
[Su98] D. Su, and W. McFarland, "An IC for Linearizing RF Power Amplifiers Using Envelope Elimination and Restoration," IEEE Journal of Solid-State Circuits, Vol. 33, No. 12, Dec. 1998, pp. 2252-2258.
[Sun00] L. Sundström, "Spectral Sensitivity of LINC Transmitters to Quadrature Modulator Misalignments," IEEE Trans. Veh. Technol., Vol. 49, pp. 1474–1487, July 2000.

[Sun94] L. Sundström, and M. Johansson, "The Effect of Modulation Scheme on LINC Transmitter Power Efficiency," Electronics Letters, Vol. 30, No. 20, pp. 1643-1645, Sept. 1994.
[Sun95a] L. Sundström, "Automatic Adjustment of Gain and Phase Imbalances in LINC Transmitters," Electronics Letters, Vol. 31, No. 3, pp. 155-156, Feb. 1995.
[Sun95b] L. Sundström, "Effects of Reconstruction Filters and Sampling Rate for a Digital Signal Component Separator on LINC Transmitter Performance," Electron. Lett., Vol. 31, No. 14, pp. 1124–1125, July 1995.
[Tom89] S. Tomisato, K. Chiba, and K. Murota, "Phase Error Free LINC Modulator," Electronic Letters, Vol. 25, No. 9, pp. 576-577, Apr. 1989.
[Var03] J. Varona, A. A. Hamoui, and K. Martin, "A Low-Voltage Fully-Monolithic ΔΣ-Based Class-D Audio Amplifier," in Proc. ESSCIRC'03, Sept. 2003, pp. 545-548.
[Yam97] T. Yamawaki, and et al., "A 2.7-V GSM RF Transceiver IC," IEEE J. Solid-State Circuits, Vol. 32, pp. 2089-2096, Dec. 1997.
[Zha00] X. Zhang, and L. E. Larson, "Gain and Phase Error Free LINC Transmitter," IEEE Trans. Veh. Technol., Vol. 49, pp. 1986–1994, Sept. 2000.
[Zha01] X. Zhang, L. E. Larson, and P. M. Asbeck, "Calibration Scheme for LINC Transmitter," Electron. Lett., Vol. 37, No. 5, pp. 317–318, Mar. 2001.

Chapter 2

2. POWER AMPLIFIER LINEARIZATION

Traditionally, constant envelope modulation schemes have been used in radio telecommunications because of their simplicity and robustness to amplitude errors. This made it possible to use high efficiency power amplifiers (PA), which are intrinsically very nonlinear devices, near the saturation region where the amplifier efficiency is at its peak.

However these modulation schemes are spectrally inefficient and the current trend is to improve the spectral efficiency or the number of bits transmitted per bandwidth by using some linear modulation scheme such as quadrature amplitude modulation (QAM). Alas, when driven through a nonlinear device the fluctuating envelope of the linear modulation schemes cause intermodulation products to appear around the signal band. This spectral spillage is effectively impossible to filter away and so can cause the amplified signal to exceed its allowed adjacent channel interference (ACI) limits.

To compensate these unwanted effects, various amplifier linearization techniques have been presented. Table 2-1 shows a comparison between three basic linearization techniques, namely, Cartesian feedback, feedforward and predistortion. The cancellation performance of the Cartesian feedback is good, but the bandwidth is narrow, making the technique unsuitable for very wideband systems. The feedforward, on the other hand, can be employed for wideband linearization, but, unfortunately, the system is extremely complicated, resulting in great power waste and large physical size. The third method, predistortion, is an optimal solution in terms of power added efficiency and physical size.

2.1 Feedforward

In the 1920s, H. S. Black invented two schemes for reducing amplifier distortion, namely, feedforward [Bla28] and negative feedback [Bla37]. Feedforward became forgotten in favor of the feedback technique, even though feedforward predated the latter by several years. Today, we are well aware of the limitations of feedback owing to the work of Nyquist and Bode. Feedback is limited by conditional stability and finite inter-modulation distortion (IMD) suppression, whereas feedforward is unconditionally stable and can, in theory, completely eliminate the IMD. However, Black himself noted that the key problem with his feedforward prototype was the primary reason for feedforward to be relegated to the background.

The feedforward prototype required perfect gain match in the different signal paths and Black reported that the gain of the amplifier had to be constantly re-adjusted. Another reason, at that time, was the simple fact that the complexity of the feedforward system compared with negative feedback was considered as a major disadvantage. However, as applications with higher frequencies and bandwidths appeared, the disadvantages of negative feedback became more apparent. This has to some extent led to a renaissance of the feedforward technique and it is nowadays considered to be one of the most established and approved methods, especially for wideband and multicarrier systems [Ken91a], [Ken91b], [Mye94]. Otherwise, it has been used in many areas spanning from low frequency audio applications [Van80] to high-frequency CATV [Pro80] and microwave [Se71b] applications.

A block diagram of the feedforward system is depicted in Figure 2-1. All blocks operate at RF. The main amplifier, a nonlinear power amplifier, is fed directly with the source signal. The distortion generated by the amplifier is isolated in the signal cancellation loop by subtracting the source signal from the amplifier output. This signal is often referred to as the error signal. In the distortion cancellation loop, the error signal is finally subtracted from the amplifier output. For perfect signal and distortion cancellation, an attenuator and auxiliary amplifier are required in the signal and distortion cancellation loops, respectively. For high frequency applications, it is evident that the performance of this scheme in terms of obtaining perfect signal and distortion cancellation is not only dependent on the amplitude match but also on the phase/delay match along the parallel signal arms. In practice, fixed delays can be inserted as noted in Figure 2-1 to balance the delays in the co-arms that are primarily dominated by the amplifiers.

The effects of delay, phase and amplitude imbalances have been treated

Table 2-1. Comparison of Three Basic Linearization Techniques [Mad99]

Technique	Cancel-lation Per-fomance	Bandwidth	Power Added Efficiency	Size	Suitability to Multicarrier
Feedback	Good	Narrow	Medium	Medium	Low
Feedforward	Good	Wide	Low	Large	High
Predistortion	Medium	Medium	High	Small	Medium

in several papers [Ste88], [Wil92a], [Par94], and [Mye94]. As an example, to obtain 25dB suppression of the amplifier distortion, an amplitude error of better than 0.5dB or a phase error of better than 0.5 degrees is required [Wil92a]. For narrowband systems, the delay mismatch can be corrected by simply adjusting the phase. However, for wideband systems, a given delay corresponds to different phase shifts at different frequencies. It is shown in [Par94] that, if the delay mismatch corresponds to one wavelength of the carrier signal, then the distortion suppression will be limited to 30dB at both ends of the bandwidth, assuming 1% bandwidth with an otherwise ideal system. Thus, to take into account the delays when designing a feedforward system, the delaying elements should be quantified by means of measurements and/ or simulations. A systematic approach based on simulation is presented in [Kon93]. The technique is based on harmonic balance simulation that takes into account the nonlinear effects in the amplifiers, provided that appropriate simulation models are available. The method was found to work quite well for a microwave feedforward amplifier. Simulation results gave 20dB suppression over 500MHz bandwidth and a 6GHz carrier frequency.

Fixed or manually controlled amplitude and phase matching networks are usually not sufficient to preserve reasonable distortion suppression. Component aging, temperature drift, change of operating frequency etc. cause variations that require automatic control of the amplitude and phase matching networks. Several solutions have appeared and three distinct techniques can be identified. One approach is based on measuring the power at some points and minimizing it [Obe91]. For example, the signal cancellation loop can be tuned by minimizing the power of the error signal and the distortion cancellation loop can be tuned by minimizing the out-of-band power at the transmitter output. Another method uses a pilot tone that is inserted at some point,

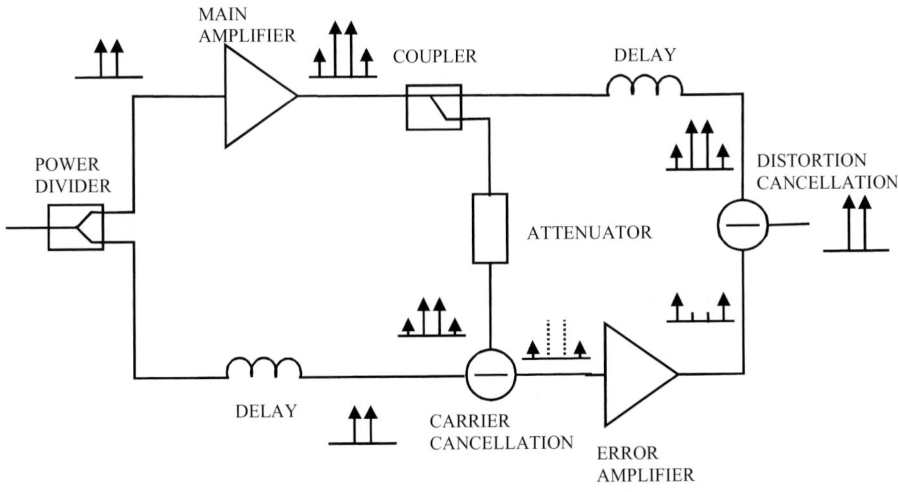

Figure 2-1. Block diagram of the feedforward system.

typically after the power amplifier, and a measure of the imbalance is obtained by detecting the pilot at another point [Nar91], [Cha91]. The level of the detected pilot tone guides a controller to adjust the amplitude and phase accordingly. The third method is based on the fact that the gradient for the function to be optimized can be calculated and used to guide the adjustment [Olv85], [Bau93], [Ken94], [Cav95], [Smi98]. Some of these solutions require a computer-based controller [Obe91], [Nar91], [Ken94], [Smi98]; others can be implemented using a continuous closed-loop system [Cha91], [Ken94]. Common for all the schemes is the fact that the total complexity of the feedforward scheme becomes quite large compared with the basic feedforward configuration shown in Figure 2-1.

Even though the main amplifier can be quite power efficient, the total efficiency of the feedforward scheme is drained due to losses in the main path delay, the couplers and the auxiliary amplifier. A high efficiency auxiliary amplifier should, of course, be used. But it must also be sufficiently linear so that no additional distortion is generated. Furthermore, the coupler that is used to subtract the error signal from the amplifier output should have a low coupling factor. With a low coupling factor, most of the power available from the main amplifier is fed to the antenna. On the other hand, the coupling factor should be high because the auxiliary amplifier must provide enough power to compensate for the losses in the coupler. Thus, an optimal coupling factor can be calculated based on the knowledge of the other components in terms of amplifier efficiency, intercept points and delay line losses [Ken92], [Dix86].

Several feedforward prototypes have been reported [Ken91b], [Se71b], [Ste88], [Mey74], [Nar91], [Dix86] with performance ranging from 20 to 40dB suppression both for narrowband and wideband systems with carrier frequencies from a couple of MHz to several GHz.

Typical applications are of narrowband type, i.e. the bandwidth is a couple of percentage points of the carrier frequency. Yet, it is interesting to note that feedforward has also been applied to systems with a bandwidth of one decade or more [Sei71a], [Mey74], even though the maximum frequency was rather low (< 300MHz). With such a large bandwidth, the designer is confronted with the problem of obtaining flat frequency responses from all components. In [Mey74], an interesting solution is presented because it combines the best sides of Black's two schemes. The main amplifier has a local negative feedback loop, while, for low frequencies, the loop-gain is large enough to make the amplifier itself sufficiently linear. However, for higher frequencies, the loop-gain is diminished and the feedforward technique, which is optimized for the higher frequencies, takes over.

2.2 Cartesian Modulation Feedback

Figure 2-2 shows the principle of the Cartesian feedback transmitter. The output of the amplifier is synchronously demodulated and compared with the source signal to obtain an error signal. The error signal is fed to the loop filter followed by upconversion in a quadrature modulator before it finally reaches the power amplifier. The Cartesian feedback was introduced by Petrovic [Pet83]. Several experimental systems have been reported operating with carrier frequencies ranging from a couple of MHz to 1.7GHz with modulation bandwidths of up to 500kHz [Joh91], [Wil92b], [Joh94], [Whi94]. Distortion suppression varies from 20dB up to 50dB with 35% to 65% amplifier power efficiency.

It is interesting to note that Cartesian feedback has been proven to work for wideband applications [Joh91]. Careful design and selection of components are required since the propagation delay will dominate the phase characteristics of the loop. The relations between loop delay, bandwidth and stability are investigated in [Joh95]. It is shown that for a 5MHz cross-over bandwidth the loop delay should not exceed 33ns (with 60° phase margin). Assuming that the loop gain is sufficiently high and that the loop contains a single pole, this corresponds to 20dB of IMD suppression at 500kHz.

A potential problem for Cartesian feedback transmitters is that the characteristics of the amplifier effectively degrade the gain and phase margins of the loop. Another problem is the phase shift that occurs in the loop when, for example, changing the carrier frequency. In practice, a phase adjuster is required to adjust the phase automatically to preserve the stability [Bro88], [Ohi92]. The phase adjuster can be placed in the loop or, as illustrated in

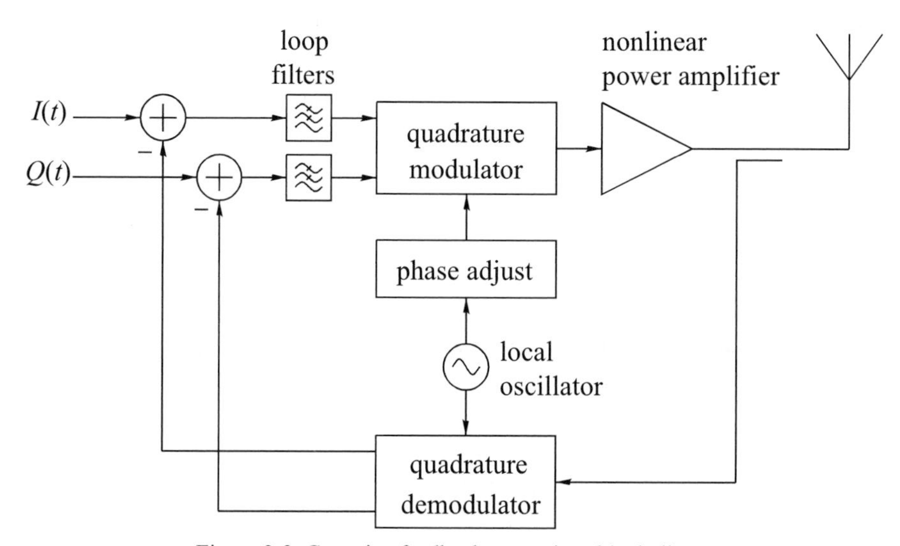

Figure 2-2. Cartesian feedback transmitter block diagram.

Figure 2-2, to differentiate the local oscillator phase between the modulator and the demodulator. This means that the complexity of a practical Cartesian feedback transmitter is much higher than the diagram in Figure 2-2 might imply.

One disadvantage of Cartesian feedback and other schemes where the amplifier is fed with a signal having a varying envelope is that the power efficiency is low for low input levels. An attempt to improve on efficiency is presented by Briffa et al. [Bri93]. Dynamic biasing is applied to the final amplifier stage. The biasing (of both base and collector) is controlled by the envelope of the input signal through mapping functions that apply predetermined biasing levels for maximum efficiency. Simulation results show that the efficiency can be increased from 50% to 60% near saturation and from 10% to 30% at low input levels. A small improvement in linearity was a welcome side effect.

To obtain low levels of ACI the dynamic range of the feedback path must be appropriately high. Upwards, the dynamic range is limited by intermodulation, especially in the quadrature demodulator, and, downwards, by the accumulated noise in the feedback path.

A derivative of Cartesian feedback was presented by Johansson et al. for multi-carrier applications [Joh93]. The principle is outlined in Figure 2-3. Several Cartesian feedback loops (Figure 2-4) operate in parallel on distinct frequency bands. Each loop can accommodate several carriers. An interesting property is that a loop can be assigned to a channel without an input sig-

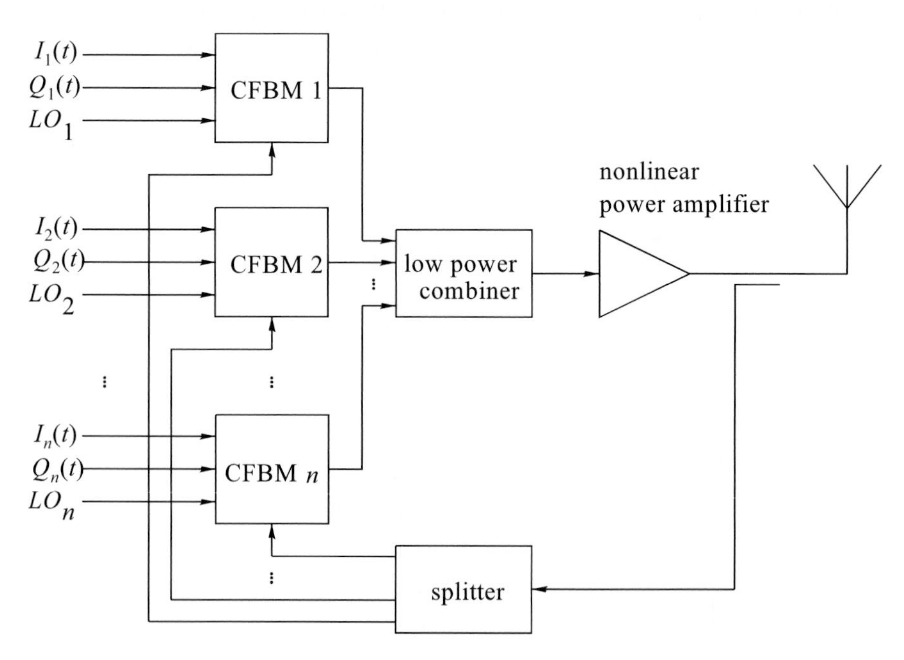

Figure 2-3. Block diagram of multi-loop Cartesian feedback system.

nal to reduce intermodulation products in that channel caused by the other channels in operation. Experimental results show that this is a viable method for broadband multi-carrier linearization with up to 30dB intermodulation suppression [Joh94]. The use of Cartesian feedback with a class-C PA amplifying an IS-136 (DAMPS) signal improves the first ACPR by 35 dB and allows the signal to be produced with an efficiency of 60% [Ken00].

2.3 Predistortion

From a mathematical point of view, predistortion is, next to feedforward, probably the most obvious technique for linearization. By preceding the nonlinear amplifier with its inverse counterpart, one-to-one mapping between the input and output can be obtained. As illustrated in Figure 2-5 predistortion in its most simple form is an open loop system. However, most solutions presented use some kind of feedback to enable adaptation of the predistorter. Several solutions have been developed to realize the predistorter, from digital baseband processing to processing the signal directly at RF using diodes as nonlinear devices.

Behavior models for PA have traditionally been developed on the basis of the AM-AM and AM-PM curves, and the PA gain is usually approximated as a complex polynomial function of instantaneous input power level. However, as the bandwidth of the signal increases, memory effects in the transmitter distort this simplified picture. Memory effects are attributed to filter group delays, the frequency response of matching networks, nonlinear capacitances of the transistors and the response of the bias networks. The performance of the predistortion algorithms that do not take these memory effects into account is severely degraded as the bandwidth of the input signal increases [Vuo01]. A nonlinear system with memory can be represented by Volterra series, which are characterized by Volterra kernels [Sch81]. How-

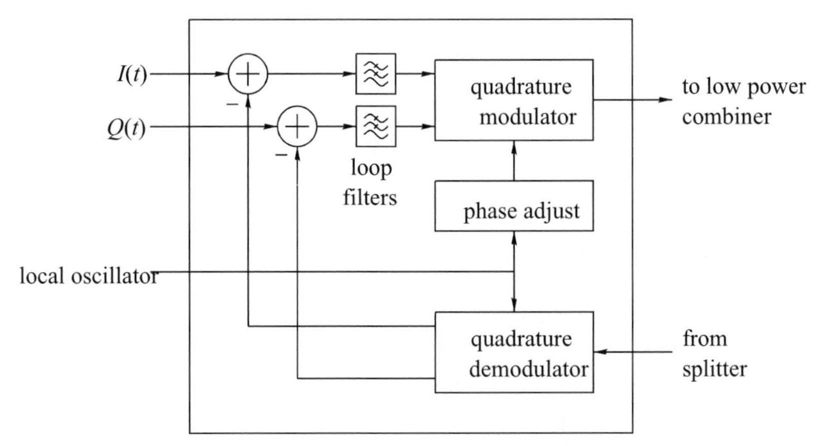

Figure 2-4. Cartesian feedback module (CFBM).

ever, the computation of the Volterra kernels for a nonlinear system is often difficult and time consuming for strongly nonlinear devices. In many applications that involve modeling of nonlinear systems, it is convenient to employ a simpler model. The Wiener model, which is a cascade connection of linear time invariant (LTI) system and memoryless nonlinear system, has been used to model nonlinear PAs with memory [Sal81]. A straightforward predistortion method is to add an adaptive filter in cascade with the memoryless predistorter [Kan98].

In applications with weak nonlinear amplifiers, or having moderate linearity requirements, coarse approximations of the wanted predistortion function can be applied using nonlinear analogue components. In fact, what this all really says is that the predistorter is essentially a technique, which can work usefully for well backed-off amplifiers showing only small amounts of compression. This is a very tough restriction for signals having high peak to average ratios.

The output of the power amplifier in Figure 2-5 is

$$v_o = v_p - a_3 v_P^3 \ (a_1 = 1). \tag{2.1}$$

A predistorter generates output v_p

$$v_o = v_{in} = v_p - a_3 v_P^3, \tag{2.2}$$

so a predistorter has to "solve" the cubic

$$v_P^3 - \frac{1}{a_3} v_p + \frac{1}{a_3} v_{in} = 0. \tag{2.3}$$

This can be solved by

$$v_P = v_{in} + b_3 v_{in}^3 + b_5 v_{in}^5 + b_7 v_{in}^7 + \ldots \text{ etc.} \tag{2.4}$$

This shows that, in general, the predistorter contains an infinite series of terms of a higher order than the amplifier distortion itself. The output signal from any useful predistorter will have a spectral bandwidth significantly greater than the distorted PA output that it strives to linearize; this has important implications for the required bandwidth of any components used in predistorter design.

The output of the power amplifier is

$$v_o = a_1 v_p - a_3 v_P^3. \tag{2.5}$$

The transfer function of the predistorter is

$$v_P = b_1 v_{in} + b_3 v_{in}^3 \ (b_1 = 1). \tag{2.6}$$

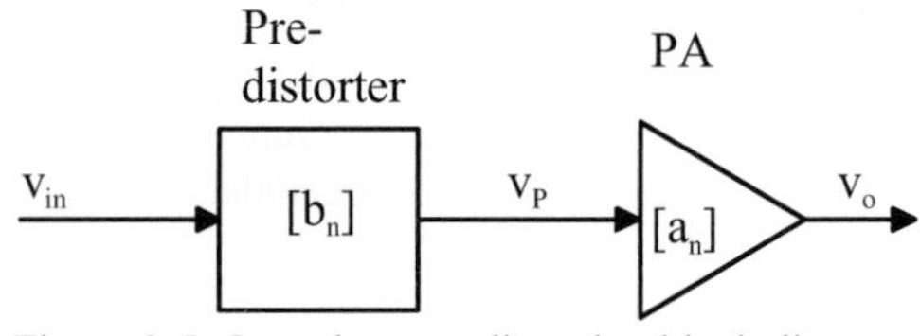

Figure 2-5. Open-loop predistortion block diagram.

Using (2.5) and (2.6), the power amplifier output is

$$\begin{aligned} v_o &= a_1(v_{in} + b_3 v_{in}^3) - a_3(v_{in} + b_3 v_{in}^3)^3 \\ &= a_1 v_{in} + (a_1 b_3 - a_3)v_{in}^3 - 3a_3 b_3 v_{in}^5 - 3a_3 b_3^2 v_{in}^7 - a_3 b_3^3 v_{in}^9. \end{aligned} \tag{2.7}$$

If $b_3 = a_3/a_1$, then the third harmonic will cancel. But we now have additional 5th, 7th and 9th order products. An important result is that a predistorter with a simple 3rd degree expansion characteristic can cancel the 3rd degree nonlinearity in an amplifier, but will create additional higher degree nonlinearities, that were absent in the basic PA itself.

2.3.1 Analog Predistortion

In [Noj84] Nojima et al. used diodes to build a third-order predistorter (see Figure 2-6) that operated at an intermediate frequency (130MHz) for a microwave system with 6GHz carrier frequency and a 30MHz bandwidth. More than 30dB suppression of the third-order IMD was obtained with a self-adjusting system that controlled the magnitude and the phase of the third-order predistorter. A continuation of this work was presented in [Noj85], [Noj85], [Nan85], where the predistorter worked at the carrier frequency instead of an intermediate frequency. One prototype was intended for mobile telephone systems operating at 800MHz and the other one for 6GHz microwave digital radio systems. For the 800MHz system, Nojima reported up to 20dB reduction of the third-order products over a bandwidth of 25MHz and, for the microwave system, about 6dB over a 500MHz bandwidth. A similar solution is presented in [Nam83], where the predistorter is realized with FET (field-effect transistors) amplifiers acting as nonlinear devices.

Third-order predistorters are not enough if we want to obtain higher accuracy or linearize amplifiers that are less linear. One solution is to use higher order polynomials. However, this requires more advanced adaptation algorithms since there will be more coefficients to adjust. A new technique for adaptation of this kind of predistortion linearizers is presented by Stapleton et al. in [Sta91], which is based on minimizing the out-of-band power. By describing both the predistorter and the amplifier with truncated 5th-order complex polynomials, the IMD power can be expressed as a function of the polynomial coefficients. From this analysis it was shown that for the dominant third order components the out-of-band power represented a quadratic surface. Although the analysis required that the input signal was a stationary Gaussian process, it was also demonstrated by means of simulations that quadratic-like surfaces were obtained for a 16-QAM signal. The fact that one global minimum exists suggests that we can choose from several powerful optimization methods. As a first step, Stapleton simulated a system using a modified version of the Hooke and Jeeves method, a direct search scheme, to prove the viability of this approach. A prototype operating with an 850MHz carrier frequency is presented in [Sta92a]. The polynomial coefficients were adjusted manually to give a 15dB improvement in the third-order and 5dB improvement in the fifth-order IMD products. Simulations estimated that the adaptation time was rather long, in the order of minutes, in fact.

The method that was used to measure the out-of-band power included the use of a mixer and a separate local oscillator followed by a filter and a rectifier. An alternative solution is given in [Sta92b], [Sta92c] that uses convolution of the RF input signal and the transmitter output signal, thereby avoiding an additional oscillator for the down conversion of the output signal. Analysis showed that the resulting signal could be used as a measure of the out-of-band power, just as in the previous approach. Prototype results gave an 11dB improvement in linearity compared with the 15dB that was predicted by means of simulation. The adaptation time was decreased to approximately 10 seconds by using more advanced optimization schemes based on surface fit.

The analysis based on complex polynomials has been extended to include quadrature modulator errors in [Hil92], [Hil94]. It is shown that the out-of-band power is a quadratic function of both the amplifier nonlinearities and the quadrature modulator errors. A prototype was reported to work excellently with up to 20dB IMD suppression and with a convergence time of below 4 seconds.

Yet another system based on polynomial predistortion is presented in [Gha93], [Gha94]. This system has a significantly larger complexity compared with the previous solutions, especially in terms of digital signal processing. The adaptation process requires synchronous detection of the output signal instead of monitoring the out-of-band power. Simulation results promise up to 45dB distortion suppression. Still no prototype results have been reported for this scheme.

The use of polynomial functions and a simple adaptation scheme based

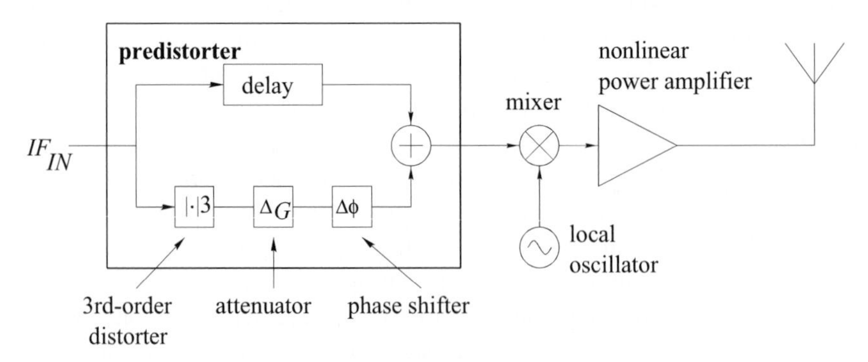

Figure 2-6. Principle of third-order predistorter.

on out-of-band power measurements offers low cost and complexity. However, a low-order polynomial is capable of canceling only weak nonlinearities. For more non-linear amplifiers, more general schemes based on DSP techniques and look-up tables have been developed.

2.3.2 Mapping Predistortion

In [Bat88], Bateman et al. suggested the use of DSP techniques and look-up tables with curve fitting to realize an adaptive predistorter. The approach required the transmission to be interrupted because of the special signal that had to be applied to characterizing the amplifier nonlinearities. However, the solutions presented below can adapt while transmitting and do not require any special signals.

A simple and "brute force" solution was presented by Nagata [Nag89], who used a huge two-dimensional table (see Figure 2-7). By using a two-dimensional table, any complex input signal represented by its Cartesian components can be mapped to a new constellation of Cartesian components. Thus, any distortion or error occurring in the conversion process can be cancelled. This even includes misalignments and nonlinearities in the quadrature modulator. For the purpose of adaptation, the amplifier output signal is synchronously demodulated and compared with the input signal. The suggested adaptation process is quite simple and is performed while transmitting. Actually, it is the time-discrete equivalent to the Cartesian feedback system described in Section 2.2. As such, the phase of the feedback signal has to be correct for stable operation. Nagata presented results from an experimental system with a 16kHz modulation bandwidth, 128kHz sampling rate and 145MHz carrier frequency. Up to 26dB distortion suppression was obtained,

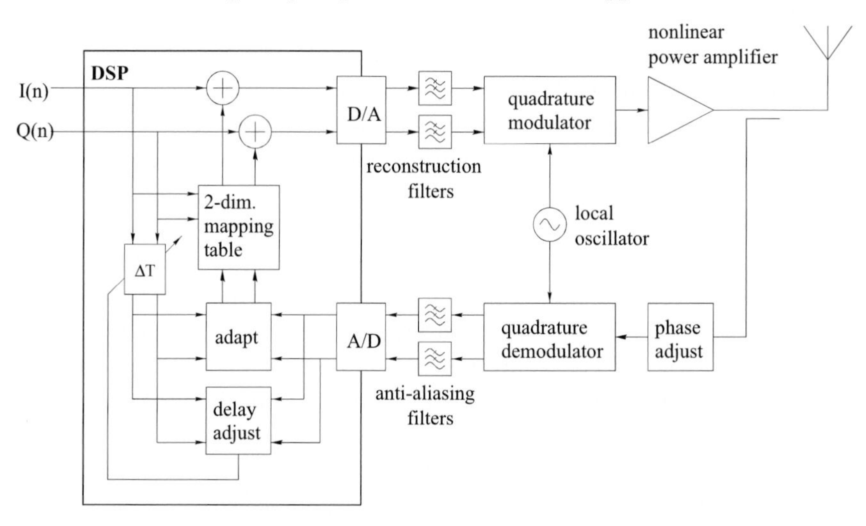

Figure 2-7. Mapping predistorter block diagram.

but the convergence time was quite long (10 seconds), which was primarily caused by the size of the table (2Mwords). The problem of correct sampling of the demodulated signal to be used in the adaptation process was addressed. It was shown by simulation that a sampling time deviation corresponding to 1% of the symbol time or more resulted in a significant degradation in performance. An automatic delay adjustment circuit that converged within 16 symbols was presented. Nagata reports that 2Mbits of memory were required in his design, along with 13000 gates in dedicated DSP hardware and up to 2W in the A/D and D/A converters. The total power consumption was therefore up to 4W, which is significantly more than the power output of the RF amplifier in most handportable equipment. It is therefore evident that the power-efficiency of an adaptive predistortion system will be poor until device technology is advanced sufficiently to enable the power consumption of the linearizer to become a small fraction of that of the RF power amplifier.

A similar system is presented by Minowa et al. [Min90], who investigated this technique in conjunction with amplifier back-off. Yet another prototype is presented in [Man94], where the computational burden was increased by using interpolation of the table entries. With this procedure, the table size could be reduced by a factor of 16 to 64 kwords. More memory efficient schemes have been developed (see Sections 2.3.3 and 2.3.4). In contrast to the mapping predistorter, these techniques can only compensate for phase-invariant non-linearities.

2.3.3 Complex Gain Predistortion

The major drawback of the mapping predistorter is the size of the two-dimensional table, which results in long adaptation times. However, if we restrict the predistorter to correct for nonlinearities in the amplifier alone, then a one-dimensional table will do, since the amplifier characteristic is a function of the input amplitude only. That is, such a table would approximate the inverse function of the amplifier nonlinearity with a finite number of table entries.

This approach has a table containing complex-valued gain factors given in Cartesian form, see Figure 2-8 [Cav90]. The address to the table is calculated as the squared magnitude of the input signal, which gives a uniform distribution of power in the table entries. Furthermore, the input signal is predistorted by a single complex multiplication. All in all, this leads to a substantially larger computational load compared with the mapping predistorter. The adaptation scheme of the complex gain predistorter is a multiplicative predistorter in contrast to the mapping predistorter, which is additive. This means that the complex gain predistorter is not sensitive to the phase of the feedback signal, as is the case for the mapping predistorter. A phase adjustment circuit in the feedback path is therefore not necessary for stable operation.

Since this solution assumes a phase-invariant characteristic it depends heavily on the use of perfect quadrature modulators and demodulators. Such modulators and demodulators are difficult and expensive to build. Methods for automatic adjustment of these errors have therefore been suggested in Chapter 3.

Cavers analyzed the effect of table size on adjacent channel interference. The effect of adaptation jitter was also investigated and it was found that the table size should be increased by 20% to account for this effect. The adaptation process was formulated as a root finding problem; the secant method was found to perform significantly faster than the linear scheme described by Nagata in [Nag89]. The convergence time was estimated to be less than 4ms for a 25kHz system with 64 table entries, provided that every table entry was accessed exactly 10 times each.

The LUT size affects linearly the speed of adaptation, so one way to increase the adaptation speed is to reduce the number of entries in LUT. The number of LUT entries, however, determines how closely the predistorter is able to follow the inverse function of the amplifier distortion as well as the maximal signal to noise ratio (SNR) available in the output of the amplifier. Both the precision and the entry number requirement can be alleviated with a nonuniform organization of the LUT entries [Cav97]. By organizing the entries in such a way that the entries do not overlap [Has01], the required precision can be reduced, while by organizing the entries according to the probability density of the amplitude values, the required number of LUT entries can be reduced [Muh00].

An experimental system based on complex gain predistortion is presented in [Wri92]. This system gave an up to 25dB improvement in linearity in nar-

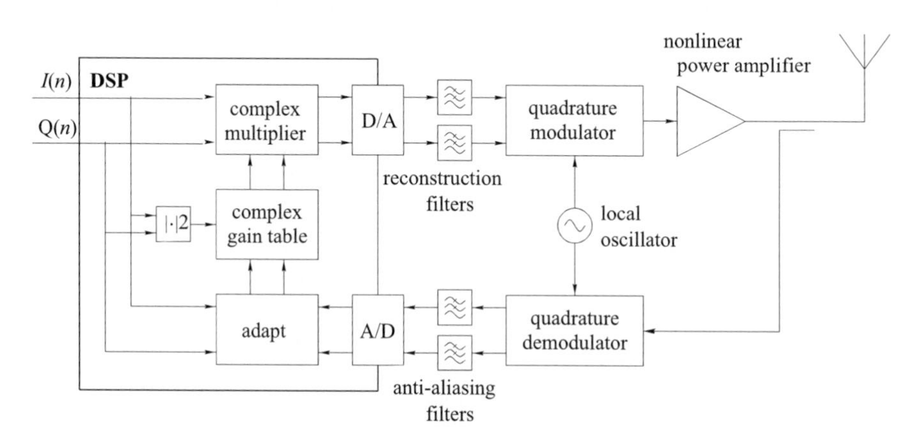

Figure 2-8. Complex gain table predistorter.

rowband operation (less than 1kHz). A recent advance in this area has been described by Sundström et al. [Sun96], [And97], in which a dedicated predistorter ASIC is outlined. The performance of this device was shown to be very good over a broad range of channel bandwidths (up to 300 kHz) and the use of an ASIC helped to reduce the linearizer power consumption to sensible levels (roughly one-tenth of that of an equivalent clock-rate DSP device, whilst providing around seven times the channel bandwidth).

2.3.4 Polar Predistortion

The approach suggested by Faulkner et al. [Fau94] uses two one-dimensional tables containing magnitude gain and phase rotation, respectively. The principle is illustrated in Figure 2-9 [Sun95]. From the input signal, given in Cartesian components, the amplitude is calculated and used as an address to the look-up table containing amplitude gain factors. The input signal is multiplied by the gain factor obtained from the table. The magnitude of the input signal is multiplied by the same gain factor and the result is used to address the second table containing the phase. Finally, phase rotation is performed by the amount obtained from the second table. This final step includes two additional look-ups to get sin() and cos() values for the rotation matrix. This technique requires more operations to predistort the signal compared with the complex gain predistorter.

Since the adaptation process is based on polar coordinates each iteration involves rectangular-to-polar (R/P) conversion of the input signal and the detected output signal. All in all, this leads to a substantially larger computational load compared with the mapping predistorter. The adaptation scheme of the polar predistorter is a multiplicative predistorter in contrast to the mapping predistorter which is additive. This means that the polar predistorter is not sensitive to the phase of the feedback signal, as is the case for the mapping predistorter. A phase adjustment circuit in the feedback path is

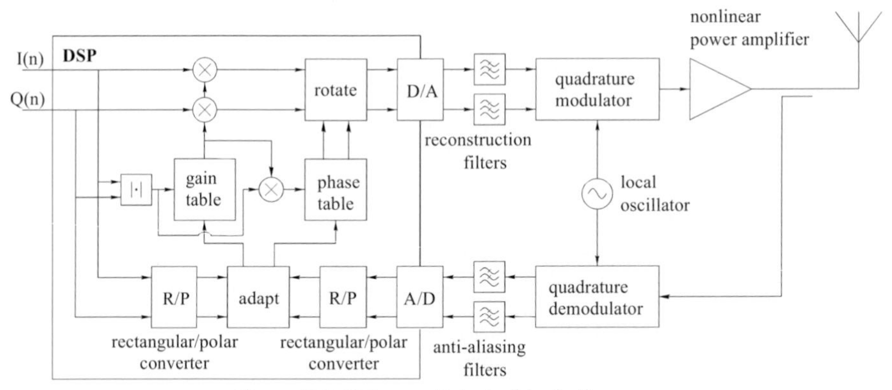

Figure 2-9. Polar predistorter block diagram.

therefore not necessary for stable operation.

This technique gives a considerable reduction in table size compared with the mapping predistorter [Fau94]. Faulkner reported that, as a compromise between convergence time and error it was found that 64 table entries gave the best result, provided that interpolation was used. This is four orders of magnitude less than the size of the mapping table. As a consequence, the adaptation time was estimated to be less than 10ms for a 25kHz system with a sampling rate eight times the symbol rate, assuming that all samples were used in the adaptation process. The experimental system gave up to 30dB IMD suppression when the system was exercised with a two-tone test [Fau94]. However, the modulation bandwidth was only 2 kHz.

The possibility of making the gain table a function of $|\cdot|^2$ (see Figure 2-8) instead of $|\cdot|$ (see Figure 2-9) was considered in [Fau94]. $|\cdot|^2$ is easy to calculate and causes the points of the table to be concentrated in the saturation region of the amplifier characteristics, while $|\cdot|$ concentrates more points into the turn-on region.

Since this solution assumes a phase-invariant characteristic, it depends heavily on the use of perfect quadrature modulators and demodulators. Therefore, methods of automatic adjustment of these errors have been suggested in Chapter 3.

2.3.5 RF-Predistortion Based on Vector Modulation

The RF-input/output predistortion operates completely independently of the circuit before the amplifier [Set00]. The RF-predistortion is implemented by transforming the analog PA input signal by analog means that are controlled by digital signals (Figure 2-10). The phase and envelope of the input and output signals are detected by analog means and A/D-converted. The result is fed to a DSP that retrieves phase and amplitude correction values corresponding to the input envelope value from a LUT. These values are D/A converted and used to control an analog AM/PM distorter that alters the RF-signal. The effect of RF-predistortion can be described with formula:

$$\begin{aligned} A_{PD}(|V_{IN}|)\, A_{PA}(|V_{PD}|) &= K \\ \phi_{PD}(|V_{IN}|) + \phi_{PA}(|V_{PD}|) &= \Delta\phi, \end{aligned} \tag{2.8}$$

where A and ϕ refer to the notation in Figure 2-10 and K and $\Delta\phi$ are constants.

The LUT is updated with a suitable algorithm on the basis of the phase and amplitude differences of the input and output signals. The signal can be detected with simple envelope and phase detectors [Ken01]. This configuration alleviates the problem of detector nonidealities that may limit the linearization ability of the baseband schemes, as both the output and input signals are detected with the same kind of detectors. Therefore, the possible error in the correction is only due to the mismatch between the detectors, which can be kept low with proper selection of the detectors. If the amplitude distortion is delayed compared to the phase distortion or vice versa, the upper and lower intermodulation sidebands become asymmetrical [Crip02]. This can be caused by, for example, different delays in the phase and amplitude feedback loops.

Another method used for the LUT update is the minimization of out of band distortion. This can be achieved by downconverting the signal to IF and bandpass filtering the signal to extract the out of band distortion. The power of the distortion is measured and the LUT is updated to minimize it. Another drawback is the slow convergence of the adaptation and complexity [Set00].

The limiting factors in the RF-predistortion mainly relate themselves to the analog parts of the circuit. The phase delay round the forward control loop reduces the effectiveness of the predistortion. The main sources for this delay are DSP latency, A/D and D/A converters latencies and the delays of the reconstruction and anti-aliasing filters. This can be alleviated to some extent by increasing the sampling rate of the digital parts and by adding an

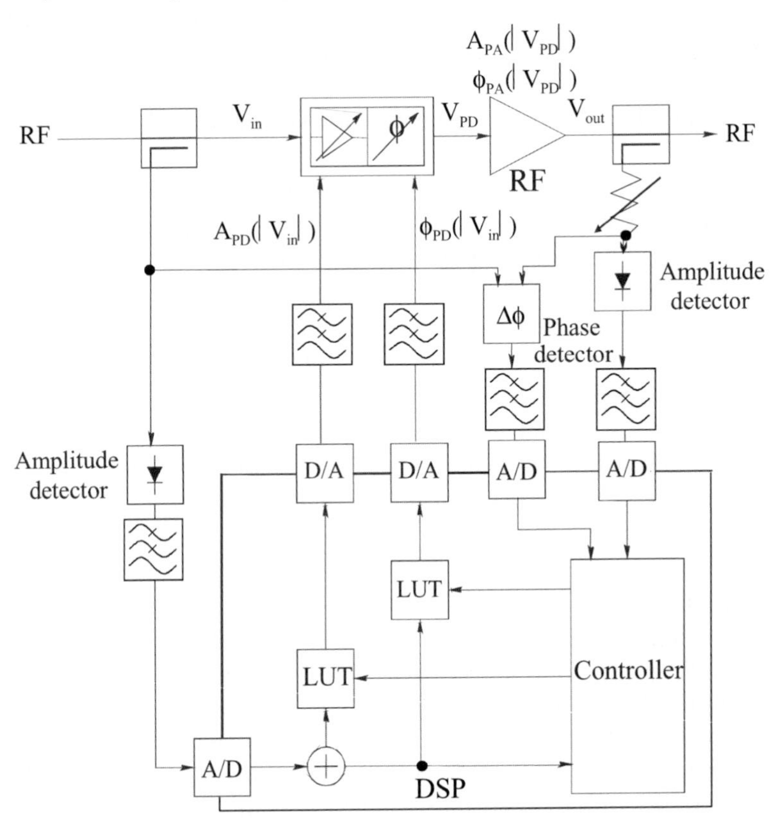

Figure 2-10. RF input/output adaptive digital linearizer.

analog delay element before the analog predistorter. The envelope detection process generates signals that have a higher bandwidth than the original modulation bandwidth of the RF signal. The RF predistortion based on vector modulation is therefore mainly suitable for narrowband systems. The wider bandwidth requires a higher sampling frequency in DSP. Furthermore, less delay for forward correction is tolerated. The operational bandwidths of the envelope detectors as well as the A/D and D/A converters limit the bandwidth of the correction signal. The use of the RF predistorter to amplify an EDGE signal improves the first ACPR by 20 dB and allows the signal to be produced with an efficiency of 30% [Ken01].

2.3.6 Data Predistorters

The technique operating at the transmitter is mainly related to distorting the data alphabet [Sal83], [Boy81]. Such predistorters compensate for the warping and clustering effects on the data constellation and therefore improve the eye openings at the maximum eye opening instants. They will thus improve the error magnitude of the amplifier output signal, but do not usually improve, intentionally, the adjacent channel performance or spectral purity of the transmitter output. They commonly employ look-up table based techniques to form the predistortion function and operate in a manner similar to their conventional equivalents. Biglieri proposed a data predistortion scheme with memory [Big88], which improves the system performance with respect to the memoryless predistorter. A key disadvantage with this form of the predistorter is that it is generally modulation format specific.

REFERENCES

[And97] P. Andreani, and L. Sundstrom, "Chip for Wideband Digital Predistortion RF Power Amplifier Linearisation," Electronics Letters," Vol. 33, No. 11, pp. 925-926, May 1997.

[Bat88] A. Bateman, D. M. Haines, and R. J. Wilkinson, "Linear Tranceiver Architectures," In Proceedings of the 38th IEEE Vehicular Technology Conference, May 1988, pp. 478-484.

[Bau93] R. M. Bauman, "Adaptive Feed-Forward System," U. S. Patent 4,389,618, June 21, 1993.

[Big88] E. Biglieri, S. Barberis, and M. Catena, "Analysis and Compensation of Nonlinearities in Digital Transmission Systems," IEEE J. Select. Areas Commun., Vol. 6, pp. 42–51, Jan. 1988.

[Bla28] H. S. Black, "Translating System," U. S. Patent 1,686,792, Oct. 9, 1928.

[Bla37] H. S. Black, "Wave Translation System," U. S. Patent 2,102,671, Dec. 21, 1937.
[Boy81] R. W. Boyd, and R. C. Davis, "Adaptive Predistortion Technique for Linearizing a Power Amplifier for Digital Data Systems," U. S. Patent 4,291,277, Sept. 22, 1981.
[Bri93] M. A. Briffa, and M. Faulkner, "Dynamically Biased Cartesian Feedback Linearization," In Proceedings of the 43th IEEE Vehicular Technology Conference, May 1993, pp. 672-675.
[Bro88] A. N. Brown, and V. Petrovic, "Phase Delay Compensation in HF Cartesian-Loop Transmitters," In Fourth International Conference on HF Communication Systems and Techniques, London, April 1988, pp. 200-204.
[Cav90] J. K. Cavers, "A Linearizing Predistorter with Fast Adaptation," IEEE Transactions on Vehicular Technology, Vol. 39, No. 4, Nov. 1990, pp 374-382.
[Cav95] J. K. Cavers, "Adaptation Behavior of a Feedforward Amplifier Linearizer," IEEE Transactions on Vehicular Technology, Vol. 44, No. 1, pp. 31-39, Feb. 1995.
[Cav97] J. Cavers, "Optimum Indexing in Predistorting Amplifier Linearizers," in Proc. of IEEE Vehicular Technology Conference, Phoenix, USA, May 1997, pp. 676–680.
[Cha91] R. H. Chapman, and W. J. Turney, "Feedforward Distortion Cancellation Circuit," U. S. Patent 5,051,704, Sept. 24, 1991.
[Crip02] S. C. Cripps, "Advanced Techniques in RF Power Amplifier Design," Norwood, MA: Artech House, 2002.
[Dix86] J. P. Dixon, "A Solid-State Amplifier with Feedforward Correction for Linear Single-Sideband Applications," In Proceedings of IEEE International Conference on Communications, Toronto, June 22-25, 1986, pp. 728-732.
[Fau94] M. Faulkner, and M. Johansson, "Adaptive Linearization using Predistortion - Experimental Results," IEEE Transactions on Vehicular Technology, Vol. 43, No. 2, pp. 323-332, May 1994.
[Gha93] M. Ghaderi, S. Kumar, and D. E. Dodds, "Fast Adaptive Predistortion Lineariser Using Polynomial Functions," Electronics Letters, Vol. 29, No. 17, pp. 1526-1528, August 1993.
[Gha94] M. Ghaderi, S. Kumar, and D. E. Dodds, "Adaptive Predistortion Lineariser Using Polynomial Functions," IEE Proceedings on Communication, Vol. 141, No. 2, pp. 49-55, Apr. 1994.
[Ham95] S.-A. El-Hamamsy, "Design of High-Efficiency RF Class-D Power Amplifier," IEEE Transactions on Power Electronics, Vol. 9, No. 3, pp. 297-308, May 1995.
[Has01] J. Hassani, and M. Kamarei, "A Flexible Method of LUT Indexing in Digital Predistortion Linearization of RF Power Amplifiers," in Proc of

IEEE International Symposium on Circuits and Systems, Sydney, Australia, Jun. 2001, pp. 53–56.

[Hil92] D. Hilborn, S. P. Stapleton, and J. K. Cavers, "An Adaptive Direct Conversion Transmitter," In Proceedings of the 42nd IEEE Vehicular Technology Conference, May 1992, pp. 764-767.

[Hil94] D. Hilborn, S. P. Stapleton, and J. K. Cavers, "An Adaptive Direct Conversion Transmitter," IEEE Transactions on Vehicular Technology, Vol. 43, No. 2, pp. 223-233, May 1994.

[Joh91] M. Johansson, and T. Mattsson, "Linearized High-Efficiency Power Amplifier for PCN," Electronics Letters, Vol. 27, No. 9, pp. 762-764, Apr. 1991.

[Joh93] M. Johansson, T. Mattsson, L. Sundström, and M. Faulkner, "Linearization of Multi-Carrier Power Amplifiers," In Proceedings of the 43rd IEEE Vehicular Technology Conference, May 1993, pp. 684-687.

[Joh94] M. Johansson, and L. Sundström, "Linearisation of RF Multicarrier Amplifiers Using Cartesian Feedback," Electronics Letters, Vol. 30, No. 14, pp. 1110-1112, July 1994.

[Joh95] M. Johansson, and M. Faulkner, "Linearization of Wideband RF Power Amplifiers," In Proceedings of Nordic Radio Symposium, Apr. 24-27, 1995, pp. 259-264.

[Kan98] H. W. Kang, Y. S. Cho, and D. H. Youn, "Adaptive Precompensation of Wiener Systems," IEEE Trans. Signal Processing, Vol. 46, No. 10, pp. 2825-2829, Oct. 1998.

[Ken00] P. B. Kenington, "High Linearity RF Amplifier Design," Norwood, MA: Artech House, 2000.

[Ken01] P. Kenington, M. Cope, R. Bennett, and J. Bishop, "A GSM-EDGE High Power Amplifier Utilising Digital Linearisation," in IEEE Microwave Symposium Digest, Phoenix, USA, May 2001, pp. 1517–1520.

[Ken91a] P. B. Kenington, R. J. Wilkinson, and J. D. Marvill, "Broadband Linear Amplifier Design for a PCN Base-Station," In Proceedings of the 41st IEEE Vehicular Technology Conference, St. Louis, May 19-22, 1991, pp. 155-160.

[Ken91b] P. B. Kenington, R. J. Wilkinson, and J. D. Marvill, "A Multi-Carrier Amplifier for Future Mobile Communications Systems," In Proceedings of the IEE 6th International Conference on Mobile Radio and Personal Communication, Warwick, Dec. 1991, pp. 151-156.

[Ken92] P. B. Kenington, "Efficiency of Feedforward Amplifiers," IEE Proceedings-G, Vol. 139, No. 5, pp. 591-593, Oct. 1992.

[Ken94] P. B. Kenington, M. A. Beach, A. Bateman, and J. P. McGeehan, "Apparatus and Method for Reducing Distortion in Amplification," U. S. Patent 5,334,946, Aug. 2, 1994.

[Kon93] K. Konstantinou, P. Gardner, and D. K. Paul, "Optimization Method for Feedforward Linearisation of Power Amplifiers," Electronics Letters, Vol. 29, No. 18, pp. 1633-1635, Sept. 1993.
[Mad99] K. Madani, "Reducing the Intermodulation Distortion in Multi-Carrier Microwave Power Amplifiers," IEEE EDMO 1999 Conference, pp. 153-157.
[Man94] A. Mansell, and A. Bateman, "Practical Implementation Issues for Adaptive Predistortion Transmitter Linearisation," In IEE Colloquium on 'Linear RF Amplifiers and Transmitters', (Digest No:1994/089), pp. 5/1-7, Apr. 1994.
[Mey74] R. G. Meyer, R. Eschenbach, and W. M. Edgerley, Jr., "A Wide-band Feed-Forward Amplifier," IEEE Journal on Solid-State Circuits, Vol. 9, No. 6, pp. 422-428, Dec. 1974.
[Min90] M. Minowa, M. Onoda, E. Fukuda, and Y. Daido, "Backoff Improvement of an 800-MHz GaAs FET Amplifier Using an Adaptive Nonlinear Distortion Canceller," In Proceedings of the 40th IEEE Vehicular Technology Conference, May 1990, pp. 542-546.
[Muh00] K. Muhonen, R. Krishnamoorthy, and M. Kahverad, "Look-up Table Techniques for Adaptive Digital Predistortion: A Development and Comparison," IEEE Trans. Vehicular Technology, Vol. 49, pp. 1995–2002, Sep. 2000.
[Mye94] D. P. Myer, "A Multicarrier Feed-Forward Amplifier Design," Microwave Journal, pp. 78-88, Oct. 1994.
[Nag89] Y. Nagata, "Linear Amplification Technique for Digital Mobile Communications," In Proceedings of the 39th IEEE Vehicular Technology Conference, May 1989, pp. 159-164.
[Nam83] J. Namiki, "An Automatically Controlled Predistorter for Multilevel Quadrature Amplitude Modulation," IEEE Trans. Commun., Vol. COM-31, No. 5, pp. 707-712, May 1983.
[Nan85] M. Nannicini, P. Magni, and F. Oggionni, "Temperature Controlled Predistortion Circuits for 64 QAM Microwave Power Amplifiers," in IEEE Microwave Symposium Digest, Vol. 85, June 1985, pp. 99-102.
[Nar91] S. Narahashi, and T. Nojima, "Extremely Low-Distortion Multicarrier Amplifier Self-Adjusting Feed-Forward (SAFF) Amplifier," In Proceedings of IEEE International Communication Conference, 1991, pp. 1485-1490.
[Noj84] T. Nojima, and Y. Okamoto, "Predistortion Nonlinear Compensator for Microwave SSB-AM System," Electronics and Communications in Japan, Vol. 67-B, No. 5, pp. 57-66, 1984.
[Noj85] T. Nojima, and T. Konno, "Cuber Predistortion Linearizer for Relay Equipment in 800 MHz band Land Mobile Telephone System," IEEE Transactions on Vehicular Technology, Vol. 34, No. 4, pp. 169-177, Nov. 1985.

[Noj85] T. Nojima, T. Murase, and N. Imai, "The Design of Predistortion Linearization Circuit for High-Level Modulation Radio Systems," In Proceedings from IEEE Global Telecommunications Conference, GLOBECOM '85, Dec. 1985, pp. 1466-1471.
[Obe91] M. G. Obermann, and J. F. Long, "Feed Forward Distortion Minimization Circuit," U. S. Patent 5,077,532, Dec. 31, 1991.
[Ohi92] Y. Ohishi, M. Minowa, E. Fukuda, and T. Takano, "Cartesian Feedback Amplifier with Soft Landing," In Proceedings of the third IEEE International Symposium on Personal, Indoor and Mobile Radio Communications, Oct. 1992, pp. 402-406.
[Olv85] T. E. Olver, "Adaptive Feedforward Cancellation Technique That is Effective in Reducing Amplifier Harmonic Distortion Products as well as Intermodulation Distortion Products," U. S. Patent 4,560,945, Dec. 24, 1985.
[Par94] K. J. Parsons, and P. B. Kenington, "The Efficiency of a Feedforward Amplifier with Delay Loss," IEEE Transactions on Vehicular Technology, Vol. 43, No. 2, pp. 407-412, May 1994.
[Pet83] V. Petrovic, "Reduction of Spurious Emission from Radio Transmitters by Means of Modulation Feedback," In IEE Conference on Radio Spectrum Conservation Techniques, Sept. 1983, pp. 44-49.
[Pro80] A. Prochazka, and R. Neumann, "Design of Wideband Feedforward Distribution Amplifier," IEEE Transactions on Cable Television, Vol. 5, No. 2, pp. 72-79, Apr. 1980.
[Sal81] A. Saleh, "Frequency-Independent and Frequency-Dependent Nonlinear Models of TWT Amplifiers," IEEE Trans. Commun., Vol. COM-29, pp.1715–1720, Nov. 1981.
[Sal83] A. Saleh, and J. Salz, "Adaptive Linearization of Power Amplifiers in Digital Radio Systems," Bell Syst. Tech. J., Vol. BSTJ-62, pp. 1019–1033, Apr. 1983.
[Sat83] G. Satoh, and T. Mizuno, "Impact of a New TWTA Linearizer upon QPSK/TDMA Transmission Performance," IEEE Journal on Selected Areas in Communications, Vol. 1, No. 1, pp. 39-45, Jan. 1983.
[Sch81] M. Schetzen, "Nonlinear System Modeling Based on the Wiener Theory," Proc. IEEE, Vol. 69, pp. 1557–1573, Dec. 1981.
[Se71b] H. Seidel, "A Microwave Feed-Forward Experiment," The Bell System Technical Journal, Vol. 50, No. 9, pp. 2879-2916, Nov. 1971.
[Sei71a] H. Seidel, "A Feedforward Experiment Applied to an L-4 Carrier System Amplifier," IEEE Transactions on Communication Technology, Vol. 19, No. 3, pp. 320-325, June 1971.
[Set00] Y. Seto, S. Mizuta, K. Oosaki, and Y. Akaiwa, "An Adaptive Predistortion Method for Linear Power Amplifiers," In Proceedings of the 51st IEEE Vehicular Technology Conference, May 2000, pp. 1889-1893.

[Smi98] A. M Smith, and J. K. Cavers, "A Wideband Architecture for Adaptive Feedforward Linearization," In Proceedings of the 48th IEEE Vehicular Technology Conference, May 1998, Vol. 3, pp. 2488-2492.
[Sta91] S. P. Stapleton, and J. K. Cavers, "A New Technique for Adaptation of Linearizing Predistorters," In Proceedings of the 41st IEEE Vehicular Technolgy Conference, May 1991, pp. 753-758.
[Sta92a] S. P. Stapleton, and F. C. Costescu, "An Adaptive Predistorter for a Power Amplifier Based on Adjacent Channel Emission," IEEE Transactions on Vehicular Technology, Vol. 41, No. 1, pp. 49-56, Feb. 1992.
[Sta92b] S. P. Stapleton, and F. C. Costescu, "An Adaptive Predistortion System," In Proceedings of the 42nd IEEE Vehicular Technology Conference, May 1992, pp. 690-693.
[Sta92c] S. P. Stapleton, G. S. Kandola, and J. K. Cavers, "Simulation and Analysis of an Adaptive Predistorter Utilizing a Complex Spectral Convolution," IEEE Transactions on Vehicular Technology, Vol. 41, No. 4, pp. 387-394, Nov. 1992.
[Ste88] R. D. Stewart, and F. F. Tusubira, "Feedforward Linearisation of 950MHz Amplifiers," IEE Proceedings, Vol. 135, Pt. H, No. 5, pp. 347-350, Oct. 1988.
[Sun95] L. Sundström, "Digital RF Power Amplifier Linearisers - Analysis and Design," PhD Thesis, Lund University, Aug. 1995.
[Sun96] L. Sundström, M. Faulkner, and M. Johansson, "Quantization Analysis and Design of a Digital Predistortion Linearizer for RF Power Amplifiers," IEEE Trans. Veh. Technol., Vol. 45, pp. 707-719, Nov. 1996.
[Van80] J. Vanderkooy, and S. P. Lipshitz, "Feedforward Error Correction in Power Amplifiers," Journal of the Audio Engineering Society, Vol. 28, No. 1/ 2, pp. 2-16, Jan./ Feb. 1980.
[Vuo01] J. H. K. Vuolevi, T. Rahkonen, and J. P. A. Manninen, "Measurement Technique for Characterizing Memory Effects in RF Power Amplifiers," IEEE Transactions on Vehicular Technology, Vol. 49, No. 8, pp. 1383-1389, Aug. 2001.
[Whi94] S. M. Whittle, "A Practical Cartesian Loop Transmitter for Narrowband Linear Modulation PMR Systems," In IEE Colloquium on 'Linear RF Amplifiers and Transmitters', (Digest No:1994/089), pp. 2 /1-5, April 1994.
[Wil92a] R. J. Wilkinson, and P. B. Kenington, "Specification of Error Amplifiers for Use in Feedforward Transmitters," IEE Proceedings-G, Vol. 139, No. 4, pp. 477-480, Aug. 1992.
[Wil92b] R. J. Wilkinson, P. B. Kenington, and J. D. Marvill, "Power Amplification Techniques for Linear TDMA Base Stations," In Proceedings from IEEE Global Communications Conference, GLOBECOM '92, part 1 (of 3), Dec. 6-9, 1992, pp. 74-78.

[Wri92] A. S. Wright, and W. G. Durtler, "Experimental Performance of an Adaptive Digital Linearized Power Amplifier," IEEE Transactions on Vehicular Technology, Vol. 41, No. 4, pp. 395-400, Nov. 1992.

Chapter 3

3. DIGITAL COMPENSATION METHODS FOR ANALOG I/Q MODULATOR ERRORS

The block diagram of the I/Q modulator and correction network is shown in Figure 3-1. The I/Q modulator has phase imbalance, gain imbalance and DC-offset errors. With a careful layout design, the errors can be minimized, but they can never be completely nulled. The phase imbalance is caused mainly by the local oscillator and phase shifter, which does not produce exactly 90 degrees of phase shift between the two channels. The gain imbalance is caused mainly by the mixers, which are not exactly balanced. The sideband suppression is a function of both the gain and phase imbalance (see Figure 1-2). The carrier suppression is a function of the DC offset between the in-phase (I) signal and the quadrature (Q) signal. This offset can be compensated by altering the DC offset of the input signals.

The effects of the modulator errors on the constellation are shown in Figure 3-2, Figure 3-3 and Figure 3-4. The phase imbalance results in rotation of the axes in the I/Q coordinates as shown in Figure 3-2. The gain imbalance results in distortion of the signal and transforms the circular constellation in the I/Q coordinates to elliptical ones, as shown in Figure 3-3. The DC-offset shifts all sample points the same amount in the same direction, as shown in Figure 3-4. Together these imbalances cause the bit-error rate of the connection to increase. In [Cav93], a more detailed discussion about analog IQ modulator errors and their effects on the communications can be found.

The effects of the analog I/Q modulator errors are modelled using matrix notation. The quadrature modulator (QM) output of Figure 3-1, when the nonidealities are taken into account, can be written as [Fau91]

$$v_q(t) = MV_m(t) + Ma, \tag{3.1}$$

where

$$M = \begin{pmatrix} \alpha\cos(\phi/2) & \beta\sin(\phi/2) \\ \alpha\sin(\phi/2) & \beta\cos(\phi/2) \end{pmatrix} \quad (3.2)$$

$$a = \begin{pmatrix} a_1 \\ a_2 \end{pmatrix}, \quad (3.3)$$

where α and β are gains of the I and Q channels, $V_m(t)$ is the quadrature input, and ϕ is the phase split between the channels. a_1 and a_2 are the dc offsets of the channels. The time invariant signals are expressed as the length of two vectors composed of the in-phase and quadrature phase parts of the signals. The representation in (3.2) is called a symmetric form of the error; (3.4) is the same error presented in asymmetric form. The symmetric model [Cav93] differs from the symmetric model presented in [Fau91] in that the phase imbalance is attributed completely to the Q channel. In this case

$$M = \begin{pmatrix} \alpha & \beta\sin(\phi) \\ 0 & \beta\cos(\phi) \end{pmatrix}. \quad (3.4)$$

To compensate the errors, correction terms should be added so that errors presented in (3.1), (3.2) and (3.3) are nulled. The corrected output for the quadrature modulator compensator (QMC) output should be (if the QMC is in series with the QM)

$$v_c(t) = Cv_d(t) + b = G\Phi v_d(t) + b, \quad (3.5)$$

where $v_d(t)$ is the compensator input, and

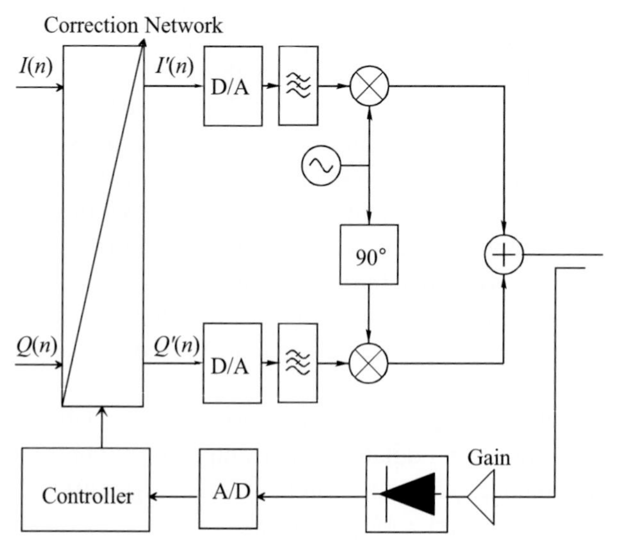

Figure 3-1. Quadrature modulator and correction network.

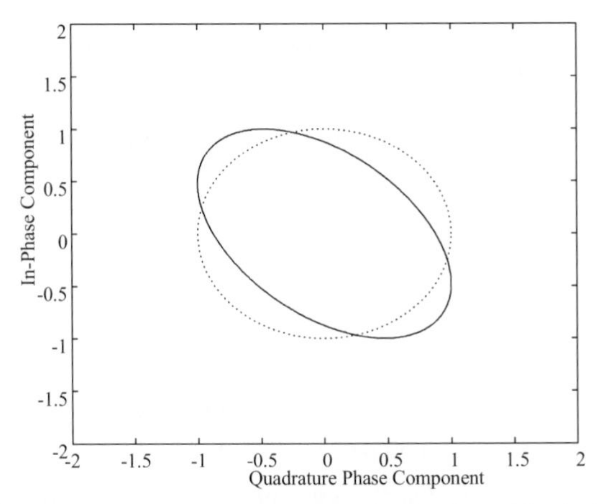

Figure 3-2. Phase imbalance causes rotation of axes in IQ plane. The dashed line represents original constellation.

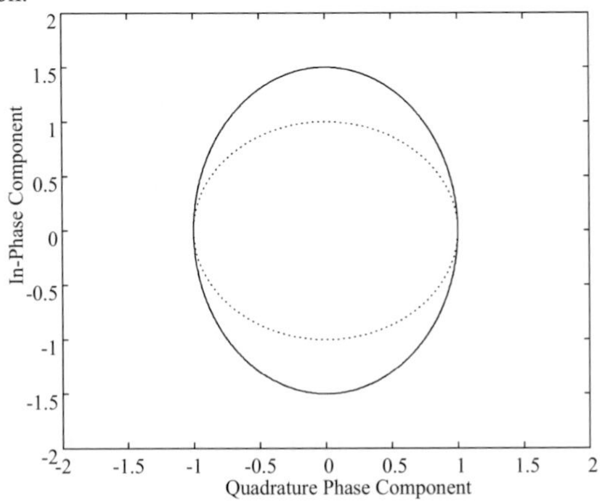

Figure 3-3. Gain imbalance transforms circular constellation into elliptical one.

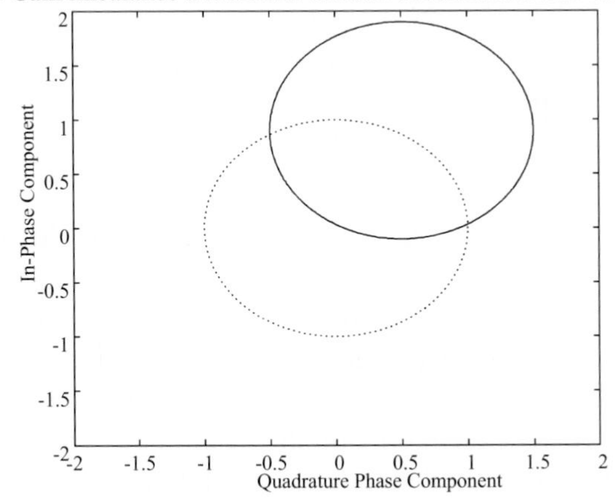

Figure 3-4. DC offset moves all points in one direction with the same amount.

$$C = G\Phi = M^{-1} = \frac{\sec\phi}{\alpha\beta}\begin{pmatrix} \beta\cos(\phi/2) & -\beta\sin(\phi/2) \\ -\alpha\sin(\phi/2) & \alpha\cos(\phi/2) \end{pmatrix}, \tag{3.6}$$

correspond to M in (3.2). The DC offset correction terms are

$$b = -a. \tag{3.7}$$

The gain imbalance is

$$G = \frac{\cos(\phi/2)}{\cos(\phi)}\begin{pmatrix} 1/\alpha & 0 \\ 0 & 1/\beta \end{pmatrix}, \tag{3.8}$$

and the phase imbalance is

$$\Phi = \begin{pmatrix} 1 & -\tan(\phi/2) \\ -\tan(\phi/2) & 1 \end{pmatrix}. \tag{3.9}$$

The compensator structure for the symmetrical model is presented in Figure 3-5. The $\cos(\phi/2)/\cos(\phi)$ in (3.8) is ignored in Figure 3-5.

With (3.4) as a starting point, C becomes

$$C = G\Phi = M^{-1} = \frac{\sec\phi}{\alpha\beta}\begin{pmatrix} \beta\cos(\phi) & -\beta\sin(\phi) \\ 0 & \alpha \end{pmatrix}. \tag{3.10}$$

The gain imbalance is

$$G = \begin{pmatrix} 1/\alpha & 0 \\ 0 & 1/\beta \end{pmatrix}, \tag{3.11}$$

and the phase imbalance is

$$\Phi = \begin{pmatrix} 1 & -\tan(\phi) \\ 0 & \sec(\phi) \end{pmatrix}. \tag{3.12}$$

The compensator structure for the asymmetrical model is presented in Figure 3-6.

In a wideband transmitter, the frequency dependence of I/Q mismatches must be taken into account. The wideband correction system is similar to the narrowband systems shown in Figure 3-5 and Figure 3-6 except that the correction terms are replaced by FIR filters, and extra delays are added due to the non-causality of the FIR filters [Pun00].

3.1 Quadrature Modulator Errors Compensation

The linearizer circuits such as mapping predistorter or Cartesian coordinate negative feedback linearizer can adapt for the quadrature modulator imperfections (see Section 2.3.2 or 2.2) [Cav90]. The linearizer types that assume the distortion to be dependent only on the magnitude of the signal and not its phase cannot be used for modulator error compensation (see Section 2.3.3).

There are analog and digital compensation methods for quadrature modulator errors. The majority of the articles suggest an adaptive way of compensating the I/Q modulator errors. There are two main ways to do this. One way is to use a particular training signal and the DSP processes the sampled information of this signal to cancel the changing nonlinearities. This method is used in [Fau91], [Fau92], [Loh93], [Cav91], [Cav93], [Jui94], [Hir96], [Cav97], [Pie01], [McV02]. The second way is to sample the data and compensate the errors as the data is transmitted or received without a test tone, as in [Hil94], [Cav97], [Mar00], [Ger92], [McV02]. A simple analog method of compensating the DC offset is presented in [Ris01], [Man9] and [Aki00].

The basic idea of adaptation is to have some samples of the data, and some sort of estimation of the imbalances. A digital signal processor calculates the correction terms to compensate for the I/Q modulator errors. In the literature, there are some algorithms used commonly when searching convergence of the correction terms. The choice of the adaptation algorithm leads to different convergence speeds and computational complexities in general, as well as to different remaining steady-state errors [Hay91]. These algorithms can be divided into gradient and surface fit (LMS, RLS) algorithms. The Newton-Rhapson algorithm used in [Loh93] is a gradient based. This method is susceptible to quantization and modeling errors [Cav97]. The least mean square (LMS) algorithm used in [Cav93], [Mar00] is simple to implement and is not computationally demanding. It is widely used because of its simple and effective nature. However, the problem with the LMS is its slow convergence. The recursive least squares (RLS) algorithm used in [Hil94] is computationally more demanding than the LMS, but the RLS method reaches convergence faster than the LMS.

The methods in [Fau91], [Fau92], [Loh93], [Cav93], [Hil94], [Cav97], [Mar00] use a feedback network as shown in Figure 3-1. At the QM (or PA) output is an envelope detector that operates at radio frequency. The detector has diode characteristics. This diode, however, has some transfer characteristics that must be taken into account. The correction terms vary with temperature and applied carrier frequency often making readjustment necessary.

3.1.1 Symmetric Compensation Method

In [Fau91], an automatic method of compensating modulator errors is proposed. The principle of the method can be seen in Figure 3-1. The method is based on baseband preconditioning, meaning that adjusting the baseband drive signals feeding the mixers compensates the errors. The output signal is measured with a diode detector and converted to digital form. Then, using this digital information, DSP searches compensation terms that are inputted to the correction network. It is possible to keep all the corrections independent of each other by applying them in the correct order [Fau91]. In [Fau91], the DC offset is cancelled first. The differential gain errors should be compensated second, and the phase errors last.

The compensation method starts by initializing the correction circuit. Then test vectors are applied to the I and Q inputs, and the adjustments begin. First, carrier leak is corrected by zeroing the drive signals. The DC compensation terms in (3.7) are adjusted by, for example, a one-dimensional search. This happens by holding a_1 constant and adjusting a_2 so that detector output is minimized. When local minimum is found, a_2 is held constant and a_1 is adjusted to find minimum. This is repeated until optimum is found [Fau91].

The gain error in (3.8) is compensated using the test vectors ($I = A$, $Q = 0$ and $I = 0$, $Q = A$) applied to the channels, and measuring the outputs. The ratio of the output measurements is the gain mismatch. An iterative approach is used in [Fau91], thus avoiding the need for a calibrated detector. The gain of the channel with the larger output is decreased until the amplitudes of both channels are equal.

When channels are balanced, the phase error in (3.9) is compensated. The system is fed with a constant amplitude phasor ($I = \cos(2\pi ft)$, $Q = \sin(2\pi ft)$). The cross-term $\tan(\phi/2)$ of Figure 3-5 is adjusted iteratively so that the I and Q amplitudes become equal, and the output circular, as it is in an ideal case.

The 900 MHz quadrature modulator consisted of a 90° phase splitter, a combiner and two diode ring mixers fed at a +7dBm RF drive level [Fau91]. The uncorrected modulator had an overall differential phase error 8°, a differential gain of 0.5 dB, and carrier leak of -20dBm. After correction the above errors improved to 0.4°, 0.02 dB and -63dBm, respectively.

The algorithm in [Fau91] can be modified so that the test vectors are not needed. The trajectory of the incoming modulation is sampled. This is not,

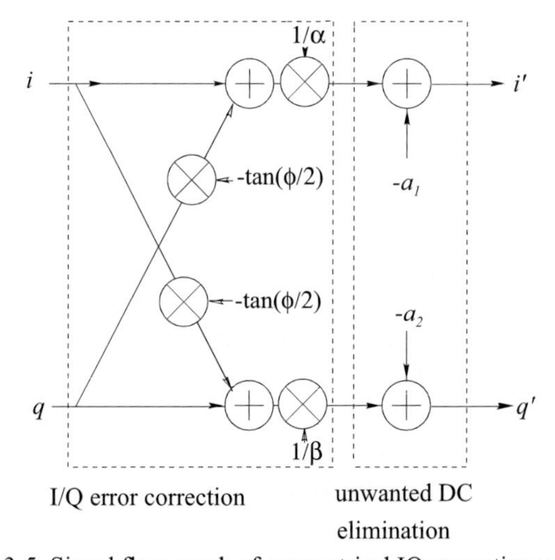

Figure 3-5. Signal flow graph of symmetrical IQ correction network.

however, tested in [Fau91].

A similar symmetric compensation method is used in [Loh93], but the Newton-Rhapson algorithm is used instead for acquiring the compensation terms. The controller stores the values inputted to the quadrature modulator and the values read from the envelope detector. Five sets of values are obtained for α, β (gains for both channels), a_1, a_2 (dc-offsets) and ϕ (phase imbalance between channels). Thus five equations are obtained for five unknown variables at different time instants. When the impairment values are known, both channels are predistorted and new values are measured; the impairment values are estimated as before. This compensation method, however, needs a perfect knowledge of the envelope detector parameters, thus limiting the usage of this method [Cav97].

In [Cav97], a symmetric model for both quadrature modulator (QM) and quadrature modulator compensator (QMC) are used. However, some improvements are made compared to previous methods. The channel imbalance and offset are measured simultaneously. The method reduces sensitivity to quantization noise in the feedback path, and thus coarse quantizers can be used.

3.1.2 Partial Correction of Mixer Nonlinearity in Quadrature Modulators

A byproduct of the baseband correction techniques [Fau91] in the previous section is that the nonlinearity effects are made worse. The differential phase correction often increases the magnitude of the drive signals. The DC correction term moves the mixer bias point, which causes the positive part of the waveform to suffer more compression than the negative part, so the system

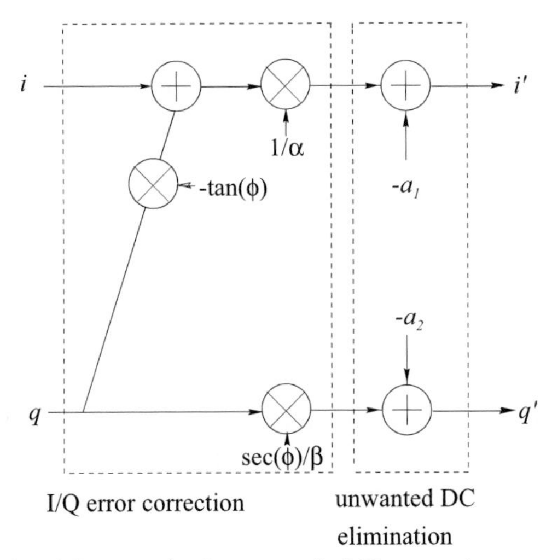

Figure 3-6. Signal flow graph of asymmetrical IQ correction network [Cav93].

is no longer balanced. This causes the growth of even order intermodulation products and the re-emergence of the carrier leak term with an amplitude that is dependent on the applied level of the signal.

In [Fau92], it was shown how the baseband correction of linear errors can be extended to include partial compensation of the third order nonlinearities in the mixer. An 8 dB improvement in mixer IM3 was obtained on a 900 MHz system.

3.1.3 Asymmetric Compensation Method

The method in [Cav93] is an improved version of the method presented in [Fau91]. The improvement suggests that a more economical asymmetric form of compensator that results in lower computational load and faster convergence should be used. The symmetrical correction network in Figure 3-5 requires 4 multipliers and 4 adders, and asymmetric in Figure 3-6 needs 3 multipliers and 3 adders [Cav93]. Moreover, a simplification is made and division with α is neglected in (3.11), since it has no substantial effect on the results, it produces only scale change of a few percentage points. This therefore reduces the number of the multipliers to two. The adaptation algorithm itself has only to acquire four remaining imbalance compensation terms in (3.10). In [Cav93], the LMS algorithm is used, providing an effective way of achieving convergence.

The DC offset is eliminated by setting $I = 0$, $Q = 0$ and adjusting a to minimize the envelope detector output. A detailed description of this can be found in [Cav93], but the principle is that the DC correction term a is obtained in two steps. First, 4 measurements are made in the output with different values of a, and a correction term is obtained from these measurements. The article suggests that the sum of squared errors between the measurements and the fitted surface of the input signal is minimized. Second, the measurements are made with the current optimal correction terms. The correction term is updated and the measurements are repeated until convergence is achieved. In practice, about eight measurements should be enough to obtain the terms.

The gain and phase imbalance compensation is achieved next. In asymmetric compensator gain and phase compensation, terms must be acquired simultaneously; [Cav93] provides an algorithm to do so. Four test signals are used with the same amplitude, but with four different phases. The adaptation is complete when all four phases give the same output from the envelope detector. The details are presented in [Cav93]. Fewer than eight iterations are needed to reach a steady state, and each iteration needs four power measurements at the envelope detector output.

The adaptation method in [Hil94] uses the same principles presented in [Cav93], but it concentrates on the adjacent channel power minimization. The QM and PA produces unwanted components on the radio frequencies and these components may affect the data transmission on adjacent channels. The method adjusts the correction coefficients in a way that the adjacent' channel power is minimized. However, the RLS algorithm used requires more memory and computational capacity to achieve good results, and thus modern digital signal processors must be used and if LUT based predistortion is used the method is far too slow [Cav97].

3.1.4 Digital Precompensation Method without Training Signal

The training signal is fed to the QM input and then read out from the QM (or PA) output, and changed again to digital form. The training signals are generated by the transmit modem and they are perfectly known. However, when using training signals, the compensation must be performed so that the training data is not transmitted. This can be achieved in, for example, the manufacturing/testing phase, when the transmitter is tested before being sold to the customers. Testing with training signals can safely be carried out when the transmitter (phone etc.) is turned on and a certain amount of time spent on the training signals before the device starts transmitting data. If the training signals are used in the middle of transmission, it must be ensured that the data is not transmitted.

[Cav97] suggests compensation methods with and without training signals. The problem with the technique without a training signal is that the loop-delay of the feedback loop is unknown. The modulator and the envelope detector, as well as the A/D converters are not infinitely fast and thus a delay is added for the signal. There are, however, compensation methods for loop-delay; [Cav97] suggests a way that does not need extra computation for the time delay parameters. Compensation from random data results in slower convergence than with training signals, but it can be accomplished in the middle of transmission without affecting the transmission. It can thus be used all the time and it can adapt to rapid changes in the temperature or other variables.

The precompensation method presented in [Mar00] does not need a training signal. The patent in [Xav01] is similar to this method. In [Mar00], for simplicity, DC, gain and phase imbalance are treated separately, but the adaptation procedure can be made concurrently. The gain of the modulator-amplifier-detector chain must be known exactly in order to achieve perfect adaptation in [Mar00]. [Mar00] uses the LMS algorithm that produces convergence in 2000-2500 iterations. However, the author accepts that the physical system is not as good as simulated, since some analog effects limit the accuracy of the compensation values.

REFERENCES

[Aki00] N. Akira, and H. Oshiroda, "Wide Band IQ Splitter Device and Proofreading Method," Japanese Patent 2000151731, April 30, 2000.
[Cav90] J. K. Cavers, "A Linearizing Predistorter with Fast Adaptation," IEEE Transactions on Vehicular Technology, Vol. 39, No. 4, Nov. 1990, pp 374-382.
[Cav91] J. K. Cavers, and M. Liao, "Adaptive Compensation for Imbalance and Offset Losses in Direct Conversion Transceivers," IEEE Vehicular Technology Conference, 1991, pp. 578-583.
[Cav93] J. K. Cavers, and M. Liao, "Adaptive Compensation Methods for Imbalance and Offset Losses in Direct Conversion Transceivers," IEEE Transactions on Vehicular Technology, Vol. 42, No. 4, Nov. 1993, pp. 581-588.
[Cav97] J. K. Cavers, "New Methods for Adaptation of Quadrature Modulators and Demodulators in Amplifier Linearization Circuits," IEEE Transactions on Vehicular Technology, Vol. 46, No. 3, Aug. 1997, pp. 707-716.
[Fau91] M. Faulkner, T. Mattsson, and W. Yates, "Automatic Adjustment of Quadrature Modulators," Electronic Letters, Vol. 27, No. 3, pp. 214-216, Jan. 1991.
[Fau92] M. Faulkner, and M. Johansson, "Correction of Mixer Nonlinearity in Quadrature Modulators," Electronic Letters, Vol. 28, No. 3, pp. 293-295, Jan. 1992.
[Ger92] K. Gerlach, "The Effect of I,Q Mismatch Errors on Adaptive Cancellation," IEEE Transactions on Aerospace and Electronic Systems, Vol. 28, No. 3, July 1992, pp. 729-740.
[Ger97] K. Gerlach, and M. J. Steiner, "An Adaptive Matched Filter that Compensates for I,Q Mismatch Errors," IEEE Transactions on Signal Processing, Vol. 45, No. 12, Dec. 1997, pp. 3104-3107.
[Gla98] J. P. F. Glas, "Digital I/Q Imbalance Compensation in a Low-IF Receiver," IEEE Global Telecommunications Conference, Vol. 3, 1998, pp. 1461-1466.
[Hay91] S. J. Haykin, "Adaptive Filter Theory," Englewood Cliffs, NJ: Prentice-Hall, 1991.
[Hid00] K. Hideaki, and N. Akira, "Wide Band IQ Splitting Apparatus and Calibration Method Therefor," European Patent Application 0984288, March 8, 2000.
[Hil92] D. Hilborn, S. P. Stapleton, and J. K. Cavers, "An Adaptive Direct Conversion Transmitter," IEEE Vehicular Technology Conference, Vol. 2, 1992, pp. 764-767.

[Hil94] D. S. Hilborn, S. P. Stapleton, and J. K. Cavers, "An Adaptive Direct Conversion Transmitter," IEEE Transactions on Vehicular Technology, Vol. 43, No. 2, May 1994, pp. 223-233.
[Hir96] M. Hironobu, "IQ Automatic Adjustment Circuit," Japanese Patent 8204771, Aug. 9, 1996.
[Jui94] N. Juichi, "PSK Modulation Signal Evaluation Device and IQ Origin Offset Detector for PSK Modulation Signal," Japanese Patent 6224952, Aug. 12, 1994.
[Loh93] A. Lohtia, P. A. Goud, and C. G. Englefield, "Digital Technique for Compensating for Analog Quadrature Modulator/Demodulator Impairments," IEEE Pacific Rim Conference on Communications Computers and Signal Processing, Vol. 2, 1993, pp. 447-450.
[Man99] H. Manabu, "Quadrature Modulation Circuit," Japanese Patent 11017752, Jan. 22, 1999.
[Man96] A. R. Mansell, and A. Bateman, "Transmitter Linearization Using Composite Modulation Feedback," Electronic Letters, Vol. 32, No. 23, Nov. 1996, pp. 2120-2121.
[Mar00] R. Marchesani, "Digital Precompensation of Imperfections in Quadrature Modulators," IEEE Transactions on Communications, Vol. 48, No. 4, Apr. 2000, pp. 552-556.
[McV02] J. D. McVey, "Modulation System Having on-line IQ Calibration," U. S. Patent 6,421,397, July 16, 2002.
[Pie01] B. Jean-Pierre, "A Method for Controlling the Transmitter Part of a Radio Transceiver and a Corresponding Radio Transceiver," European Patent Application 1154580, Nov. 14, 2001.
[Pun00] K-.P Pun, J. E. Franca, and C. Azeredo-Leme, "Wideband Digital Correction of I and Q Mismatch in Quadrature Radio Receivers," in Proc. IEEE International Symposium on Circuits and Systems (ISCAS), May 2000, pp. 661-664.
[Ris01] M. Rishi, and M. P. Stroet, "Quadrature Modulator with Set-and-Forget Carrier Leakage Compensation," U. S. Patent 6,169,463, Jan. 2, 2001.
[Xav01] C. Jean-Xavier, and M. Rossano, "Quadrature Modulator Imbalance Estimator and Modulator Stage Using It," U. S. Patent 6,208,698, Mar. 27, 2001.

Chapter 4

4. DIRECT DIGITAL SYNTHESIZERS

In this chapter, the operation of the direct digital synthesizer is described first. It is simple to add modulation capabilities to the DDS, because the DDS is a digital signal processing device. The digital quadrature modulator architectures, where the DDS generates in-phase and quadrature carriers, are also reviewed.

4.1 Conventional Direct Digital Synthesizer

The direct digital synthesizer (DDS) is shown in a simplified form in Figure 4-1. The direct digital frequency synthesizer (DDFS) or numerically controlled oscillator (NCO) is also widely used to define this circuit. The DDS has the following basic blocks: a phase accumulator, a phase to amplitude converter (conventionally a sine ROM), a digital to analog converter and a filter [Tie71], [Web72], [Gor75], [Bra81]. The phase accumulator consists of a j-bit frequency register that stores a digital phase increment word followed by a j-bit full adder and a phase register. The digital input phase increment word is entered in the frequency register. At each clock pulse, this data is added to the data previously held in the phase register. The phase increment word represents a phase angle step that is added to the previous value at each $1/f_s$ seconds to produce a linearly increasing digital value. The phase value is generated using the modulo 2^j overflowing property of a j-bit phase accumulator. The rate of the overflows is the output frequency

$$f_{out} = \frac{\Delta P f_s}{2^j} \quad \forall \quad f_{out} \le \frac{f_s}{2}, \tag{4.1}$$

where ΔP is the phase increment word, j is the number of phase accumulator bits, f_s is the sampling frequency and f_{out} is the output frequency. The constraint in (4.1) comes from the sampling theorem. The phase increment word in (4.1) is an integer, therefore the frequency resolution is found by setting $\Delta P = 1$

$$\Delta f = \frac{f_s}{2^j}. \tag{4.2}$$

The read only memory (ROM) is a sine look-up table, which converts the digital phase information into the values of a sine wave. In the ideal case with no phase and amplitude quantization, the output sequence is given by

$$\sin(2\pi \frac{P(n)}{2^j}), \tag{4.3}$$

where $P(n)$ is a (the j-bit) phase register value (at the nth clock period). The numerical period of the phase accumulator output sequence is defined as the minimum value of Pe for which $P(n) = P(n+Pe)$ for all n. The numerical period of the phase accumulator output sequence (in clock cycles) is

$$Pe = \frac{2^j}{\mathrm{GCD}(\Delta P, 2^j)}, \tag{4.4}$$

where GCD $(\Delta P, 2^j)$ represents the greatest common divisor of ΔP and 2^j. The numerical period of the sequence samples recalled from the sine ROM

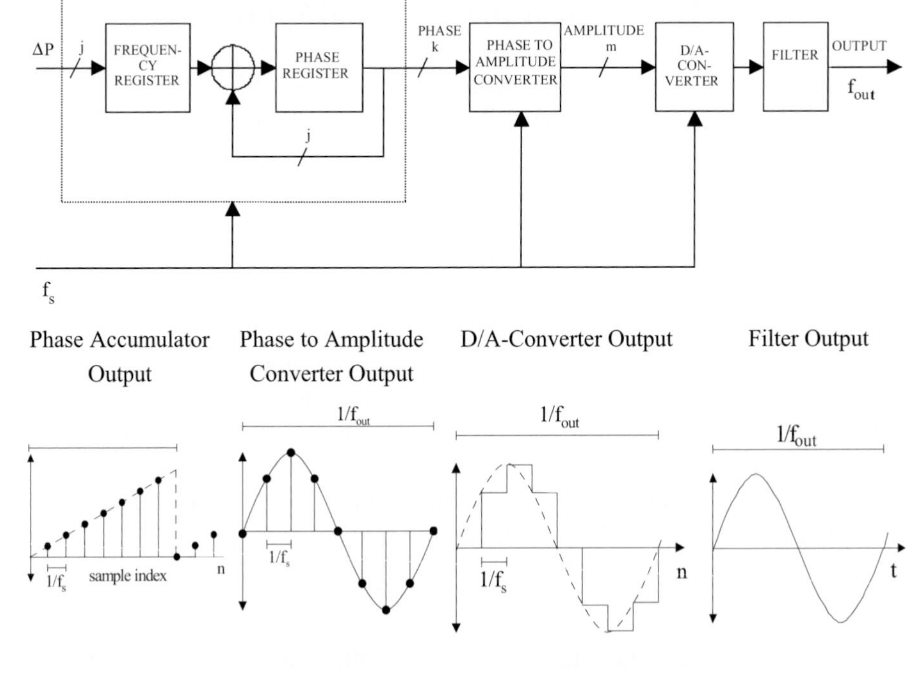

Figure 4-1. Simplified block diagram of the direct digital synthesizer, and the signal flow in the DDS.

will have the same value as the numerical period of the sequence generated by the phase accumulator [Dut78], [Nic87]. The spectrum of the output waveform of the DDS prior to a digital-to-analog conversion is therefore characterized by a discrete spectrum consisting of *Pe* points. The ROM output is presented to the D/A-converter, which develops a quantitized analog sine wave. The D/A-converter output spectrum contains frequencies $nf_s \pm f_{out}$, where $n = 0, 1, \ldots$etc. The amplitudes of these components are weighted by a function

$$\sin c(\pi \frac{f}{f_s}). \qquad (4.5)$$

This effect can be corrected by an inverse $\mathrm{sinc}(\pi f/f_s)$ filter. The filter that is after the D/A converter removes the high frequency sampling components and provides a pure sine wave output. As the DDS generates frequencies close to $f_s/2$, the first image $(f_s - f_{out})$ becomes more difficult to filter. This results in a narrower transition band for the filter. The complexity of the filter is determined by the width of the transition band. In order to keep the filter simple, the DDS operation is therefore limited to less than 35 percent of the sampling frequency.

4.2 Pulse Output DDS

The pulse output DDS is the simplest DDS type. It has only a phase accumulator. The MSB or carry output signal of the phase accumulator is used as an output. The average frequency of the DDS is obtained from (4.1). As long as ΔP divides into 2^j, the output is periodic and smooth (see column 3 in Table 4-1), but all other cases create jitter. The output can change its state only at the clock rate. If the desired output frequency is not a factor (a divider) of 2^j,

Table 4-1. For an accumulator of 3 bits (j=3) controlled with an input of $\Delta P = 3$ and $\Delta P = 2$.

Accumulator output $\Delta P = 3$ and $j = 3$	Carry output	Accumulator output $\Delta P = 2$ and $j = 3$	Carry output
000 (0)	1 Cycle begins	000 (0)	1 Cycle begins
011 (3)	0	010 (2)	0
110 (6)	0	100 (4)	0
001 (1)	1	110 (6)	0
100 (4)	0	000 (0)	1
111 (7)	0	010 (2)	0
010 (2)	1	100 (4)	0
101 (5)	0	110 (6)	0
000 (0)	1	000 (0)	1

then a phase error is created between the ideal and the actual output. This phase error will increase (or decrease) until it reaches a full clock period, at which time it returns to zero and starts to build up again (see column 1 in Table 4-1). Ideally, we would like to generate a transition every 8/3 = 2.6667 cycles (see column 1 in Table 4-1), but this is not possible because the phase accumulator can generate a transition only at integer multiples of the clock period. After the first transition the error is -1/3 clock period (we should transit after 2.6667 clocks, and we transit after 3), and after the second it is -2/3 clock period (we should transit after 5.33, and we do after 6). There is a clear relation between the error and the parameters ΔP (phase increment word) and C (phase accumulator output at the moment of carry generation). The error is exactly $-C/\Delta P$.

By using a digital delay generator (see Figure 4-2), the carry output is first connected to a logic circuit that calculates first the ratio $-C/\Delta P$ and delays the carry signal [Nuy90], [Gol96], [Rah01], [Nos01a], [Nos01b], [Ric01]. The negative delay must be converted into a positive delay, which is $1 - C/\Delta P$, $\Delta P > C$ in all cases (the carry overflow error can never be as large as ΔP).

It is assumed that the delay-time of the whole delay line meets exactly $T_s = 1/f_s$. For the delay components inside the delay line there are B outputs with delay times

$$T_{cv} = \frac{yT_s}{B}, \quad \text{where } y = 1,...,B, \tag{4.6}$$

and where $B = 2^b$ in this case. The applied delay (yT_s/B) is a multiple of the delay components inside the delay line, and the positive delay time is

$$T_s - \frac{CT_s}{\Delta P}. \tag{4.7}$$

From these two equations, it is easy to solve y (digital delay generator input)

$$y = \left[B(1 - \frac{C}{\Delta P}) \right], \tag{4.8}$$

where [] denotes truncation to integer values. The division $C/\Delta P$ requires a lot of hardware.

The delay compensation could also be implemented with other tech-

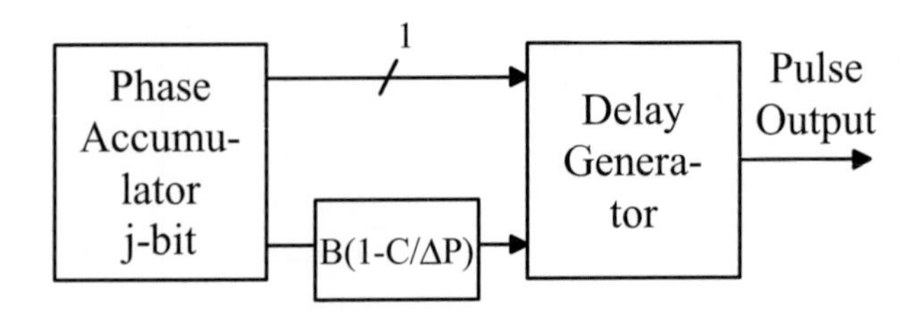

Figure 4-2. Single bit DDS with a digital delay generator.

niques [Nak97], [Mey98], [Nie98].

4.3 DDS Architecture for Modulation Capability

It is simple to add modulation capabilities to the DDS, because the DDS is a digital signal processing device. In the DDS it is possible to modulate numerically all three waveform parameters

$$s(n) = A(n)\sin(2\pi(\Delta P(n) + P(n))), \tag{4.9}$$

where $A(n)$ is the amplitude modulation, $\Delta P(n)$ is the frequency modulation, and $P(n)$ is the phase modulation. All known modulation techniques use one, two or all three basic modulation types simultaneously. Consequently, any known waveform can be synthesized from these three basic types within the Nyquist band limitations in the DDS. Figure 4-3 shows a block diagram of a basic DDS system with all three basic modulations in place [Zav88a], [McC88]. The frequency modulation is made possible by placing an adder before the phase accumulator. The phase modulation requires an adder between the phase accumulator and the phase to amplitude converter. The amplitude modulation is implemented by inserting a multiplier between the phase to amplitude converter and the D/A-converter. The multiplier adjusts the digital amplitude word applied to the D/A-converter. Also, with some D/A-converters, it is possible to provide an accurate analog amplitude control by varying a control voltage [Sta94].

4.4 QAM Modulator

The block diagram of the conventional QAM modulator with quadrature outputs is shown in Figure 4-4. The output of the QAM modulator is

$$\begin{aligned} I_{out}(n) &= I(n)\cos(\omega_{out}\, n) + Q(n)\sin(\omega_{out}\, n) \\ Q_{out}(n) &= Q(n)\cos(\omega_{out}\, n) - I(n)\sin(\omega_{out}\, n), \end{aligned} \tag{4.10}$$

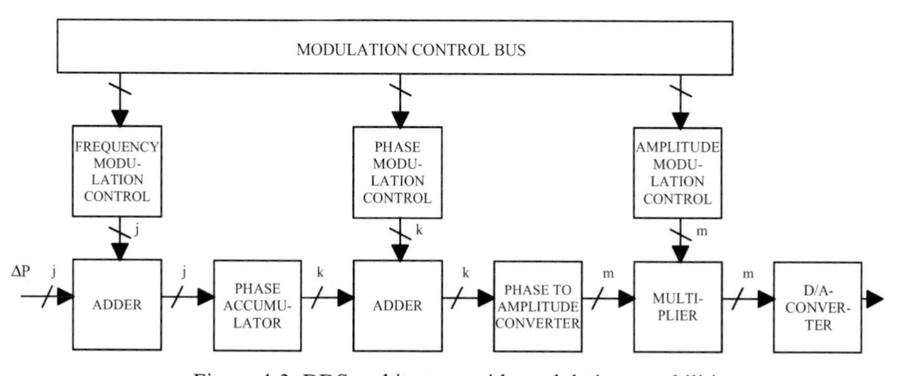

Figure 4-3. DDS architecture with modulation capabilities.

where ω_{out} is the quadrature direct digital synthesizer (QDDS) output frequency from (4.1), and $I(n)$, $Q(n)$ are pulse shaped and interpolated quadrature data symbols [Tan95a]. The direct implementation of (4.10) requires a total of four real multiplications and two real additions, as shown in Figure 4-4. However, we can reformulate (4.10) as [Wen95]

$$\begin{aligned} I_{out}(n) &= I(n)(\cos(\omega_{out} n) + \sin(\omega_{out} n)) + \sin(\omega_{out} n)(Q(n) - I(n)) \\ Q_{out}(n) &= Q(n)(\cos(\omega_{out} n) - \sin(\omega_{out} n)) + \sin(\omega_{out} n)(Q(n) - I(n)). \end{aligned} \quad (4.11)$$

The term $\sin(\omega_{out})$ $(Q(n) - I(n))$ appears in the both outputs. Therefore, the total number of real multiplications is reduced to three. This, however, is at the expense of having five real additions.

The quadrature amplitude modulation (QAM) could be also performed

$$\begin{aligned} I_{out}(n) &= I(n)\cos(\omega_{out} n) + Q(n)\sin(\omega_{out} n) \\ &= A(n)\cos(\omega_{out} n - P(n)), \\ Q_{out}(n) &= Q(n)\cos(\omega_{out} n) - I(n)\sin(\omega_{out} n) \\ &= A(n)\sin(\omega_{out} n - P(n)), \end{aligned} \quad (4.12)$$

and $A(n) = \sqrt{I(n)^2 + Q(n)^2}$, $P(n) = \arctan(Q(n)/I(n))$,

where arctan is the four quadrant arctangent of the quadrature phase data ($Q(n)$) and in-phase ($I(n)$). If the in-phase output only is needed, this requires one adder for phase modulation ($P(n)$) before and a multiplier for amplitude modulation ($A(n)$) after the phase to amplitude converter (sine) according to (4.12) (Figure 14-1) If the quadrature output is needed, this requires one adder before and two multipliers after the phase to amplitude converter (quadrature output) according to (4.12) (Figure 14-2).

The CORDIC algorithm also performs quadrature modulation (see Chap-

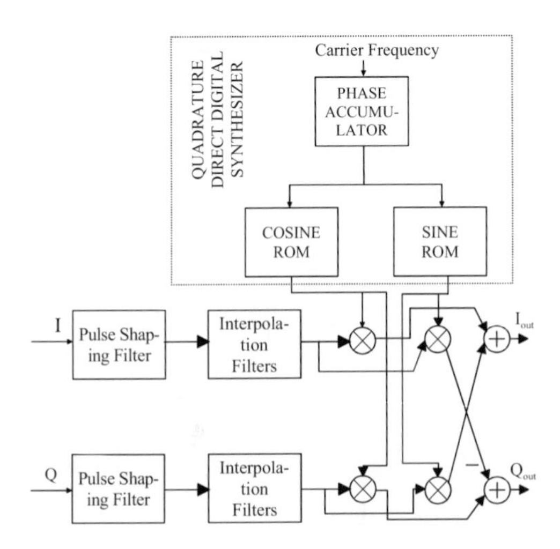

Figure 4-4. QAM modulator with quadrature outputs.

ter 6). The QAM modulation can be performed by two cascaded rotation stages, a coarse rotation stage and a fine rotation stage (see Section 6.4) [Ahm89], [Cur01], [Tor03], [Son03].

If the quadrature output is not needed, then the complex multiplier can be replaced with the two multipliers and an adder, as shown in Figure 4-5. The output of the quadrature modulator is shown in Figure 4-5

$$O(n) = I(n)\cos(2\pi n f_{out} / f_s) + Q(n)\sin(2\pi n f_{out} / f_s), \tag{4.13}$$

where f_s and f_{out} are the sampling and output frequency, and $I(n)$, $Q(n)$ are interpolated in-phase and quadrature-phase carriers, respectively. Solutions for the quadrature modulation, when f_{out}/f_s is equal to 0, 1/2, 1/3, 1/4, -1/4, 1/6 and 1/8, are listed in Table 4-2. In order to implement a multiplier-free quadrature modulation, we must therefore require that for all values of n, the quadrature modulator output (4.13) leads either to +1, -1, or 0 for sine and cosine terms in Table 4-2. It should, however, be noted that the multiplications by 0.5, -0.5, $1/\sqrt{2}$, $-1/\sqrt{2}$, $\sqrt{3}/2$, $-\sqrt{3}/2$ in other cases are also relatively simple to implement with hardware shifts and canonic signed digit (CSD) multipliers (see section 11.6). The multiplications could also be included in the filter coefficients of I and Q filter branches. Furthermore, if we require that either the sine or cosine term is zero in (4.13) at every n, then, for each output value, one of the in-phase or quadrature-phase part needs to be processed, which reduces hardware [Dar70], [Won91]. This results in oscillator samples of $\cos(n\pi/2)$ and $\sin(n\pi/2)$ that have the trivial values of 1, 0, -1, 0, …, thus eliminating the need for high-speed digital multipliers and adders to implement the mixing functions. Furthermore, since half of the

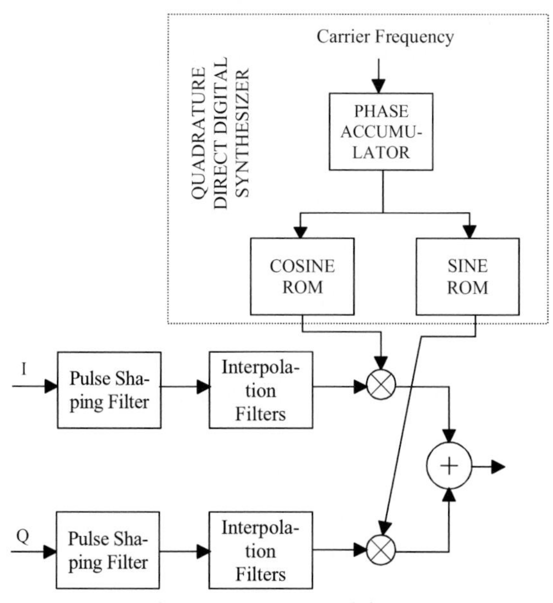

Figure 4-5. QAM modulator.

cosine and sine oscillator samples are zero, only a single interpolate-by-4 transmit filter (over-sampling factor of 4) can be used to process the data in both I and Q rails of the modulator, as shown in Figure 4-6. Thus the only hardware operating at $4/T_{sym}$ in the modulator is a 4:1 multiplexer at the output.

The pulse shaping filter in Figure 4-4 and Figure 4-5 reduces the transmitted signal bandwidth, which results in an increase in the number of available channels, and at the same time it maintains low adjacent channel interferences. Furthermore, it minimizes the inter-symbol interference (ISI). The interpolation filters increase the sampling rate and reject the extra images of the signal spectrum resulting from the interpolation operations. The quadrature DDS and the complex multiplier translate the signal spectrum from the baseband into the IF.

All the waveshaping is performed by the lower sample rate in the pulse shaping filter, and it is essential that the interpolation filters do not introduce any additional magnitude and phase distortion [Cro83]. The interpolation filters are usually implemented with multirate FIR structures. There exists a well-known multirate architecture for implementing very narrow-band FIR filters, which consists of a programmable coefficient FIR filter, and half band filters (see Section 11.9) followed by the cascaded-integrator-comb (CIC) structure (see Section 11.10) [Hog81]. A re-sampler allows the use of sampling rates that are not multiples of the symbol rates (see Chapter 12). It

Table 4-2
Solutions for Multiplier-Free Quadrature Modulation

f_{out}/f_s	$\cos(2\pi n\, f_{out}/f_s)$	$\sin(2\pi n\, f_{out}/f_s)$	$\cos(2\pi n\, f_{out}/f_s)\, I(n) + \sin(2\pi n\, f_{out}/f_s)\, Q(n)$
0	··· 1 ···	··· 0 ···	··· $I(n)$ ···
½	··· 1, -1 ···	··· 0, 0 ···	··· $I(n)$, $-I(n)$ ···
1/3	··· 1, -0.5, -0.5 ···	··· 0, $\frac{\sqrt{3}}{2}$, $-\frac{\sqrt{3}}{2}$ ···	··· $I(n)$, $-I(n)/2+Q(n)\frac{\sqrt{3}}{2}$, $-I(n)/2-Q(n)\frac{\sqrt{3}}{2}$ ···
¼	··· 1, 0, -1, 0 ···	··· 0, 1, 0, -1 ···	··· $I(n)$, $Q(n)$, $-I(n)$, $-Q(n)$ ···
-1/4	··· 1, 0, -1, 0 ···	··· 0, -1, 0, 1 ···	··· $I(n)$, $-Q(n)$, $-I(n)$, $Q(n)$ ···
1/6	··· 1, 0.5, -0.5, -1, -0.5, 0.5 ···	··· 0, $\frac{\sqrt{3}}{2}$, $\frac{\sqrt{3}}{2}$, 0, $-\frac{\sqrt{3}}{2}$, $-\frac{\sqrt{3}}{2}$ ···	··· $I(n)$, $I(n)/2+Q(n)\frac{\sqrt{3}}{2}$, $-I(n)/2+Q(n)\frac{\sqrt{3}}{2}$, $-I(n)$, $-I(n)/2-Q(n)\frac{\sqrt{3}}{2}$, $I(n)/2-Q(n)\frac{\sqrt{3}}{2}$ ···
1/8	··· 1, $\frac{1}{\sqrt{2}}$, 0, $-\frac{1}{\sqrt{2}}$, -1, $-\frac{1}{\sqrt{2}}$, 0, $\frac{1}{\sqrt{2}}$ ···	··· 0, $\frac{1}{\sqrt{2}}$, 1, $\frac{1}{\sqrt{2}}$, 0, $-\frac{1}{\sqrt{2}}$, -1, $-\frac{1}{\sqrt{2}}$ ···	··· $I(n)$, $I(n)/\sqrt{2}+Q(n)/\sqrt{2}$, $Q(n)$, $-I(n)/\sqrt{2}+Q(n)/\sqrt{2}$, $-I(n)$, $-I(n)/\sqrt{2}-Q(n)/\sqrt{2}$, $-Q(n)$, $I(n)/\sqrt{2}-Q(n)/\sqrt{2}$, ···

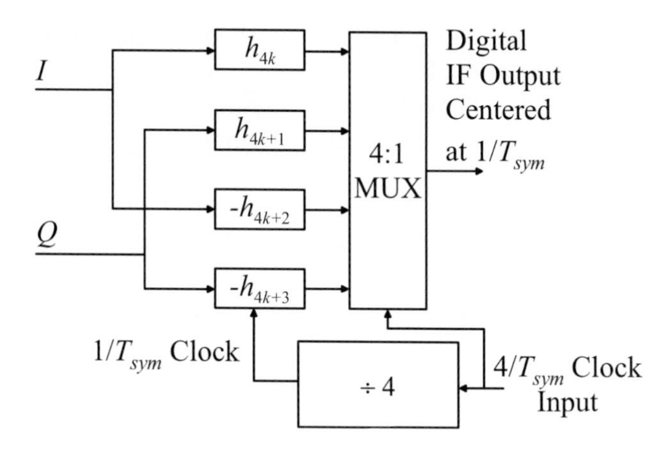

Figure 4-6. Simplified digital modulator.

enables the transmission of signals having different symbol rates.

REFERENCES

[Ahm89] H. M. Ahmed, "Efficient Elementary Function Generation with Multipliers," in Proc. 9th Symposium on Computer Arithmetic, Santa Monica, CA, USA, Sept. 1989, pp. 52-59.

[Bra81] A. L. Bramble, "Direct Digital Frequency Synthesis," in Proc. 35th Annu. Frequency Contr. Symp., USERACOM (Ft. Monmouth, NJ), May 1981, pp. 406-414.

[Cro83] R. E. Crochiere, and L. R. Rabiner, "Multirate Digital Signal Processing," Englewood Cliffs, NJ: Prentice-Hall, 1983.

[Cur01] F. Curticăpean, and J. Niittylahti, "An Improved Digital Quadrature Frequency Down-Converter Architecture," Asilomar Conf. on Signals, Syst. and Comput., Nov. 2001, pp. II-1318-1321.

[Dar70] S. Darlington, "On Digital Single-Sideband Modulators," IEEE Trans. Circuit Theory, Vol. 17, No. 3, pp. 409-414, Aug. 1970.

[Dut78] D. L. Duttweiler, and D. G. Messerschmitt, "Analysis of Digitally Generated Sinusoids with Application to A/D and D/A Converter Testing," IEEE Trans. on Commun., Vol. COM-26, pp. 669-675, May 1978.

[Gol96] B. G. Goldberg, "Digital Techniques in Frequency Synthesis," McGraw-Hill, 1996.

[Gor75] J. Gorski-Popiel, "Frequency-Synthesis: Techniques and Applications," IEEE Press, New York, USA, 1975.

[Hog81] E. B. Hogenauer, "An Economical Class of Digital Filters for Decimation and Interpolation," IEEE Trans. Acoust., Speech, Signal Process., Vol. ASSP-29, No. 2, pp. 155-162, Apr. 1981.

[McC88] E. W. McCune, Jr., "Number Controlled Modulated Oscillator," U. S. Patent 4,746,880, May 24, 1988.

[Mey98] U. Meyer-Bäse, S. Wolf, and F. Taylor, "Accumulator-Synthesizer with Error-Compensation," IEEE Trans. Circuits and Systems II, Vol. 45, No. 7, pp. 885 - 890, July 1998.

[Nak97] T. Nakagawa, and H. Nosaka, "A Direct Digital Synthesizer with Interpolation Circuits," IEEE J. of Solid State Circuits, Vol. 32, No. 5, pp. 766-770, May 1997.

[Nic87] H. T. Nicholas, and H. Samueli, "An Analysis of the Output Spectrum of Direct Digital Frequency Synthesizers in the Presence of Phase-Accumulator Truncation," in Proc. 41st Annu. Frequency Contr. Symp., June 1987, pp. 495-502.

[Nie98] J. Nieznanski, "An Alternative Approach to the ROM-less Direct Digital Synthesis," IEEE J. of Solid State Circuits, Vol. 33, No. 1, pp. 169 - 170, Jan. 1998.

[Nos01a] H. Nosaka, Y. Yamaguchi, and M. Muraguchi, "A Non-Binary Direct Digital Synthesizer with an Extended Phase Accumulator," IEEE Transactions on Ultrasonics, Ferroelectrics and Frequency Control, Vol. 48, pp. 293-298, Jan. 2001.

[Nos01b] H. Nosaka, Y. Yamaguchi, A. Yamagishi, H. Fukuyama, and M. Muraguchi, "A Low-Power Direct Digital Synthesizer Using a Self-Adjusting Phase-Interpolation Technique," IEEE J. Solid-State Circuits, Vol. 36, No. 8, pp. 1281-1285, Aug. 2001.

[Nuy90] P. Nuytkens, and P. V. Broekhoven, "Digital Frequency Synthesizer," U. S. Pat. 4,933,890, June 12, 1990.

[Rah01] T. Rahkonen, H. Eksymä, A. Mäntyniemi, and H. Repo, "A DDS Synthesizer with Digital Time Domain Interpolator," Analog Integrated Circuits and Signal Processing, Vol. 27, No. 1-2, pp. 111-118, Apr. 2001.

[Ric01] R. Richter, and H. J Jentschel, "A Virtual Clock Enhancement Method for DDS Using an Analog Delay Line," IEEE J. Solid-State Circuits, Vol. 36, No. 7, pp. 1158-1161, July 2001.

[Son03] Y. Song and B. Kim, "A 330-MHz 15-b Quadrature Digital Synthesizer/Mixer in 0.25 μm CMOS," in Proc. 29th European Solid-State Circuits Conference, Estoril, Portugal, Sept. 2003, pp. 513-516.

[Sta94] Stanford Telecom STEL-2173 Data Sheet, the DDS Handbook, Fourth Edition, and Triquint Semiconductor TQ6122 Data Sheet, TQS Digital Communications and Signal Processing, 1994.

[Tan95a] L. K. Tan, and H. Samueli, "A 200 MHz Quadrature Digital Synthesizer/Mixer in 0.8 μm CMOS," IEEE J. of Solid State Circuits, Vol. 30, No. 3, pp. 193-200, Mar. 1995.

[Tie71] J. Tierney, C. Rader, and B. Gold, "A Digital Frequency Synthesizer," IEEE Trans. Audio and Electroacoust., Vol. AU-19, pp. 48-57, Mar. 1971.
[Tor03] A. Torosyan, D. Fu, and A. N. Willson, Jr., "A 300-MHz Quadrature Direct Digital Synthesizer/Mixer in 0.25µm CMOS," IEEE J. of Solid State Circuits, Vol. 38, No. 6, pp. 875-887, June 2003.
[Web72] J. A. Webb, "Digital Signal Generator Synthesizer," U. S. Patent 3,654,450, Apr. 4, 1972.
[Wen95] A. Wenzler, and E. Lüder, "New Structures for Complex Multipliers and their Noise Analysis," in Proc. IEEE International Symposium on Circuits and Systems (ISCAS), 1995, pp. 1,432-1,435.
[Won91] B. C. Wong, and H. Samueli, "A 200-MHz All-Digital QAM Modulator and Demodulator in 1.2-µm CMOS for Digital Radio Applications," IEEE J. Solid-State Circuits, Vol. 26, No. 12, pp. 1970-1979, Dec. 1991.
[Zav88a] R. J. Zavrel, Jr., "Digital Modulation Using the NCMO," RF Design, pp. 27-32, Mar. 1988.
[Zav88b] R. Zavrel, and E. W. McCune, "Low Spurious Techniques & Measurements for DDS Systems," in RF Expo East Proc., 1988, pp. 75-79.

Chapter 5

5. RECURSIVE OSCILLATORS

In this chapter, the operation of the digital recursive oscillators is described first. It is shown that it produces spurs as well as the desired output frequency. A coupled-form complex oscillator is also presented.

5.1 Direct-Form Oscillator

Figure 5.1 shows the signal flow graph of the well-known second-order direct-form feedback structure with state variables $x_1(n)$ and $x_2(n)$ [Gol69], [Fur75], [Har83], [Abu86a], [Gor85], [Asg93], [Pre94], [Ali97], [Ali01], [Tur03]. The corresponding difference equation for this system is given by

$$x_2(n+2) = \alpha\, x_2(n+1) - x_2(n). \tag{5.1}$$

The two state variables are related by

$$x_1(n) = x_2(n+1). \tag{5.2}$$

Solving the one-sided z transform of (5.1) for $x_2(n)$ leads to

$$X_2(z) = \frac{(z^2 - \alpha z)\, x_2(0) + z\, x_1(0)}{z^2 - \alpha z + 1}, \tag{5.3}$$

where $x_1(0)$ and $x_2(0)$ are the initial values of the state variables. Identifying the second state variable as the output variable

$$y(n) = x_2(n), \tag{5.4}$$

as shown in Figure 5.1, and choosing the denominator coefficient α to be

$$\alpha = 2\cos\theta_{out}, \qquad \theta_{out} = \omega_{out} T = 2\pi\, f_{out} / f_s, \tag{5.5}$$

with f_{out} being the oscillator frequency and f_s the sampling frequency, then, on choosing the initial values of the state variables to be

$$x_1(0) = A\cos\theta_{out}, \qquad x_2(0) = A, \tag{5.6}$$

we obtain from (5.3) a discrete-time sinusoidal function as the output signal:

$$Y(z) = \frac{A(z^2 - \cos\theta_{out}\, z)}{z^2 - 2\cos\theta_{out}\, z + 1}. \tag{5.7}$$

It has complex-conjugate poles at $p = \exp(\pm j\theta_{out})$, and a unit sample response

$$y(n) = A\cos(n\theta_{out}), \qquad n \geq 0. \tag{5.8}$$

Thus the impulse response of the second-order system with complex-conjugate poles on the unit circle is a sinusoidal waveform.

An arbitrary initial phase offset φ_0 can be realized [Fli92], namely,

$$y(n) = A\cos(\theta_{out} n + \varphi_0), \tag{5.9}$$

by choosing the initial values:

$$x_1(0) = A\cos(\theta_{out} + \varphi_0), \tag{5.10}$$

$$x_2(0) = A\cos(\varphi_0). \tag{5.11}$$

Thus, any real-valued sinusoidal oscillator signal can be generated by the second-order structure shown in Figure 5.1.

In fact, the difference equation in (5.1) can be obtained directly from the trigonometric identity

$$A\cos(\theta_{out}(n+2)) = A2\cos(\theta_{out})\cos(\theta_{out}(n+1)) - A\cos(\theta_{out} n), \tag{5.12}$$

where, by definition, $x_2(n + 2) = A\cos(\cos\theta_{out}(n + 2))$.

The output sequence $y(n)$ of the ideal oscillator is the sampled version of a pure sine wave. The angle θ_{out} represented by the oscillator coefficient is given by

$$\theta_{out} = 2\pi f_{out} / f_s, \tag{5.13}$$

where f_{out} is the desired output frequency. In an actual implementation, the multiplier coefficient $2\cos\theta_{out}$ is assumed to have $b + 2$ bits. In particular, 1

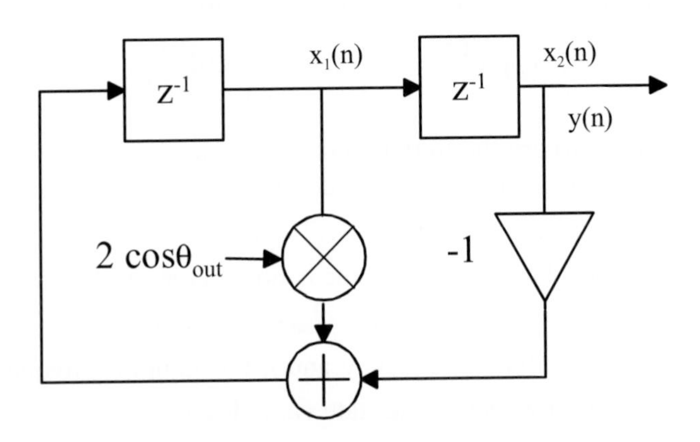

Figure 5.1. Recursive digital oscillator structure.

bit is for the sign, 1 bit for the integer part and b bits for the remaining fractional part in the fixed-point number representation. Then the largest value of the coefficient $2\cos\theta_{out}$ that can be represented is $(2 - 2^{-b})$. This value of the coefficient gives the smallest value of θ_{min} that can be implemented by the direct form digital oscillator using b bits

$$\theta_{min} = \cos^{-1}\left[\frac{1}{2}(2-2^{-b})\right]. \tag{5.14}$$

Therefore, the smallest frequency that the oscillator can generate is

$$f_{min} = \frac{\theta_{min}}{2\pi} f_s, \tag{5.15}$$

where f_s is the sampling frequency. It should be emphasized that the frequency resolution of the direct-form oscillator is a function of the wordlength of the multiplier coefficient. Therefore, to increase the frequency resolution, the wordlength of the multiplier coefficient must be increased. Because the poles generated by a quantized coefficient ($2\cos\theta_{out}$) are not uniformly distributed on the unit circle, the distance between two adjacent generated frequencies varies (the distance decreases as the generated frequency increases). The inability of the direct form digital oscillator to generate equally spaced frequencies is a disadvantage from the point of view of a communication system. A modified direct-form digital oscillator capable of generating almost equally spaced frequencies has been proposed in [Ali97]. However, it comes at a higher hardware cost as the solution in [Ali97] requires one additional multiplier and two integrators.

Every time when the frequency is changed, a new set of parameters have to be computed. For the direct-form digital oscillator it is rather difficult to keep the phase continuous when the frequency is changed. This is a disadvantage if compared with the DDS (Figure 4-1).

In this digital oscillator, besides the zero-input response $y(n)$ of the second-order system we get a zero-state response $y_{err}(n)$ due to the random sequence $e_2(n)$ acting as an input signal. From (5.1) we obtain

$$y(n+2) = \alpha\, y(n+1) - y(n) + e_2(n), \tag{5.16}$$

and by the z transformation

$$Y(z) = Y_{ideal}(z) + Y_{err}(z), \tag{5.17}$$

with $Y_{ideal}(z)$ derived from (5.7). The z transform of the output error $y_{err}(n)$ is given by

$$Y_{err}(z) = \frac{z^2 E_2(z) - z^2 e_2(0) - z e_2(1)}{z^2 - 2\cos\theta_{out}\, z + 1}, \tag{5.18}$$

with $E_2(z)$ being the z transform of the quantization error signal $e_2(n)$. Transforming $Y_{err}(z)$ back into the time domain results in an output error sequence

$$y_{err}(n) = \frac{1}{\sin\theta_{out}} \sum_{k=2}^{n} e_2(k)\sin(\theta_{out}(n-k+1)), \quad \text{for } n \geq 2, \tag{5.19}$$

where $e_2(0)$ and $e_2(1)$ are assumed to be zero. Equation (5.19) shows that the output error is inversely proportional to $\sin(\theta_{out})$, thus the output error increases with the decreasing digital oscillator frequency. The accumulation of the quantization errors degrades the spectral purity of the generated wave and can produce computation overflows. Such computation overflows can easily make the oscillator unstable.

The output quantization error can be reduced by an appropriate error feedback [Abu86b]. In addition to the error feedback, a periodic oscillator reset could be applied. In order to eliminate an infinite accumulation of errors, the direct-form oscillator could be reset to its initial states after N samples (K cycles) if the normalized frequency $\theta_{out}/2\pi$ equals the rational number K/N [Fur75]. This method is not effective for a large N, because the accumulated error will also be large. As a result, the signal spectrum will be degraded. Therefore, it is possible to reset the oscillator many times over one period (N) in [Fur75].

5.2 Coupled-Form Complex Oscillator

The coupled-form digital oscillator generates simultaneously sine and the cosine waves ($A\sin\theta_{out}n$ and $A\cos\theta_{out}n$) [Gol69], [Gor85], [Fli92], [Kro96], [Pro97], [Gra98], [Pal99], [Pal00], [Tur03] and can be obtained from the trigonometric formulae

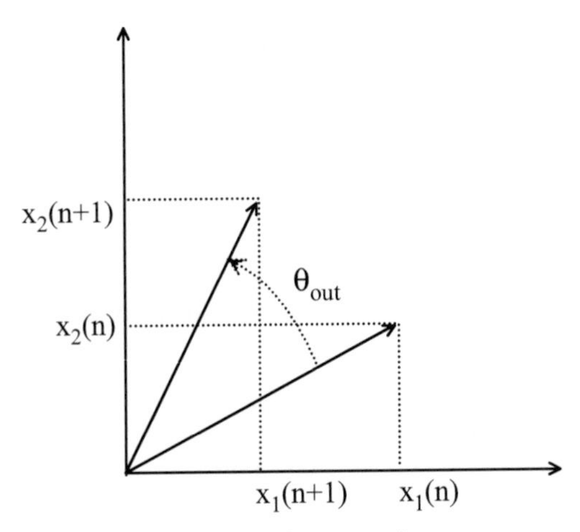

Figure 5.2. Vector rotation.

$$\begin{aligned} A\cos(\alpha+\beta) &= A\cos(\alpha)\cos(\beta) - A\sin(\alpha)\sin(\beta) \\ A\sin(\alpha+\beta) &= A\cos(\alpha)\sin(\beta) + A\sin(\alpha)\cos(\beta), \end{aligned} \tag{5.20}$$

where, by definition, $\alpha = n\theta_{out}$, $\beta = \theta_{out}$, and

$$\begin{aligned} x_1(n+1) &= A\cos((n+1)\theta_{out}) \\ x_2(n+1) &= A\sin((n+1)\theta_{out}). \end{aligned} \tag{5.21}$$

Thus we obtain the two coupled equations

$$\begin{aligned} x_1(n+1) &= x_1(n)\cos(\theta_{out}) - x_2(n)\sin(\theta_{out}) \\ x_2(n+1) &= x_1(n)\sin(\theta_{out}) + x_2(n)\cos(\theta_{out}), \end{aligned} \tag{5.22}$$

that perform a general rotational transform anti-clockwise with angle θ_{out}; the coordinates of a vector in Figure 5.2 transform from (x_1(n), $x_2(n)$) to (x_1(n+1), $x_2(n+1)$). The structure for the realization of the coupled-form oscillator is illustrated in Figure 5.3. This is a two-output system which is not driven by any input, but which requires the initial conditions $x_1(0) = A\cos(\theta_{out})$ and $x_2(0) = A\sin(\theta_{out})$ in order to begin its self-sustaining oscillations.

Obviously, the total hardware cost of the coupled-form digital oscillator (Figure 5.3) is significantly higher than the cost of a direct-form oscillator (Figure 5.1). The coupled-form complex oscillator in Figure 5.3 could be implemented by a complex multiplier [Kro96], [Gra98], [Pal99], [Pal00]. The drawback of this method is low maximum clock frequency, which is limited by the speed of the complex multiplier. Due to the recursion, regular pipelining cannot be utilized [Pal00]. The design in [Gra98] uses CORDIC algorithm to compute the initial values.

Changing the frequency of the sinusoids generated by the coupled-form oscillator is simpler than in the case of the direct-form oscillator, because

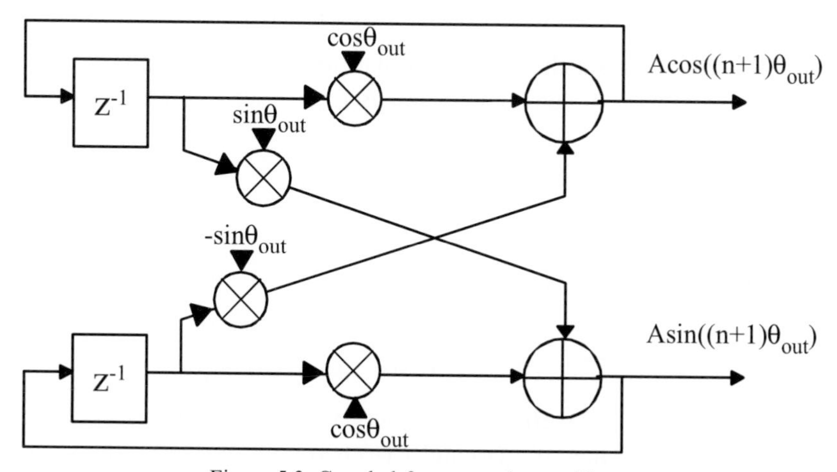

Figure 5.3. Coupled-form complex oscillator.

only the oscillator's coefficients have to be updated. Moreover, the change of frequency is phase continuous.

From the z transform of the state equations (5.22)

$$z\,X = V\,X, \quad X = [X_1(z)\,X_2(z)]^T \tag{5.23}$$

the characteristic equation

$$\begin{aligned} D(z) = |z\,I - V| &= \begin{vmatrix} z - \cos\theta_{out} & \sin\theta_{out} \\ -\sin\theta_{out} & z - \cos\theta_{out} \end{vmatrix} \\ &= z^2 - 2\cos\theta_{out}\,z + \cos^2\theta_{out} + \sin^2\theta_{out} \end{aligned} \tag{5.24}$$

can be derived. From (5.24), it is obvious that the eigenvalues (poles) are lying on the unit circle of the z plane. In a finite register length arithmetic, however, the eigenvalues almost never have exactly unit magnitude because the two coefficients $\cos(\Theta_{out})$ and $\sin(\Theta_{out})$ are realized separately. Thus we observe no stable limit cycle but a waveform with an increasing or decreasing amplitude. The stability of the oscillator is also influenced by the accumulation of the roundoff errors caused by finite wordlength arithmetic. Clearly, these negative effects can be reduced by increasing the internal wordlength of the oscillator. Unfortunately, this solution has a high negative impact on the total hardware cost. Moreover, the oscillator may become instable even if it happens after the higher number of oscillating cycles.

Because the poles of the coupled-form digital oscillator are conditioned by two coefficients, ($\cos(\Theta_{out})$ and $\sin(\Theta_{out})$), they are uniformly distributed on a rectangular grid. As a result, the frequency resolution is improved in comparison to that of a direct-form oscillator (especially at low frequencies).

Equation (5.22) shows that x_1 and x_2 will both be sinusoidal oscillations that are always in exact phase quadrature. Furthermore, if quantization effects are ignored, then, for any time n, the equality

$$x_1^2(n) + x_2^2(n) = x_1^2(0) + x_2^2(0) \tag{5.25}$$

holds. In order to reset the system so that, after every k iterations, the variables $x_1(n)$ and $x_2(n)$ are changed to satisfy (5.25), we can multiply both $x_1(n)$ and $x_2(n)$ by the factor

$$f(n) = \sqrt{\frac{x_1^2(0) + x_2^2(0)}{x_1^2(n) + x_2^2(n)}}. \tag{5.26}$$

Thus, for each k iterations of (5.22), we perform once the non-linear iteration

$$\begin{aligned} x_1(n+1) &= f(n)\left[x_1(n)\cos(\theta_{out}) - x_2(n)\sin(\theta_{out})\right] \\ x_2(n+1) &= f(n)\left[x_1(n)\cos(\theta_{out}) + x_2(n)\sin(\theta_{out})\right]. \end{aligned} \tag{5.27}$$

Execution of (5.27) effectively resets $x_1(n+1)$ and $x_2(n+1)$ so that (5.25) is satisfied. Thus, if $x_1(n)$ and $x_2(n)$ had both drifted by the same relative amount to a lower value, both would be raised, in one iteration cycle, to the value they would have had if no noise were present. If, however, $x_2(n)$ had drifted up and $x_1(n)$ down so that the sum of the squares were satisfied (5.25), then (5.27) would have no effect. Thus the drifts of phase are not compensated by (5.27). In [Fli92], the amplitude gain is controlled by changing the oscillator coefficients. If $f(n)$ is larger than one, then the coefficient-set generating poles outside the unit circle are utilized. If $f(n)$ is smaller than one, then the coefficient-set generating poles inside the unit circle are used for computing the next sample. By these means, the amplitude of the generated waves is kept within given limits. Other amplitude correction methods are presented in [Gra98], [Pal99], [Pal00]. In these methods, the feedback signal is saturated if overflows or underflows occur. The most efficient method of eliminating the infinite accumulation of errors of the coupled-form complex oscillator is to reset its initial states after N samples (K cycles) if the normalized frequency $\theta_{out}/2\pi$ equals the rational number K/N.

An increase in the frequency resolution requires that the word length of the whole complex oscillator is widened; however, in the case of the conventional direct digital synthesizer, it is only necessary to increase the phase accumulator word length (see (4.2)).

REFERENCES

[Abu86a] A. I. Abu-El-Haija, and M. M. Al-Ibrahim, "Digital Oscillator Having Low Sensitivity and Roundoff Errors," IEEE Trans. on Aerospace and Electronic Systems, Vol. AES-22, No. 1, pp. 23-32, Jan. 1986.

[Abu86b] A. I. Abu-El-Haija, and M. M. Al-Ibrahim, "Improving Performance of Digital Sinusoidal Oscillators by Means of Error-Feedback Circuits," IEEE Trans. Circuits Syst., Vol. CAS-33, pp. 373-380, Apr. 1986.

[Ali01] M. M. Al-Ibrahim, "A Multifrequency Range Digital Sinusoidal Oscillator with High Resolution and Uniform Frequency Spacing," IEEE Trans. on Circuits and Systems II, Vol. 48, No. 9, pp. 872-876, Sept. 2001.

[Ali97] M. M. Al-Ibrahim and A. M. Al-Khateeb, "Digital Sinusoidal Oscillator with Low and Uniform Frequency Spacing," IEE Proceedings on Circuits, Devices and Systems, Vol. 144, No. 3, pp. 185-189, June. 1997.

[Asg93] S. M. Asghar, and A. R. Linz, "Frequency Controlled Recursive Oscillator Having Sinusoidal Output," U. S. Patent 5,204,624, Apr. 20, 1993.

[Fli92] N. J. Fliege, and J. Wintermantel, "Complex Digital Oscillators and FSK Modulators," IEEE Trans. on Signal Processing, Vol. SP-40, No. 2, pp. 333-342, Feb. 1992.

[Fur75] K. Furuno, S. K. Mitra, K. Hirano, and Y. Ito, "Design of Digital Sinusoidal Oscillators with Absolute Periodicity," IEEE Trans. on Aerospace and Electronic Systems, Vol. AES-11, No. 6, pp. 1286-1298, Nov. 1975.

[Gol69] B. Gold, and C. M. Rader, "Digital Processing of Signals," New York: McGraw-Hill, 1969.
[Gor85] J. W. Gordon and J. O. Smith, "A Sine Generation Algorithm for VLSI Applications," in Proc. 1985 International Computer Music Conference (ICMC), Vancouver, 1985, pp. 165-168.
[Gra98] E. Grayver, and B. Daneshrad, "Direct Digital Frequency Synthesis Using a Modified CORDIC," Proc. IEEE Int. Symp. Circuits and Systems (ISCAS), Vol. 5, pp. 241–244, 1998.
[Har83] I. Hartimo, "Self-Sustained Stable Oscillations of Second Order Recursive Algorithms," in Proc. IEEE Int. Conference on Acoustics, Speech, and Signal Processing, USA, Apr. 1983, pp. 635-638.
[Kro96] B. W. Kroeger, and J. S. Baird "Numerically Controlled Oscillator with Complex Expotential Outputs Using Recursion Technique," U. S. Patent 5,517,535, May. 14, 1996.
[Pal00] K. I. Palomäki, J. Niittylahti, and V. Lehtinen, "A Pipelined Digital Frequency Synthesizer Based on Feedback," Proceedings of the 43rd IEEE Midwest Symposium on Circuits and Systems, Vol. 2, Aug. 2000 pp. 814-817.
[Pal99] K. I. Palomäki, J. Niittylahti, and M. Renfors, "Numerical Sine and Cosine Synthesis Using a Complex Multiplier," Proceedings of the 1999 IEEE International Symposium on Circuits and Systems, Vol. 4, June 1999, pp. 356 -359.
[Pre94] L. Presti, and G. Cardamone, "A Direct Digital Frequency Synthesizer Using an IIR Filter Implemented with a DSP Microprocessor," IEEE Int. Conf. Acoustics, Speech, and Signal Processing (ICASSP), Vol. 3, pp. 201–204, 1994.
[Pro97] J. G. Proakis, and D. G. Manolakis, "Digital Signal Processing, Principles," Macmillan Publishing Company, 1998, pp. 365.
[Tur03] C. S. Turner, "Recursive Discrete-Time Sinusoidal Oscillators," IEEE Signal Processing Magazine, Vol. 20, No. 3, pp. 103–111, May 2003.

Chapter 6

6. CORDIC ALGORITHM

Algorithms used in communication technology require the computation of trigonometric functions, coordinate transformations, vector rotations, or hyperbolic rotations. The CORDIC, an acronym for COordinate Rotation DIgital Computer, algorithm offers an opportunity to calculate the desired functions in a rather simple and elegant way. The CORDIC algorithm was first introduced by Volder [Vol59]. Walter [Wal71] later developed it into a unified algorithm to compute a variety of transcendental functions. Two basic CORDIC modes leading to the computation functions exist, the rotation mode and the vectoring mode. For both modes the algorithm can be realized as an iterative sequence of additions/subtractions and shift operations, which are rotations by a fixed rotation angle, but with a variable rotation direction. Due to the simplicity of the operations involved, the CORDIC is very well suited for a VLSI realization ([Sch86], [Dur87], [Lee89], [Not88], [Bu88], [Cav88a], [Cav88b], [Lan88], [Sar98], [Kun90], [Lee92], [Hu92b], [Fre95], [Hsi95], [Phi95], [Ahn98], [Dac98], [Mad99]). It has been implemented in pocket calculators like Hewlett Packard's HP-35 [Coc92], and in arithmetic coprocessors like Intel 8087.

In this book, the interest is in the rotation mode, because the QAM modulator (our application) performs a circular rotation (see (4.10)). The basic task performed in the CORDIC algorithm is to rotate a 2 by 1 vector through an angle using a linear, circular or hyperbolic coordinate system [Wal71]. This is accomplished in the CORDIC by rotating the vector through a sequence of elementary angles whose algebraic sum approximates the desired rotation angle.

The CORDIC algorithm provides an iterative method of performing vector rotations by arbitrary angles using only shifts and adds. The algorithm is derived from the general rotation transformation. In Figure 6-1, a pair of rectangular axes is rotated clockwise through the angle *Ang* by the CORDIC algorithm where the coordinates of a vector transform (I,Q) to (I',Q')

$$\begin{aligned} I' &= I\cos(Ang) + Q\sin(Ang) \\ Q' &= Q\cos(Ang) - I\sin(Ang). \end{aligned} \tag{6.1}$$

which rotates a vector clockwise in a Cartesian plane through the angle *Ang*, as shown in Figure 6-1. These equations can be rearranged so that

$$\begin{aligned} I' &= \cos(Ang)\left[I + Q\tan(Ang)\right] \\ Q' &= \cos(Ang)\left[Q - I\tan(Ang)\right]. \end{aligned} \tag{6.2}$$

If the rotation angles are restricted to $\tan(Ang_i) = \pm 2^{-i}$, the multiplication by the tangent term is reduced to a simple shift operation. Arbitrary angles of rotation are obtainable by performing a series of successively smaller elementary rotations. If the decision at each iteration, i, is which direction to rotate rather than whether or not to rotate, then the term $\cos(Ang_i)$ becomes a constant, because $\cos(Ang_i) = \cos(-Ang_i)$. The iterative rotation can now be expressed as

$$\begin{aligned} I_{i+1} &= K_i\left[I_i + Q_i d_i 2^{-i}\right] \\ Q_{i+1} &= K_i\left[Q_i - I_i d_i 2^{-i}\right], \end{aligned} \tag{6.3}$$

where $d_i = \pm 1$ and

$$K_i = \cos(\tan^{-1}(2^{-i})) = 1/\sqrt{(1+2^{-2i})}. \tag{6.4}$$

Removing the scale constant from the iterative equations yields a shift-add algorithm for the vector rotation. The product of the K_i's approaches 0.6073 as the number of iterations goes to infinity. The exact gain depends on the number of iterations, and obeys the relation

$$G_N = \prod_{i=0}^{N-1}\sqrt{(1+2^{-2i})}. \tag{6.5}$$

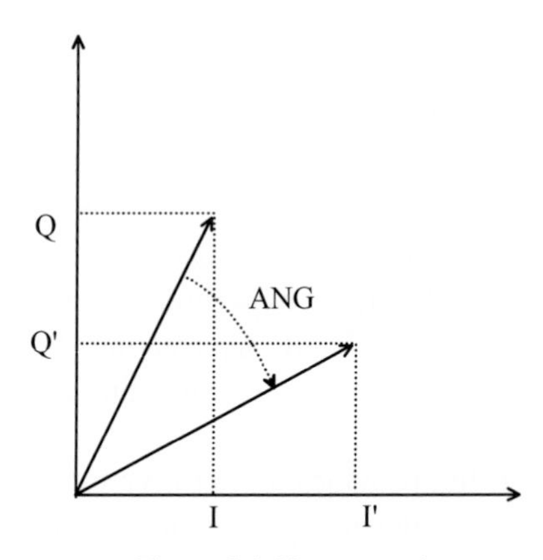

Figure 6-1. Vector rotation.

The CORDIC rotation algorithm has a gain, G_N, of approximately 1.647 as the number of iterations goes to infinity.

If both vector component inputs are set to the full scale simultaneously, the magnitude of the resultant vector is 1.414 times the full scale. This, combined with the CORDIC gain, yields a maximum output of 2.33 times the full-scale input.

The angle of a composite rotation is uniquely defined by the sequence of the directions of the elementary rotations. This sequence can be represented by a decision vector. The set of all possible decision vectors is an angular measurement system based on binary arctangents. Conversions between this angular system and others can be accomplished using an additional adder/subtractor that accumulates the elementary rotation angles at each iteration. The angle computation block adds a third equation to the CORDIC algorithm

$$z_{i+1} = z_i - d_i \tan^{-1}(2^{-i}). \tag{6.6}$$

The CORDIC algorithm can be operated in one of two modes. The first one, called rotation by Volder [Vol59], rotates the input vector by a specified angle (given as an argument). The second mode, called vectoring, rotates the input vector to the I axis while recording the angle required to make that rotation. The CORDIC circular rotator operates in the rotation mode. In this mode, the angle computation block is initialized with the desired rotation angle. The rotation decision at each iteration is made in order to decrease the magnitude of the residual angle in the angle computation block. The decision at each iteration is therefore based on the sign of the residual angle after each step. The CORDIC equations for the rotation mode are

$$\begin{aligned} I_{i+1} &= I_i + Q_i d_i 2^{-i} \\ Q_{i+1} &= Q_i - I_i d_i 2^{-i} \\ z_{i+1} &= z_i - d_i \tan^{-1}(2^{-i}), \end{aligned} \tag{6.7}$$

where d_i = -1 if $z_i < 0$, and +1 otherwise, so that z is iterated to zero. These equations provide the following result, after n iterations

$$\begin{aligned} I_n &= G_n \left[I_0 \cos(A) + Q_0 \sin(A)\right] \\ Q_n &= G_n \left[Q_0 \cos(A) - I_0 \sin(A)\right] \\ G_n &= \prod_{i=0}^{n-1} \sqrt{1 + 2^{-2i}} \\ A &= Ang - z_n, \end{aligned} \tag{6.8}$$

where A is the rotated angle

$$A = \sum_{i=0}^{n-1} d_i \tan^{-1}(2^{-i}). \tag{6.9}$$

The CORDIC rotation algorithm as stated is limited to rotation angles between -π/2 and π/2, because of the use of 2^0 for the tangent in the first iteration. For composite rotation angles larger than π/2, an initializing rotation is required. For example, if it is desired to perform rotations with rotation angles between -π and π, it is necessary to make an initial rotation of ±π/2

$$\begin{aligned} I_0 &= d\, Q_{in}, \\ Q_0 &= -d\, I_{in}, \\ z_0 &= z_{in} - d\, 2 \tan^{-1}(2^0), \end{aligned} \tag{6.10}$$

where $d = -1$ if $z_{in} < 0$, and +1 otherwise.

6.1 Scaling of I_n and Q_n

The results from the CORDIC operation have to be corrected because of the inherent magnitude expansion in the circular mode. This increases the latency and requires a lot of hardware. Many articles have dealt with this problem and have suggested different methods to reduce the cost of the scaling. Timmermann et al. compares the different approaches in [Tim91].

Each iteration of the CORDIC algorithm extends the vector $[I_i\ Q_i]$ in the rotational mode. Because of this, the resulting vector $[I_n\ Q_n]$ has to be scaled with the scaling factor G_n given in (6.5). The correction due to the scaling factor can be performed in three different ways:

1. Post-multiplying the result I_n, Q_n or pre-multiplying the input I_0, Q_0 with $1/G_n$. This is the straightforward way to compensate for the scaling factor; it increases the latency with one multiplication.
2. Separate scaling iterations can be included in the CORDIC algorithm, or CORDIC iterations can be repeated, such that the scaling factor becomes a power of two, thus reducing the final scaling operation to a shift operation [Ahm82], [Hav80].
3. The CORDIC iterations can be merged with the scaling factor compensation (as described in [Bu88]).

The conclusion in [Tim91] is that the third type of scaling tends to increase the overall latency. Therefore, to minimize the latency, the normal iterations and the scaling should be separated.

In our design, the scaling factor is constant because the number of the iterations is constant. The scaling factor is simply factored into an aggregate processing gain attributed to the filter chain in the QAM modulator (see Figure 16-6).

6.2 Quantization Errors in CORDIC Algorithm

Hu [Hu92a] provided an accurate description of the errors encountered in all modes of the CORDIC operation. In [Hu92a], two major sources of error are identified: the (angle) approximation error and the rounding error. The first type of error is due to the quantized representation of a CORDIC rotation angle by a finite number of elementary angles. The second one is due to the finite precision arithmetic used in a practical implementation. However, the bound for the approximation error has been set without taking into account the effects of the quantization of the angles (the inverse tangents). In [Kot93], the study of the numerical accuracy in the CORDIC includes the accuracy problem with the inverse tangent calculations. The error analysis in [Hu92a] and [Kot93] is based on the assumption that an error reaches its maximum value at each quantization step. This gives quite pessimistic results especially in QAM modulator applications, where the I/Q inputs are random signals.

6.2.1 Approximation Error

The equations (6.7) can be rewritten as

$$v_{i+1} = p_i \cdot v_i , \tag{6.11}$$

where $v_i = [I_i\ Q_i]^T$ is the rotation vector at the ith iteration, and

$$p_i = \begin{bmatrix} 1 & d_i 2^{-i} \\ -d_i 2^{-i} & 1 \end{bmatrix} = \sqrt{1+2^{-2i}} \begin{bmatrix} \cos a_i & d_i \sin a_i \\ -d_i \sin a_i & \cos a_i \end{bmatrix} \tag{6.12}$$

is an unnormalized rotation matrix. The magnitude of the elementary angle rotated in the ith iteration is $a_i = \tan^{-1}(2^{-i})$.

In the CORDIC algorithm, each rotation angle A is represented by a restricted linear combination of the n elementary angles a_i (n is the number of iterations), as follows:

$$Ang = \sum_{i=0}^{n-1} d_i\ a_i + z_n = A + z_n , \tag{6.13}$$

where z_n is the error due to this angle quantization. The CORDIC computation error in v_n due to the presence of "z_n" is defined as the approximation error. In the following derivations, infinite precision arithmetic will be applied in order to suppress the effect due to the rounding error.

Conventionally, in the CORDIC algorithm, two convergence conditions will be set [Hu92a]. The first condition states that the rotation angle must be bounded

$$|A| \leq \sum_{i=0}^{n-1} a_i \equiv A_{max}. \tag{6.14}$$

The second condition is set to ensure that if the rotation angle A satisfies (6.14), its angle approximation error will be bounded by the smallest elementary rotation angle a_{n-1}. That is,

$$|z_n| \leq a_{n-1}. \tag{6.15}$$

To satisfy this condition, the elementary angle sequence (a_i; $i = 0$ to $i = n-1$) must be chosen so that [Wal71]

$$a_i - \sum_{j=i+1}^{n-1} a_j \leq a_{n-1}. \tag{6.16}$$

Based on the above result, it is quite obvious, that in order to minimize the approximation error, the smallest elementary rotation angle a_{n-1} must be made small. This can be achieved by increasing the number of the CORDIC iterations.

6.2.2 Rounding Error of Inverse Tangents

If the numerically controlled oscillator (NCO) output in Figure 16-6 has a long period (from (4.4)), then the approximation errors are uncorrelated and uniformly distributed within each quantization step

$$-a_{n-1} \leq z_n \leq a_{n-1}. \tag{6.17}$$

The quantization of the angles is defined as

$$e_i = a_i - Q[a_i], \tag{6.18}$$

where $Q[.]$ denotes the quantization operator and the rounding error is

$$-\frac{2\pi}{2^{ba+1}} \leq e_i \leq \frac{2\pi}{2^{ba+1}}, \tag{6.19}$$

for a fixed-point angle computation data path with ba bits, which is assumed to be greater than the number of iteration stages (n). This is a reasonable assumption because the size of the residual angle becomes smaller in the successive iteration stages, approximately by one bit after each iteration.

6.2.3 Rounding Error of I_n and Q_n

The rounding error of z_n is quite straightforward as it involves only the inner product operation. Hence, the focus will be on the rounding error in I_n and Q_n. The quantization of the error $v_i = [I_i\ Q_i]^T$ is defined as

$$e_i = \begin{bmatrix} I_i \\ Q_i \end{bmatrix} - \begin{bmatrix} Q[I_i] \\ Q[Q_i] \end{bmatrix}, \tag{6.20}$$

where $e_i = [e_i^I\ e_i^Q]^T$ is an error vector due to rounding. For the fixed-point arithmetic, the absolute rounding error will be bounded by

$$\left|e_i^I\right| \le \frac{2^{-bb}}{2}, \quad \left|e_i^Q\right| \le \frac{2^{-bb}}{2}, \tag{6.21}$$

where bb is the number of fractional bits in the angle rotation data path. The variance of the error is (assuming that the error is a white noise process and probability distribution of the error sample values is uniform over the range of the quantization error)

$$\delta_I^2 = \delta_Q^2 = \frac{2^{-2bb}}{12}. \tag{6.22}$$

The variance of the rounding error of I_i and Q_i is

$$\delta_{IQ}^2 = \delta_I^2 + \delta_Q^2 = \frac{2^{-2bb}}{6}. \tag{6.23}$$

In each CORDIC iteration, the rounding error consists of two components: the rounding error propagated from the previous iterations and the rounding error introduced in the present iteration. The variance due to the rounding error of I_n and Q_n at the CORDIC rotator output is therefore

$$\delta_{\text{tot2}}^2 = \delta_{IQ}^2 \left\{ 1 + \sum_{j=0}^{n-1} \left\| \prod_{i=j}^{n-1} K_i^2 \right\| \right\}, \tag{6.24}$$

where K_i^2 is $(1 + 2^{-2i})$ from equation (6.12).

6.3 Redundant Implementations of CORDIC Rotator

The computation time and the achievable throughput of CORDIC processors using conventional arithmetic are determined by the carry propagation involved with the additions/subtractions, since the direction of the CORDIC microrotation is steered by the sign of the previous iteration results. This sign is not known prior to the computation of the MSB. The use of redundant arithmetic is well known to speed up additions/subtractions, because a carry-free or limited carry-propagation operation becomes possible. However, the application of redundant arithmetic in the CORDIC is not straightforward, because a complete word level carry-propagation is still required in order to determine the sign of a redundant number (this also holds for generalized signed digit numbers as described in [Par93]).

In order to overcome this problem, several authors have proposed techniques for estimating the sign of the redundant intermediate results from a number of MSDs (most significant digits) ([Erc90], [Erc88], [Tak87]). If the sign, and therefore the rotation direction, cannot be estimated reliably from the MSDs, no microrotation occurs at all. However, the scaling factor involved in the CORDIC algorithm depends on the actual rotations. Therefore, here, the scaling factor is variable, and has to be calculated in parallel to the usual CORDIC iteration. Additionally, a division by the variable scaling factor has to be implemented following the CORDIC iteration.

A number of publications dealing with constant scale factor redundant (CSFR) CORDIC implementations of the rotation mode ([Erc90], [Erc88], [Tak87], [Kun90], [Nol91], [Tak91], [Lin90], [Nol90], [Yos89]) describe sign estimation techniques, where every iteration is reactually performed, in order to overcome this problem. However, either a considerable increase (about 50 per cent) in the complexity of the iterations (double rotation method [Tak91]) or a 50 percent increase in the number of iterations (correcting iteration method [Tak91], [Nol90], [Kun90], [Nol91]) occurs.

In [Dup93], a different CSFR algorithm is proposed for the rotation mode. Using this "branching CORDIC", two iterations are performed in parallel if the sign cannot be estimated reliably, each assuming one of the possible choices for the rotation direction. It is shown in [Dup93] that, at most, two parallel branches can occur. However, this is equivalent to an almost twofold effort in terms of implementation complexity of the CORDIC rotation engine. In contrast to the above mentioned approaches, in [Daw96], transformations of the usual CORDIC iteration are developed, resulting in a constant scale factor redundant implementation without additional or branching iterations. It is shown in [Daw96] that this "Differential CORDIC (DCORDIC)" method compares favorably to the sign estimation methods.

However, the architecture described in Section 16.5 does not use any of these techniques. The adder/subtractors used in the CORDIC rotator unit allow the operation frequency to be reached with carry-ripple arithmetic, and therefore the problem of sign estimation is avoided.

6.4 Hybrid CORDIC

Ahmed has developed a hybrid approach to elementary function generation that combines memory, multipliers and the CORDIC [Ahm82], [Ahm89].

The vector rotation by angle (*Ang*) could be partitioned into two parts, a coarse rotation stage and fine rotation stage [Ahm89]. The basic principle is general, although Ahmed has restricted his attenuation to the CORDIC rotation algorithm in [Ahm89]. Similar results were published in [Tim89], [Hwa03] but the effect on the vectoring mode were considered. Wang generalized Ahmed's work and presented two Hybrid CORDIC algorithms: Mixed Hybrid algorithm and Partioned-Hybrid algorithm based on [Wan97]. The hybrid CORDIC algorithms offer a considerable latency time reduction and chip area savings when compared with the original CORDIC algorithm [Wan97].

6.4.1 Mixed-Hybrid CORDIC Algorithm

The architectural approach for the case of the Mixed-Hybrid CORDIC algorithm is given in Figure 6-2. The CORDIC iterations related to the coarse part are performed as in the conventional CORDIC algorithm. In order to avoid errors, each rotation direction can be evaluated only after the previous iteration has been completed. The iterations related to the fine part can be simplified [Ahm89], [Tim89]. Applying the Taylor series expansions to the z data path in (6.6), and taking only the first terms, the z data path becomes

$$z_{i+1} \approx z_i - d_i 2^{-i}. \tag{6.25}$$

In [Wan97], it has been showed that the rotation directions for approximately 2/3 of the iterations (*N*) can be derived using (6.25) in the rotation mode. The error is negligible in finite length arithmetic [Wan97]. All rotation directions associated with such iterations are available in parallel, since each of them is given by the corresponding bit of the residue rotation angle (1 and 0 identify positive and negative rotation, respectively). It is worthwhile noting that application of the CORDIC iterations related to the most

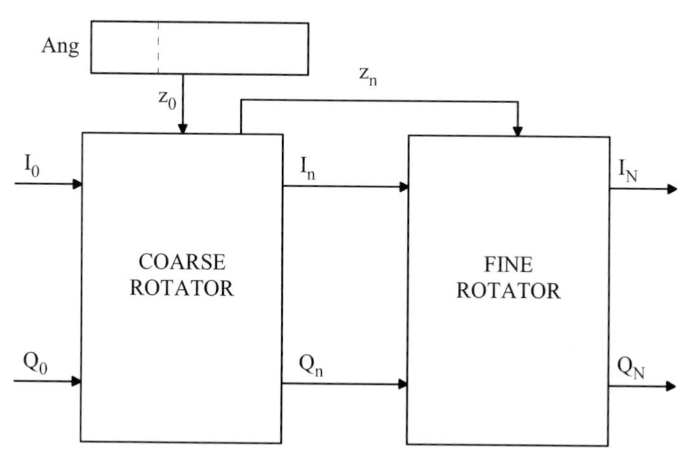

Figure 6-2. The architecture for Mixed-Hybrid CORDIC algorithms.

significant part of *Ang* may change the value of the bits of the least significant part with respect to their initial value in *Ang*. As a consequence, the parallel evaluation of the directions associated with the fine part can be performed only after rotations of the coarse one. The coarse rotator uses the whole rotation angle *Ang* and the initial vector coordinates I_0 and Q_0 to generate z_n, I_n, and Q_n at the end of the first n iterations ($n \approx 1/3\ N$) [Wan97]. The rotation directions are generated sequentially, as in the conventional CORDIC algorithm. At the end of these iterations, the most significant part of z_n is zero. The fine rotator operates starting from z_n, I_n, and Q_n to generate I_n and Q_n at the end of the remaining N - n iterations.

A method in [Ahm89] implements a coarse stage with CORDIC for the first n rotations and a fine stage computing as a single rotation via two multiplications and without a trigonometric lookup table. A coarse stage is implemented with CORDIC rotator and the fine stage by a lookup table and multipliers in [Cur01]. Applying the Taylor series expansions to the trigonometric functions in (6.1), and taking only the first terms, (6.1) becomes

$$\begin{aligned} I_N &= I_n + Q_n\, AngL \\ Q_N &= Q_n - I_n\, AngL. \end{aligned} \tag{6.26}$$

Therefore, no look-up tables for sine and cosine terms are needed to complete the fine rotation according to (6.26) [Ahm89].

6.4.2 Partitioned-Hybrid CORDIC Algorithm

To increase the performance, the computation of the coarse rotation must be completely separated from the least significant one while preserving full accuracy. To achieve this goal, the *Ang* is partitioned in two parts: the n most significant bits (*AngH*) and the remaining N - n bits (*AngL*). The rotation *Ang* is partitioned in two rotations; *AngH* is the most relevant, while *AngL* adjusts the position of the rotated vector to achieve the final expected (or

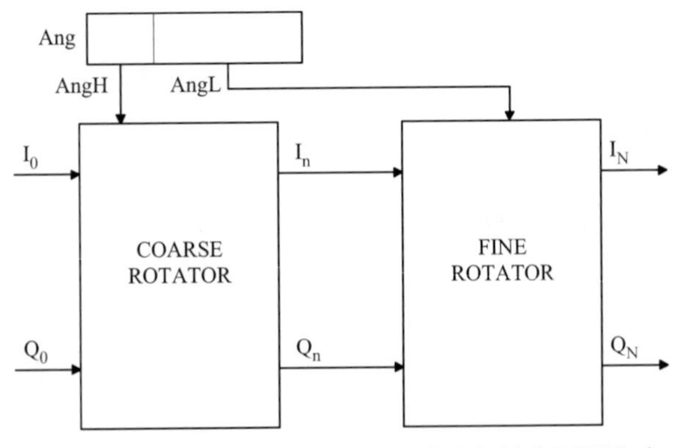

Figure 6-3. The architecture for Partitioned-Hybrid CORDIC algorithms.

nearby) position. *AngL* always has zeros in the n most significant bits. The angle *AngH* is completely independent from *AngL* only if it does not change the N - n least significant bits of the residue rotation angle with respect to *AngL*. To achieve this goal, the exact rotation *AngH* must be performed separately from *AngL*. To avoid a possible increase in the number of the rotation iterations and to speed up the whole algorithm, the traditional CORDIC algorithm is modified by assuming that the whole rotation *AngH* is performed in only one iteration [Wan97]. This is not an actual limit since, when n is not too large, the iteration can be realized by using a ROM-based architecture.

The architectural approach to the case of the Partitioned-Hybrid CORDIC algorithm is given in Figure 6-3. The coarse rotator operates a rotation of *AngH* on I_0 and Q_0 and generates I_n and Q_n. The fine rotator receives these intermediate rotated coordinates and applies the rotation *AngL* to generate the final I_N and Q_N. The coarse rotator could be implemented by the look-up table, where the *AngH* is used as an address. This solution is feasible in terms of circuit complexity only for relatively small values of n, as occurs in several applications [Mad99], [Jan02]. A coarse stage is implemented with lookup tables and multipliers in [Ahm89], [Tor03] or by a lookup table [Jan02]. The fine rotator could be implemented as a single rotation using approximation (6.26) in [Ahm89], [Tor03] or by CORDIC iterations using [Ahm89], [Mad99], [Jan02]. The rotation directions for the CORDIC iterations are obtained from (6.25), which reduces the latency time.

The coarse rotation stage performs the rotation with the *AngH* using a complex multiplier in [Son03], [Tor03]. The sin(*AngH*) and cos(*AngH*) required for this rotation are stored in a small lookup table. To reduce the hardware cost of the complex multiplication, sin(*AngH*) and cos(*AngH*) are quantized on a small number of bits in [Son03]. Naturally, this brute quantization introduces a significant angle rotation error. To avoid this situation, the angle rotation error of the coarse rotation stage is stored in a lookup table and taken into consideration by the fine rotation block [Son03]. The fine rotation is also performed using a complex multiplier, but the sin(*AngL*) and cos(*AngL*) are computed using Taylor series expansion (6.26) [Tor03] or linear interpolation [Son03]. Even if these algorithms require two complex multipliers, the total area is smaller than the one of a classic mixer due to the fact that the wordlength of the multipliers is small [Tor03]. However, in order to reach high operation speed, both designs have to rely on heavy pipelining. For instance, the design in [Tor03] has 13 pipeline stages for a 300 MHz clock, whereas the design in [Son03] requires 17 pipeline stages for a 330 MHz clock. Therefore, these architectures have the same drawback as the CORDIC algorithm, i.e., high tuning latency.

REFERENCES

[Ahm82] H. M. Ahmed, "Signal Processing Algorithms and Architectures," Ph. D. dissertation, Department of Electrical Engineering, Stanford University, CA. Jun. 1982.

[Ahm89] H. M. Ahmed, "Efficient Elementary Function Generation with Multipliers," in Proc. 9th Symposium on Computer Arithmetic, Santa Monica, CA, USA, Sept. 1989, pp. 52-59.

[Ahn98] Y. Ahn, and S. Nahm, "VLSI Design of a CORDIC-based Derotator," in Proc. ISCAS'98, Vol. 2, June 1998, pp. 449-452.

[Bu88] J. Bu, E. F. Deprettere, and F. du Lange, "On the Optimization of Pipelined Silicon CORDIC Algorithm," in Proc. European Signal Processing Conference (EUSIPCO), Sep. 1988, pp. 1,227-1,230.

[Cav88a] J. R. Cavallaro, and F. T. Luk, "Floating Point CORDIC for Matrix Computations," in Proc. IEEE International Conference on Computer Design, Oct. 1988, pp. 40-42.

[Cav88b] J. R. Cavallaro, and F. T. Luk, "CORDIC Arithmetic for a SVD processor," Journal of Parallel and Distributed Computing, Vol. 5, pp. 271-290, June 1988.

[Coc92] D. Cochran, "Algorithms and Accuracy in the HP-35," Hewlett Packard Journal, pp. 10-11, Jun. 1992.

[Cur01] F. Curticăpean, and J. Niittylahti, "An Improved Digital Quadrature Frequency Down-Converter Architecture," Asilomar Conf. on Signals, Syst. and Comput., Nov. 2001, pp. II-1318-1321.

[Dac98] M. Dachroth, B. Hoppe, H. Meuth, and U. W. Steiger, "High-Speed Architecture and Hardware Implementation of a 16-bit 100 MHz Numerically Controlled Oscillator," in Proc. ESSCIRC'98, Sept. 1998, pp. 456-459.

[Daw96] H. Dawid, and H. Meyr, "The Differential CORDIC Algorithm: Constant Scale Factor Redundant Implementation without Correcting Iterations," IEEE Trans. on Computers, Vol. 45, No. 3, pp. 307-318, Mar. 1996.

[Dup93] J. Duprat, and J. M. Muller, "The CORDIC Algorithm: New Results for Fast VLSI Implementation," IEEE Trans. on Computers, Vol. 42, No. 2, pp. 168-178, Feb. 1993.

[Dur87] R. A. Duryea, and C. Pottle, "Finite Precision Arithmetic Units in Jacobi SVD Architectures," School of Electrical Engineering, Cornell University, Ithacaa, NY, Technical Report EE-CEG-87-11, Mar. 1987.

[Erc88] M. D. Ercegovac, and T. Lang, "Implementation of Fast Angle Calculation and Rotation Using On-Line CORDIC," in Proc. IEEE International Symposium on Circuits and Systems (ISCAS), June 1988, pp. 2,703-2,706.

[Erc90] M. D. Ercegovac, and T. Lang, "Redundant and On-Line CORDIC Application to Matrix Triangularization and SVD," IEEE Trans. on Computers, Vol. 38, No. 6, pp. 725-740, June 1990.

[Fre95] S. Freeman, and M. O'Donnell, "A Complex Arithmetic Digital Signal Processor Using Cordic Rotators," in Proc. ICASSP-95, Vol. 5, pp. 3191-3194, May 1995.

[Hav80] G. L. Haviland, and A. A. Tuszynski, "A CORDIC Arithmetic Processor Chip," IEEE Trans. on Computers, Vol. 29, No. 2, pp. 68-79, Feb. 1980.

[Hsi95] S. F. Hsiao, and J. M. Delosme, "Householder CORDIC Algorithms," IEEE Trans. Comput., Vol. 44, No. 8, pp. 990-1001, Aug. 1995.

[Hu92a] Y. H. Hu, "The Quantization Effects of the CORDIC Algorithm," IEEE Trans. Signal Processing, Vol. 40, No. 4, pp. 834-844, April 1992.

[Hu92b] H. Y. Hu, "CORDIC-Based VLSI Architectures for Digital Signal Processing," IEEE Signal Processing Magazine, pp. 16-35, July 1992.

[Hwa03] D. D. Hwang, D. Fu, and A. N. Willson, Jr, "A 400-MHz Processor for the Conversion of Rectangular to Polar Coordinates in 0.25-μm CMOS," IEEE J. Solid-State Circuits, Vol. 38, , No. 10, pp. 1771-1775, Oct. 2003.

[Jan02] I. Janiszewski, B. Hoppe, and H. Meuth, "Numerically Controlled Oscillators with Hybrid Function Generators," IEEE Transactions on Ultrasonics, Ferroelectrics and Frequency Control, Vol. 49, pp. 995-1004, July 2002.

[Kot93] K. Kota, and J. R. Cavallaro, "Numerical Accuracy and Hardware Tradeoffs for CORDIC Arithmetic for Special-Purpose Processors," IEEE Trans. Comput., Vol. 42, No. 7, pp. 769-779, July 1993.

[Kun90] H. Kunemund, S. Soldner, S. Wohlleben, and T. Noll, "CORDIC Processor with Carry Save Architecture," in Proc. ESSCIRC'90, Sep. 1990, pp.193-196.

[Lan88] A. A. de Lange, A. J. van der Hoeven, E. F. Deprettere, and J. Bu, "An Optimal Floating-Point Pipeline CMOS CORDIC Processor," in Proc. IEEE International Symposium on Circuits and Systems (ISCAS), June 1988, pp. 2,043-2,047.

[Lee89] J. Lee, and T. Lang, "On-Line CORDIC for Generalized Singular Value Decomposition," SPIE High Speed Computing II Vol 1058, pp.235-247, 1989.

[Lee92] J. Lee, and T. Lang, "Constant-Factor Redundant CORDIC for Angle Calculation and Rotation," IEEE Trans. Comput., Vol. 41, No. 8, pp. 1016-1025, Aug. 1992.

[Lin90] H. X. Lin, and H. J. Sips, "On-Line CORDIC Algorithms," IEEE Trans. on Computers, Vol. 38, No. 8, pp. 1,038-1,052, Aug. 1990.

[Mad99] A. Madisetti, A. Kwentus, and A. N. Wilson, Jr., "A 100-MHz, 16-b, Direct Digital Frequency Synthesizer with a 100-dBc Spurious-Free Dynamic Range," IEEE J. of Solid State Circuits, Vol. 34, No. 8, pp. 1034-1044, Jan. 1999.

[Nol90] T. Noll, "Carry-Save Arithmetic for High-Speed Digital Signal Processing," in Proc. IEEE International Symposium on Circuits and Systems (ISCAS), Vol. 2, May 1990, pp. 982-986.
[Nol91] T. Noll, "Carry-Save Architectures for High-Speed Digital Signal Processing," Journal of VLSI Signal Processing, Vol. 3, pp. 121-140, June 1991.
[Not88] S. Note, J. van Meerbergen, Catthoor, and H. de Man, "Automated Synthesis of a High Speed CORDIC Algorithm with the Cathedral-III Compilation System," in Proc. IEEE International Symposium on Circuits and Systems (ISCAS), June 1988, pp. 581-584.
[Par93] B. Parhami, "On the Implementation of Arithmetic Support Functions for Generalized Signed-Digit Number Systems," IEEE Trans. on Computers, Vol. 42, No. 3, pp. 379-384, Mar. 1993.
[Phi95] L. Philips, I. Bolsens, and H. D. Man, "A Programmable CDMA IF Transceiver ASIC for Wireless Communications," in Proc. IEEE Custom Integrated Circuits Conf., 1995, pp. 307-310.
[Sar98] R. Sarmiento, and et al., "A CORDIC Processor for FFT Computation and Its Implementation Using Gallium Arsenide Technology," IEEE Trans. on VLSI Systems, Vol. 6, No. 1, pp. 18-30, Mar. 1998.
[Sch86] G. Schmidt, D. Timmermann, J. F. Bohme, and H. Hahn, "Parameter Optimization of the CORDIC Algorithm and Implementation in a CMOS Chip," in Proc. European Signal Processing Conference (EUSIPCO), Sep. 1986, pp. 1,291-1,222.
[Son03] Y. Song and B. Kim, "A 330-MHz 15-b Quadrature Digital Synthesizer/Mixer in 0.25 μm CMOS," in Proc. 29th European Solid-State Circuits Conference, Estoril, Portugal, Sept. 2003, pp. 513-516.
[Tak87] N. Takagi, T. Asada, and S. A. Yajima, "Hardware Algorithm for Computing Sine and Cosine using Redundant Binary Representation," Systems and Computers in Japan, Vol. 18, No. 9, pp. 1-9, 1987.
[Tak91] N. Takagi, T. Asada, and S. A. Yajima, "Redundant CORDIC Methods with a Constant Scale Factor for a Sine and Cosine Computation," IEEE Trans. on Computers, Vol. 40, No. 9, pp. 989-995, Sep. 1991.
[Tim89] D. Timmermann, H. Hahn, and B. Hosticka, "Modified CORDIC Algorithm with Reduced Iterations," Electronics Letters, Vol. 25, No. 15, pp. 950-951, July 1989.
[Tim91] D. Timmermann, H. Hahn, B. J. Hostica, and B. Rix, "A New Addition Scheme and Fast Scaling Factor Compensation Methods for CORDIC Algorithms," INTEGRATION, the VLSI Journal, Vol. 11, pp. 85-100, 1991.
[Tor03] A. Torosyan, D. Fu, and A. N. Willson, Jr., "A 300-MHz Quadrature Direct Digital Synthesizer/Mixer in 0.25μm CMOS," IEEE J. of Solid State Circuits, Vol. 38, No. 6, pp. 875-887, June 2003.
[Vol59] J. E. Volder, "The CORDIC Trigonometric Computing Technique," IRE Trans. on Electron. Comput., Vol. C-8, pp. 330–334, Sept. 1959.

[Wal71] J. S. Walther, "A Unified Algorithm for Elementary Functions," in Proc. Spring Joint Computer Conference, May 1971, pp. 379-385.
[Wan97] S. Wang, V. Piuri, and E. E. Swartzlander, "Hybrid CORDIC Algorithms," IEEE Trans. on Computers, Vol. 46, No. 11, pp. 1202-1207, Nov. 1997.
[Yos89]H. Yoshimura, T. Nakanishi, and H. Yamauchi, "A 50 MHz CMOS Geometrical Mapping Processor," IEEE Trans. on Circuits and Systems, Vol. 36, No. 10, pp. 1,360-1,363, 1989.

[19] S. [illegible], "A Unified Algorithm for Elementary Functions," [illegible] Proc. Spring Joint Computer Conference, Vol. [illegible], pp. [illegible], [illegible].

[20] [illegible] Wong, [illegible] and E. E. Swartzlander, "Hybrid CORDIC Algorithms," IEEE Trans. on Computers, Vol. 46, No. 11, pp. 1202-1207, [illegible] 1997.

[21] [illegible] and H. Yamauchi, "A [illegible] CMOS Computational [illegible]," IEEE Journal of Solid-State Circuits, Vol. 30, No. [illegible], pp. [illegible], [illegible] 1995.

Chapter 7

7. SOURCES OF NOISE AND SPURS IN DDS

The model of the noise and spurs in the DDS has six sources. These sources are depicted symbolically in Figure 7-1. The sources are: the truncation of the phase accumulator bits addressing the sine LUT (e_P), a distortion from compressing the sine LUT (e_{COM}), the finite precision of the sine samples stored in the LUT (e_A), the digital-to-analog conversion (e_{DA}) (see Chapter 10), a post-filter (e_F), the phase noise of the clock (n_{clk}), and the frequency error (Δf). The frequency error (Δf) causes a frequency offset (4.2), but not noise and spurs.

7.1 Phase Truncation Related Spurious Effects

In the ideal case, with no phase and amplitude truncation, the output sample sequence of the DDS is given by

$$s(n) = \sin(2\pi \frac{\Delta P}{2^j} n). \tag{7.1}$$

Since the amount of memory required to encode the entire width of the phase accumulator would usually be prohibitive, only k of the most significant bits of the accumulator output are generally used to calculate the sine-wave samples. If the phase accumulator value is truncated to k bits prior to performing the look-up operation, the output sequence must be modified as

$$s(n) = \sin(\frac{2\pi}{2^k} \left[\frac{\Delta P}{2^{j-k}} n \right]), \tag{7.2}$$

where [] denotes truncation to integer values. This may be rewritten as

$$s(n) = \sin(\frac{2\pi}{2^j} (\Delta P n - e_P(n))), \tag{7.3}$$

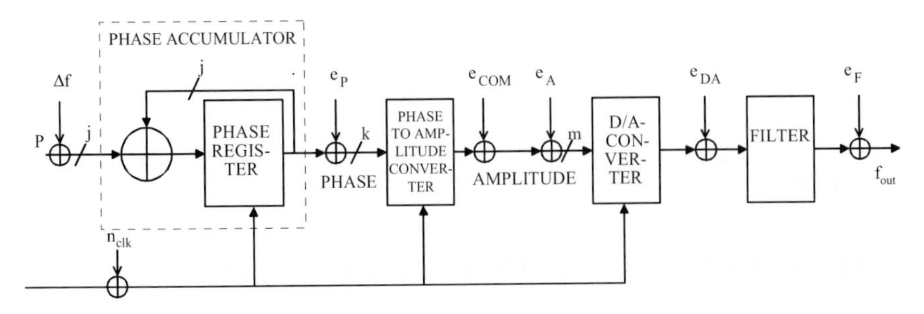

Figure 7-1. Block diagram of the sources of noise and spurs.

where $e_P(n)$ is the error associated with the phase truncation. The phase error sample sequence is also restricted in magnitude as

$$e_P(n) < 2^{j-k}, \tag{7.4}$$

and is also periodic with some period. The phase truncation occurs only when GCD (ΔP, 2^j) is smaller than 2^{j-k}. If GCD (ΔP, 2^j) is equal or greater than 2^{j-k}, then the phase increment bits are zeros below 2^{j-k} and no phase error occurs.

This sawtooth waveform (see Figure 7-2) is identical to the waveform that would be generated by a phase accumulator of word length (j-k) with an input phase increment word of

$$(n\,\Delta P) \bmod 2^{j-k}. \tag{7.5}$$

A complete derivation of the phase accumulator truncation effects on the output spectrum is given in [Meh83], [Nic87], [Jen88b], [Kro00], [Tor01]. References [Meh83] and [Nic87] describe very similar approaches and base their analyses on the phase-error sequence due to phase truncation and, using properties of this error sequence along with the assistance of small-angle approximations, derive a rather complex procedure for the characterization of phase-truncation spurs. Kroupa [Kro00] used an approach similar to that of [Meh83] and [Nic87] and presented an algorithm for the estimation of phase-truncation spurs with the introduction of more approximations. Jenq [Jen88b] used a more elegant approach, one for analyzing a class of non-

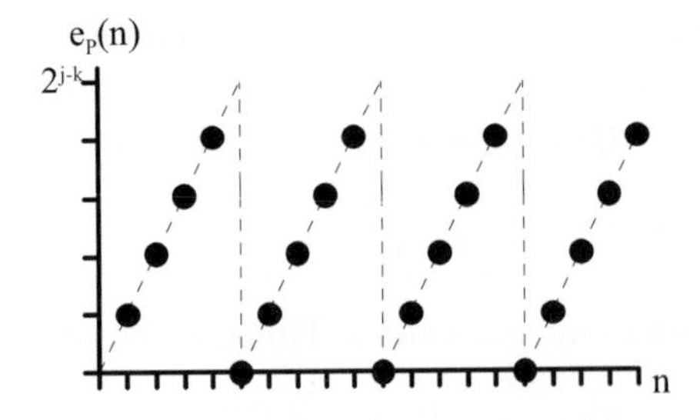

Figure 7-2. Phase accumulator error sequence.

uniformly sampled signals, to model the phase error due to phase truncation without approximations. References [Jen88b] and [Tor01] describe similar approaches. The difference between them is that, in reference [Jen88b], the results are given in Fourier spectrum, while in [Tor01], the result are given in discrete Fourier transform, which comprise just the samples of the Fourier transform.

The process of phase truncation occurs in a periodic pattern due to the periodic characteristics of the DDS. Jenq obtains the equivalence of the phase truncation with a non-uniform sampling process [Jen88b]. The phase increment (ΔP) is a number with an integer part W and a fractional part L/M, i.e.

$$\Delta P = W + L / M, \tag{7.6}$$

where L and M have no common factor. The integer part of the address increment register should be set to W, and its fractional part to L/M. Only the integer part of the phase accumulator is supplied to the addressing circuit of the sine LUT; data points sent to the D/A converter are offset from the intended uniform sampling instances, except for those where the fractional part of the phase accumulator is zero. Since the ratio of M to L is prime, M is the smallest integer to make $M\Delta P = M(W + L/M)$ an integer. Therefore, the output data sequence is obtained by sampling the sine wave stored in the sine LUT non-uniformly, but having an overall period MT_s, where M is

$$M = \frac{2^{j-k}}{\mathrm{GCD}(\Delta P, 2^{j-k})}, \tag{7.7}$$

and where GCD (ΔP, 2^{j-k}) denotes the greatest common divisor of ΔP and 2^{j-k}. The number of spurs due to the phase truncation is [Nic87]

$$Y = \frac{2^{j-k}}{\mathrm{GCD}(\Delta P, 2^{j-k})} - 1 = M - 1. \tag{7.8}$$

It has been shown in [Jen88a] that if one samples a sinusoidal $e^{j\omega_{out} t}$ non-uniformly with sampling advancement offsets (i.e. sampling earlier than it should be) $\{t_m T_s,\ m = 0, 1, 2, \ldots M\text{-}1\}$, then the digital spectrum of the sampled waveform is given by

$$G(\omega) = \frac{1}{T_s} \sum_{r=-\infty}^{\infty} A(r)\, 2\pi\, \delta[\omega - \omega_{out} - r(2\pi / MT_s)], \tag{7.9}$$

where the coefficient $A(r)$ is given by

$$A(r) = \sum_{m=0}^{M-1} \left[\frac{1}{M} e^{-j2\pi t_m f_{out} / f_s} \right] e^{-jrm(2\pi / M)}, \tag{7.10}$$

and $f_s = 1/T_s$ and $f_{out} = \omega_{out}/2\pi$.

To utilize (7.9) and (7.10) for this situation, let Δ be the time duration corresponding to

$$(W + L/M)\Delta = T_s \tag{7.11}$$

and let $[x]_{\text{frac}}$ be the fractional part of x, then we then have

$$\begin{aligned} t_m / f_s &= t_m T_s \\ &= [m(W + L/M)]_{\text{frac}} \Delta \\ &= [mL/M]_{\text{frac}} \Delta \end{aligned} \tag{7.12}$$

$$f_{out} = (W + L/M)(\frac{1}{NT_s}), \tag{7.13}$$

where N is 2^k (k is the number of bits used to calculate the sine-wave samples).

Hence

$$\begin{aligned} 2\pi t_m f_{out} / f_s &= 2\pi [mL/M]_{\text{frac}} / N \\ &= \frac{2\pi \langle mL \rangle_M}{MN}, \end{aligned} \tag{7.14}$$

where $\langle mL \rangle_M$ stands for mL modulo M. Substituting (7.14) into (7.10), we then have

$$A(r, L, M, N) = \sum_{m=0}^{M-1} \left[\frac{1}{M} e^{-j2\pi \langle mL \rangle_M /(MN)} \right] e^{-jrm2\pi/M}. \tag{7.15}$$

It is noted from (7.15) that the finite sequence [A(r, L, M, N), r = 0, 1,…, M - 1] is the discrete Fourier transform (DFT) of the sequence [(1/M) $e^{-j2\pi t_m f_{out}/f_s}$, m = 0, 1,…, M - 1]; therefore, by Parseval's theorem, the sum of the squares of $|A(r, L, M, N)|$ for r = 0, 1,…, M – 1 is equal to M times the sum of the squares of $|(1/M)\, e^{-j2\pi t_m f_{out}/f_s}|$ which is unity, i.e.

$$\sum_{r=0}^{M-1} |A(r, L, M, N)|^2 = 1. \tag{7.16}$$

This result is used to calculate the S/N, which is defined as the ratio of the power of the desirable harmonic component to the sum of the powers of the spurious harmonic components, i.e.

$$S/N = 10 \log_{10} \left[\frac{|A(0, L, M, N)|^2}{1 - |A(0, L, M, N)|^2} \right], \tag{7.17}$$

where $|A(0, L, M, N)|^2$ can be readily obtained from (7.15)

$$|A(0,L,M,N)|^2 = \left[\frac{\sin^2(\pi/N)}{(\pi/N)^2}\frac{(\pi/MN)^2}{\sin^2(\pi/MN)}\right]. \tag{7.18}$$

There are three interesting properties of $|A(0, L, M, N)|^2$ worth mentioning:

1) For $M = 1$, $|A(0, L, 1, N)|^2 = 1$, hence there is no spurious harmonic component due to the phase truncation.
2) For a fixed N, $|A(0, L, M, N)|^2$ is a decreasing function of M. Therefore, the S/N is also decreasing on M.
3) For a fixed M, $|A(0, L, M, N)|^2$ is an increasing function of N. Hence, the S/N can be made arbitrarily large by choosing a sufficiently large N.

From the properties listed above, we can have closed-form expressions for both the maximum and the minimum S/N for a fixed N, by making $M = 2$ and ∞, respectively, as follows:

$$S/N(\max) = 20\log_{10}[\cot(\pi/2N)] \tag{7.19}$$

and

$$S/N(\min) = 10\log_{10}\left[\frac{[\sin(\pi/N)/(\pi/N)]^2}{1-[\sin(\pi/N)/(\pi/N)]^2}\right]. \tag{7.20}$$

For a reasonably large N, say $N > 10$ (in practice, N is larger than 1000), (7.19) and (7.20) can be simplified by expanding the arguments of the log function in (7.19) and (7.20) in the Taylor's series form, and retaining only the first significant term. By doing so, we obtain

$$\begin{aligned} S/N(\max) &\approx 20\log_{10}(N) - 10\log_{10}(\pi/2) \\ &\approx 6.02k - 3.92 \text{ dB}, \end{aligned} \tag{7.21}$$

and

$$\begin{aligned} S/N(\min) &\approx 20\log_{10}(N) - 10\log_{10}(\pi^2/3) \\ &\approx 6.02k - 5.17 \text{ dB}. \end{aligned} \tag{7.22}$$

Equations (7.21) and (7.22) give very handy and accurate estimates of the S/N as a function of the size of the sine LUT [Jen88b].

The *SP* is defined as the ratio of the power of the desirable harmonic component to the power of the spurious harmonic components

$$SP(r) = 10\log_{10}\left[\frac{|A(0,L,M,N)|^2}{|A(r,L,M,N)|^2}\right], \tag{7.23}$$

where $|A(0, L, M, N)|^2$ and $|A(r, L, M, N)|^2$ can be readily obtained from (7.15)

$$SP(r) = 10\log_{10}\left(\frac{\text{sinc}(1/N)^2\,\text{sinc}(Nr/(NM)+1/(NM))^2}{\text{sinc}(1/(NM))^2\,\text{sinc}(r+1/N)^2}\right), \quad (7.24)$$

$$r = 1,\ldots,M-1,$$

where sinc(x) = sin(πx)/πx.

The corresponding spur locations for quadrature DDS (complex in-input to Discrete Fourier Transform (DFT)) are given by

$$F(r) = (\Delta P/\text{GCD}(\Delta P, 2^j) + rN\Delta P/\text{GCD}(\Delta P, 2^j))\text{mod}(Pe), \quad (7.25)$$

where $r = 1,\ldots M-1$ and $0 \le F(r) \le Pe-1$, the DFT size is NM, which is equal to the period of the DDS output (Pe) from (4.4). If the location of a spur were calculated using (7.25), and if the resulting spur number, $F(r)$, was larger than $Pe/2$, then the aliased positions of the spurs for cosine DDS output (real input to the DFT) are

$$F(r) = Pe - F(r), \quad (7.26)$$

where $0 \le F(r) \le Pe/2$. The worst case signal to spur power ratio (minimum ratio) occurs when $r = M-1$ in (7.24) [Jen88a]. The signal to spur power ratios from minimum to maximum are in the following order $SP(M-1)$, $SP(1)$, $SP(M-2)$, $SP(2)$, $SP(M-3)$… in (7.24). The worst-case carrier to the spur ratio due to the phase truncation occurs when $M = 2$ ($r = 1$)

$$SP(1)_{\min} = 10\log_{10}\left[\frac{A(0,L,2,N)^2}{A(1,L,2,N)^2}\right] = 20\log_{10}\left[\cot\left(\frac{\pi}{2N}\right)\right]. \quad (7.27)$$

The carrier to spur ratio due to the phase truncation when $r = 1$ and $M \approx \infty$ ($2^{j-k} >> \text{GCD}(\Delta P, 2^{j-k})$ in (7.7) is given by

$$SP(1)_{\max} = 20\log_{10}\left[\frac{A(0,L,\infty,N)}{A(1,L,\infty,N)}\right] = 20\log_{10}[N+1]. \quad (7.28)$$

For a reasonably large N, say $N > 10$ (in practice, N is larger than 1000), (7.27) can be simplified by expanding the argument of the log function in (7.27) in the Taylor's series form and retaining only the first significant term. By doing so, we obtain the worst-case carrier to spur ratio

$$SP(1)_{\min} \approx 20\log_{10}(N) - 20\log_{10}\left[\frac{\pi}{2}\right] \approx 6.02k - 3.92\text{ dB}. \quad (7.29)$$

The carrier to spur ratio due to the phase truncation when $r = 1$ and $M \approx \infty$ (from (7.28) is

$$SP(1)_{\max} \approx 20\log_{10}(N) = 6.02k. \quad (7.30)$$

The phase truncation error analysis in [Jen88b] is extended here so that it includes the worst-case carrier to spur ratio bounds (7.29) and (7.30). The spur power is concentrated in one peak in Figure 8-2, because M is 2 (7.8). The worst-case carrier-to-spur level due to the phase truncation appears to be 44.24 dBc. The expect worst-case carrier-to-spur value is 44.17 dBc (7.29), which agrees closely. The frequency bin of the worst case spur in Figure 8-2 is 1784 ($8\times F(1)$), where $F(1)$ is from (7.25) and (7.26) and the DFT is calculated over eight DDS output periods (Pe). If M is larger than 2, the spur power is spread over many peaks (see Figure 8-3). The number of spurs is 15 from (7.8) in Figure 8-3. Since $M = 16$ for this case, the expected worst-case carrier-to-spur value is approximately 48.16 dBc (7.30). The worst-case carrier-to-spur level due to the phase truncation appears to be 48.08 dBc.

7.2 Finite Precision of Sine Samples Stored in LUT

Finite quantization in the sine LUT values also leads to the DDS output spectrum impairments. If it is assumed that the phase truncation does not exist, then the output of the DDS is given by

$$\sin(\frac{2\pi}{2^j}(\Delta P\, n)) - e_A(n), \tag{7.31}$$

where $e_A(n)$ is the quantization error due to the finite sine LUT data word. The sequence of the LUT quantization errors is periodic, repeating every Pe samples (4.4). There are two limiting cases, i.e. cases where the numerical period of the output sequence (Pe) is either long or short, to consider.

In the first case, the quantization error results in what appears to be a white noise floor, but is actually a "sea" of very finely spaced discrete spurs. The amplitude quantization errors can be assumed to be totally uncorrelated and uniformly distributed within each quantization step,

$$-\frac{\Delta_A}{2} \le e_A \le \frac{\Delta_A}{2}, \tag{7.32}$$

where the quantization step size is

$$\Delta_A = \frac{1}{2^m}, \tag{7.33}$$

and where m is the word length of the sine values stored in the sine LUT. Then the amplitude error power is [Ben48]

$$E\{e_A^2\} = \frac{1}{\Delta_A} \int_{-\frac{\Delta_A}{2}}^{\frac{\Delta_A}{2}} e_A^2 \, de_A = \frac{\Delta_A^2}{12}. \tag{7.34}$$

The signal power of the sine wave is

$$P_A = \frac{A^2}{2}, \tag{7.35}$$

where A is the amplitude of the sine wave. In the DDS literature, there is well-known formula for the signal to noise ratio due to the amplitude quantization (rounding)

$$\left(\frac{S}{N}\right) = 10\ \log_{10}\left(\frac{P_A}{E\{ea^2\}}\right) = 10\ \log_{10}\left(\frac{P_{CA}}{E\{cea^2\}}\right) \approx (1.76 + 6.02m)\ \text{dB}, \tag{7.36}$$

where P_A is $A^2/2$ (sine wave power), P_{CA} is A^2 (quadrature wave power), A is 0.5 in (7.36) and the quantization power of the sine wave

$$E\{ea^2\} = \frac{\Delta_m^2}{12} = \frac{2^{-2m}}{12}, \tag{7.37}$$

and the error power for the quadrature output signal is

$$E\{cea^2\} = \frac{\Delta_m^2}{6} = \frac{2^{-2m}}{6}, \tag{7.38}$$

where Δ_m is the phase to amplitude converter step size and m is the phase to amplitude converter wordlength.

Comparing (7.22) and (7.36), if $k \leq m + 2$, then the phase truncation dominates S/N, otherwise amplitude quantization. Due to the sine/cosine wave symmetry the error is also symmetric, therefore the spurs located at the even bin positions are zero for all phase increment words [Tie71] ($Pe/2$ spurs for cosine output and $Pe/4$ spurs for quadrature output, where the period of the DDS output (Pe) is from (4.4)). The cosine generated is real, so its power is equally divided into negative and positive frequency components. The quadrature output is complex, so its power is in one frequency component. Therefore, the amplitude quantization noise floor is from (7.36) for the cosine output and quadrature output

$$NF \approx (1.76 + 6.02m + 10\ \log_{10}\left(\frac{Pe}{4}\right))\ \text{dBc, when } Pe >> 1. \tag{7.39}$$

In the second case, there will be no quantization errors if the samples match exactly the quantization levels, e. g., $f_{out} = f_s/4$. The assumption that the error is evenly distributed in one period is really not valid due to the shortness of the period. Assuming that the amplitude error gets its maximum absolute value ($\Delta_A/2$) at every sampling instance, and that all the energy is in one spur, the carrier-to-spur ratio is

$$\left(\frac{C}{S}\right) = 10 \times \log_{10}\left(\frac{4P_A}{\Delta_A^2}\right) = (-3.01 + 6.02m)\ \text{dBc}. \tag{7.40}$$

However, simulations indicate that in the worst-case the sum of the discrete spurs is approximately equal to

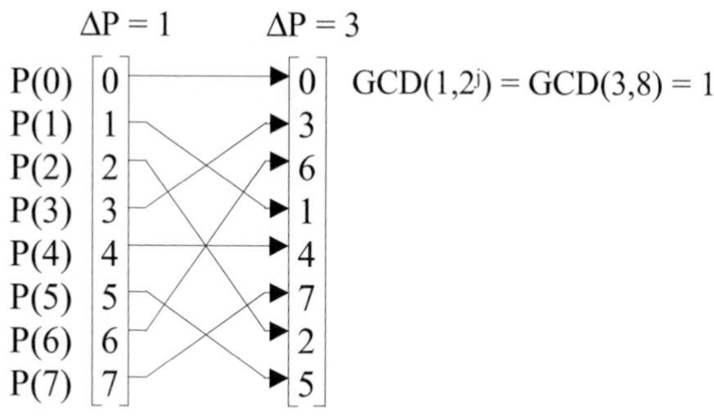

Figure 7-3. Time series vectors for a 3-bit phase accumulator for $\Delta P = 1$ and $\Delta P = 3$. The column vector for $\Delta P = 3$ can be formed from a permutation of the values of the $\Delta P = 1$ vector, regardless of the initial phase accumulator contents.

$$\left(\frac{C}{S_{sum}}\right) = \left(\frac{P_A}{E\{e_A^2\}}\right) = (1.76 + 6.02m) \text{ dBc.} \tag{7.41}$$

7.3 Distribution of Spurs

One of the advantages of a DDS is its ability to provide a continuous phase when changing phase increment words; the phase accumulator need not be reset when a new ΔP is applied. The state of the phase accumulator at the point in time when a new ΔP is applied provides a natural phase-offset for the subsequent DDS output, thereby providing continuous-phase frequency switching. The initial phase of the phase accumulator at the time when a new ΔP is applied could, however, also be a factor in determining the output spurs.

The j-bit phase accumulator can be considered as a permutation generator, where each value of ΔP provides a different permutation of the values from 0 to $2^j - 1$ given by

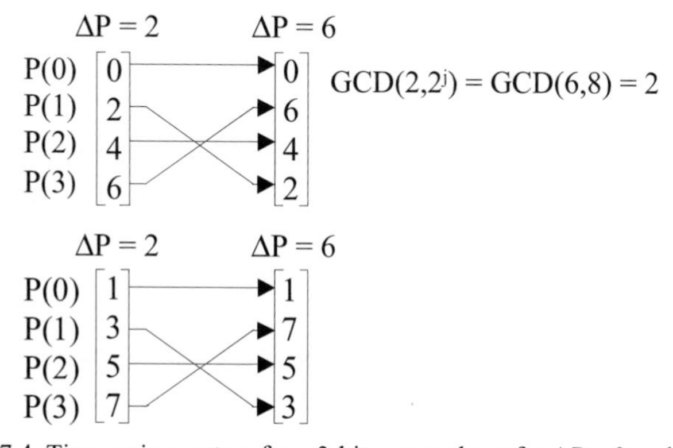

Figure 7.4. Time series vectors for a 3-bit accumulator for $\Delta P = 2$ and $\Delta P = 6$.

$$ {}_{\Delta P}P(n) = (n\Delta P + p) \bmod 2^j. \tag{7.42}$$

The output vector generated by the phase accumulator depends on the phase increment word and on the initial accumulator value, i.e., the initial phase (p) in (7.42). In Figure 7-3, any phase accumulator output vector can be formed from the permutation of another output vector, regardless of the initial phase accumulator contents, when GCD(ΔP, 2^j) = 1 for all values of ΔP. In Figure 7.4, the time vectors are formed from phase increment values that have the property GCD(ΔP, 2^j) = 2^1 when the initial phase is not same (0 and 1). From this figure it is evident that the phase accumulator is now characterized by having two different sets of possible output vectors, depending on the initial contents of the phase accumulator.

The number of the least significant bits (i), which are zero in the phase increment word, can be obtained from

$$\mathrm{GCD}(\Delta P, 2^j) = 2^i, \quad \text{where } i = (0, 1, ..., j-1), \tag{7.43}$$

and the phase value of the least i significant bits of the initial phase (p) is

$$p_i = (p) \bmod 2^i, \quad \text{where } p_i = (0, ..., 2^i - 1). \tag{7.44}$$

The most (j-i) significant initial phase bits do not affect output vector values but cause a constant phase shift to the output vector. Consequently, for the phase increment words with the property of GCD(ΔP, 2^j) = 2^i, we have to evaluate only for initial phases (0, ..., 2^i - 1). Therefore, by evaluating the output vector for $\Delta P = 2^i$ and every initial phase p_i ε (0, ..., $2^i - 1$) with i ε (0, ..., $j - 1$) we know the output vector for any ΔP and initial phase p.

The time output vector generated by the phase increment words with the property of GCD(ΔP, 2^j) = 2^i can be formed from a permutation of the individual elements of the vector for $\Delta P = 2^i$ (assuming the initial phase $p = 0$),

$$ {}_{\Delta P}P(n) = {}_{2^i}P((n\Delta P / 2^i) \bmod Pe), \tag{7.45}$$

where Pe is the period of the phase accumulator output (2^{j-i}) from (4.4), and $\Delta P/2^i$ and Pe are relatively prime. As in (7.45), output time vectors may be formed from a permutation of another time vector by permuting the indices using ($n\Delta P/2^i$) mod Pe. The converse follows from the existence of a unique integer $0 \le J < Pe$ satisfying the relation

$$(\Delta P / 2^i)\, J \bmod Pe = 1. \tag{7.46}$$

This is a fundamental result of number theory that requires that $\Delta P/2^i$ and Pe are relatively prime [McC79]. In a sense, J is the multiplicative inverse of $\Delta P/2^i$. From the above equation it follows that $\Delta P/2^i$ and J must be odd because Pe is even. Therefore J and Pe are relatively prime, too.

The DDS with a sinusoidal output operates by applying some memoryless non-linear function $s\{\}$ to the phase accumulator output to produce the sine function. The DFT of the phase to amplitude converter output using (7.45) is

$$S\{_{\Delta P} P(m)\} = \sum_{n=0}^{Pe-1} s\{_{2^i} P((n\Delta P/2^i) \bmod Pe)\} W_{Pe}^{mn} \quad m = 0,1,..., Pe-1, \tag{7.47}$$

where $W_{Pe} = e^{-j2\pi/Pe}$,

and Pe is the period of the phase accumulator output (4.4). (7.46) can be used to show that permutation samples in the time domain produce the same type of permutation in the frequency domain by defining the new index

$$q = (n\Delta P/2^i) \bmod Pe, \tag{7.48}$$

and noting that

$$\begin{aligned} qJ \bmod Pe &= J((n\Delta P/2^i) \bmod Pe) \bmod Pe \\ &= n(\Delta P/2^i) J \bmod Pe. \end{aligned} \tag{7.49}$$

Substituting from (7.46), (7.49) becomes

$$n = qJ \bmod Pe. \tag{7.50}$$

Re-indexing (7.47) using (7.48) and (7.50), then

$$\begin{aligned} S\{_{\Delta P} P(m)\} &= \sum_{q=0}^{Pe-1} s\{_{2^i} P(q)\} W_{Pe}^{m(qJ \bmod Pe)} \\ &= \sum_{q=0}^{Pe-1} s\{_{2^i} P(q)\} W_{Pe}^{q(mJ \bmod Pe)} \\ &= S\{_{2^i} P((mJ) \bmod Pe)\} \quad m = 0,1,..., Pe-1. \end{aligned} \tag{7.51}$$

The above equation establishes that the permutation of the samples in the time domain results in the same type of permutation as the DFT samples in the frequency domain, because J and Pe are relatively prime. This means that the spurious spectrum due to all system non-linearities generated by the phase increment words with the property of GCD(ΔP, 2^j) = 2^i is a permutation of the spectrum generated by $\Delta P = 2^i$ (assuming the same initial phase, p = 0 in this case), because each spectrum will differ only in the position of the spurs and not in the magnitudes. Therefore, by evaluating the DDS spectrum for $\Delta P = 2^i$ and every initial phase $p_i \in (0, ..., 2^i - 1)$ with $i \in (0, ..., j-1)$, we know the DDS spectrum magnitude for any ΔP and initial phase p. For phase increment words with GCD(ΔP, 2^j) = 1, the DDS spectrum magnitude does not depend on the initial phase and the evaluation of the spectrum magnitude has to be made just for $p = 0$ (Figure 7-3).

Moreover, when the output of the phase accumulator is truncated to k bits (see Section 7.1), for phase increment words with GCD(ΔP, 2^j) = $2^i \leq 2^{j-k}$, the least i significant bits of the initial phase p does not affect the output sequence. For the spectrum evaluation procedure, it means that for these phase increment words the DDS spectrum magnitude does not depend on the initial phase and the evaluation of the spectrum has to be made just for $p = 0$.

Two ΔP values 619 and 1121 are considered in the example. The DFT of the DDS output sequence for ΔP = 619 is shown in Figure 7-5, where the worst-case carrier-to-spur level due to the phase truncation appears to be 48.08 dBc. Since $M >> 2$ for this case, the expected worst-case carrier-to-spur value is 48.16 dBc (7.30), which agrees closely. The number of spurs in the figures is 15, from (7.8). The frequency bins of the spurs in Figure 7-5 could be calculated using (7.25) and (7.26). The DFT spectrum for the second frequency case of ΔP = 1121 is shown in Figure 7-6. As predicted, since GCD (ΔP, 2^j) = 1 for this case as well, the worst-case carrier-to-spur level is unchanged and only the position of the spurs has been permutated.

7.4 Phase Noise of DDS Output

Leeson has developed a model that describes the origins of phase noise in oscillators [Lee66]; since it closely fits experimental data, the model is widely used in describing the phase noise of the oscillators [Roh83], [Man87]. In the model, the sampling clock signal (oscillator output) is phase modulated by a sine wave of frequency f_m

$$y_s(t) = \cos(\omega_s t + \beta \sin \omega_m t), \tag{7.52}$$

where ω_s is the sampling clock frequency of DDS, β is the maximum value of the phase deviation, ω_m is the offset frequency. The spectrum of the sampling clock signal is shown in Figure 7-7.

The frequency of the sampling clock signal is

$$f_s(t) = \frac{1}{2\pi}\frac{d(\theta_s(t))}{dt} = \frac{1}{2\pi}(\omega_s + \beta\omega_m \cos \omega_m t). \tag{7.53}$$

The DDS could be described as a frequency divider, and so the output frequency of the DDS is

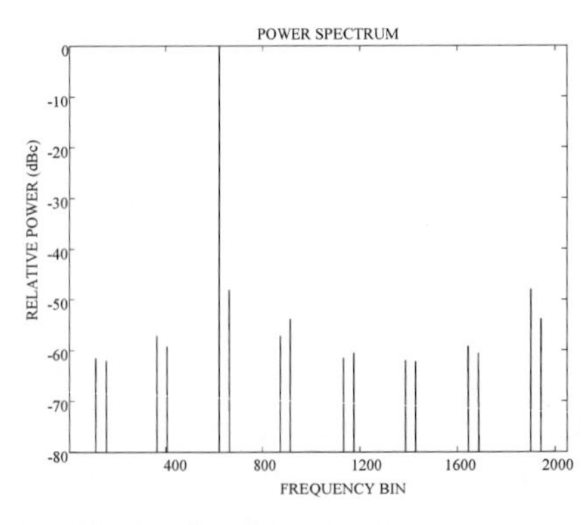

Figure 7-5. Discrete Fourier transform of the DDS output sequence for j = 12, k = 8 and ΔP = 619.

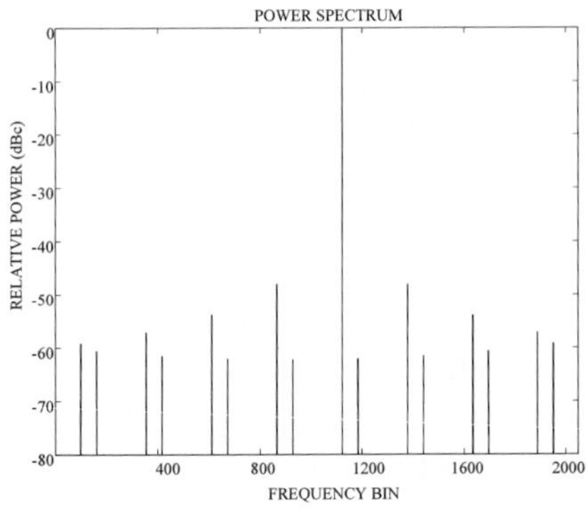

Figure 7-6. Discrete Fourier transform of the DDS output sequence for j = 12, k = 8 and ΔP = 1121.

$$f_{out}(t) = \frac{\Delta P f_s(t)}{2^j} = \frac{f_s(t)}{N} = \frac{1}{2\pi}(\omega_{out} + \frac{\beta}{N}\omega_m \cos\omega_m t), \tag{7.54}$$

where j is the word length of the DDS phase accumulator, ΔP is the phase increment word, N is the division ratio. The phase of the DDS output is

$$\theta_{out}(t) = (\omega_{out} t + \frac{\beta}{N}\sin\omega_m t), \tag{7.55}$$

and the DDS output is

$$y_{out}(t) = \cos(\omega_{out} t + \frac{\beta}{N}\sin\omega_m t). \tag{7.56}$$

Comparing (7.52) and (7.56), the modulation index is changed from β to β/N, but the offset frequency is not changed. The spectrum of the DDS sampling clock is given by inspection from the equivalent relationship

$$\begin{aligned} y_s(t) &= \cos(\omega_s t + \beta \sin\omega_m t) = \mathrm{Re}\left\{e^{j\omega_s t} e^{j\beta \sin\omega_m t}\right\} \\ &= \mathrm{Re}\left\{e^{j\omega_s t} \sum_{i=-\infty}^{\infty} J_i(\beta) e^{ji\omega_m t}\right\} = \sum_{i=-\infty}^{\infty} J_i(\beta)\cos(\omega_s t + i\omega_m t), \end{aligned} \tag{7.57}$$

where $J_i(\beta)$ are Bessel functions of the first kind. The spectrum of the DDS output is given by inspection from the equivalent relationship

$$\begin{aligned} y_{out}(t) &= \cos(\omega_{out} t + \frac{\beta}{N}\sin\omega_m t) = \mathrm{Re}\left\{e^{j\omega_{out} t} e^{j\frac{\beta}{N}\sin\omega_m t}\right\} \\ &= \mathrm{Re}\left\{e^{j\omega_{out} t} \sum_{i=-\infty}^{\infty} J_i(\frac{\beta}{N}) e^{ji\omega_m t}\right\} = \sum_{i=-\infty}^{\infty} J_i(\frac{\beta}{N}) \cos(\omega_{out} t + i\omega_m t). \end{aligned} \tag{7.58}$$

The relative power of the DDS output phase noise at offset $i\omega_m$ is from (7.57) and (7.58)

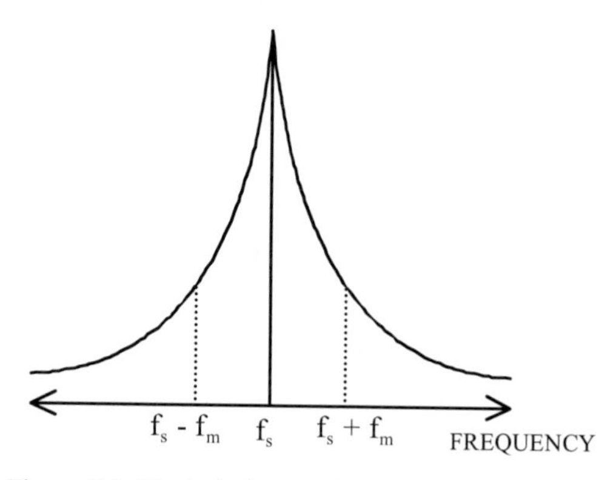

Figure 7-7. Typical phase noise sidebands of an oscillator.

$$\frac{P_{out_i}}{P_{s_i}} = \left(\frac{J_i(\frac{\beta}{N})}{J_i(\beta)} \right)^2 . \tag{7.59}$$

If $\beta << 1$, then $J_0(\beta) \approx 1$, $J_0(\beta/N) \approx 1$, $J_1(\beta) \approx \beta/2$, $J_1(\beta/N) \approx \beta/(2N)$ and $J_i(\beta) \approx 0$ (i = 2, 3...), and

$$\left(\frac{P_{out_1}}{P_{s_1}} \right)_{\text{dB}} \approx -20 \times \log_{10}(N)\ [\text{dB}]. \tag{7.60}$$

From the above equation, it can be seen that the relative power level of the DDS output phase noise depends on the ratio between the output frequency and sampling clock frequency. The output signal will exhibit the improved phase noise performance [Jen97], [Ana99]

$$n_{clk} - 20 \times \log_{10}(\frac{f_s}{f_{out}}). \tag{7.61}$$

The DDS circuitry has a noise floor, which, at some point, will limit this improvement. An output phase noise floor of -160 dBc/Hz is possible, depending on the logic family used to implement the DDS [Qua90]. The frequency accuracy of the sampling clock is propagated through the DDS [Qua90]. Therefore, if the sampling clock frequency is 0.1 PPM higher than desired, the output frequency will be also higher by 0.1 PPM.

7.5 Post-Filter Errors

The sixth source of noise at the DDS output is the post-filter, e_F, which is needed to remove the high frequency sampling components. Since this post-filter is an energy storage device, the problem of the response time arises. The filter must have a very flat amplitude response and a constant group delay across the bandwidth of interest so that the perfectly linear digital modulation and frequency synthesis advantages are not lost. The output filter also affects the switching time of the DDS output.

REFERENCES

[Ana99] Analog Devices, "A Technical Tutorial on Digital Signal Synthesis," Application Note, 1999.

[Ben48] W. R. Bennett, "Spectra of Quantized Signals," Bell Sys. Tech. J., Vol. 27, pp. 446-472, July 1948.

[Jen88a] Y. C. Jenq, "Digital Spectra of Nonuniformly Sampled Signals: Fundamentals and High-Speed Waveform Digitizers," IEEE Trans. Inst. and Meas., Vol. 37, pp. 245-251, June 1988.

[Jen88b] Y. C. Jenq, "Digital Spectra of Nonuniformly Sampled Signals - Digital Look-Up Tunable Sinusoidal Oscillators," IEEE Trans. on Inst. and Meas., Vol. 37, No. 3, pp. 358-362, Sept. 1988.

[Jen97] Y. C. Jenq, "Direct Digital Synthesizer with Jittered Clock," IEEE Trans. on Instrumentation and Measurement, Vol. 46, No. 3, pp. 653–655, June 1997.

[Kro00] V. F. Kroupa, V. Cizek, J. Stursa, and H. Svandova, "Spurious Signals in Direct Digital Frequency Synthesizers due to the Phase Truncation," IEEE Transactions on Ultrasonics Ferroelectrics and Frequency Control, Vol. 47, No. 5, pp. 1166-1172, Sept. 2000.

[Lee66] D. B. Leeson, "A Simple Model of Feedback Oscillator Noise Spectrum," IEEE Proc., Vol. 54, pp. 329-330, Feb. 1966.

[Man87] V. Manassewitsch, "Frequency Synthesizers Theory and Design," 3nd Edition, New York: Wiley, 1980.

[McC79] J. H. McClellan, and C. M. Rader, "Number Theory in Digital Signal Processing," Englewood Cliffs, NJ: Prentice-Hall, 1979.

[Meh83] S. Mehrgardt, "Noise Spectra of Digital Sine-Generators Using the Table-Lookup Method," IEEE Trans. Acoust., Speech, Signal Process., Vol. ASSP-33, No. 4, pp. 1037-1039, Aug. 1983.

[Nic87] H. T. Nicholas, and H. Samueli, "An Analysis of the Output Spectrum of Direct Digital Frequency Synthesizers in the Presence of Phase-Accumulator Truncation," in Proc. 41st Annu. Frequency Contr. Symp., June 1987, pp. 495-502.

[Qua90] Qualcomm, "Hybrid PLL/DDS Frequency Synthesiser," AN2334-4, 1990.

[Roh83] U. L. Rohde, "Digital PLL Frequency Synthesizers Theory and Design," Prentice-Hall Inc., 1983.

[Tie71] J. Tierney, C. Rader, and B. Gold "A Digital Frequency Synthesizer," IEEE Trans. Audio and Electroacoust., Vol. AU-19, pp. 48-57, Mar. 1971.

[Tor01] A. Torosyan, and A. N. Willson, Jr., "Analysis of the Output Spectrum for Direct Digital Frequency Synthesizers in the Presence of Phase Truncation and Finite Arithmetic Precision," Proceedings of the 2nd International Symposium on Image and Signal Processing and Analysis, 2001, pp. 458-463.

Chapter 8

8. SPUR REDUCTION TECHNIQUES IN SINE OUTPUT DIRECT DIGITAL SYNTHESIZER

The drawback of the direct digital synthesizer (DDS) is the high level of spurious frequencies [Rei93]. In this chapter we concentrate only on the spurs that are caused by the finite word length representation of phase and amplitude samples. The number of words in the LUT (phase to amplitude converter) will determine the phase quantization error, while the number of bits in the digital-to-analog converter (D/A-converter) will affect amplitude quantization. Therefore, it is desirable to increase the resolution of the LUT and D/A-converter. Unfortunately, larger LUT and D/A-converter resolutions mean higher power consumption, lower speed, and greatly increased costs. Memory compression techniques could be used to alleviate the problem, but the cost of different techniques is an increase in circuit complexity and distortions (see Section 9.2).

Additional digital techniques may be incorporated in the DDS in order to reduce the presence of spurious signals at the DDS output. The Nicholas modified phase accumulator does not destroy the periodicity of the error sequences, but it spreads the spur power into many spur peaks [Nic88]. Non-subtractive dither is used to reduce the undesired spurious components, but the penalty is that the broadband noise level is quite high after dithering [Rei93], [Fla95]. To alleviate the increase in noise, subtractive dither can be used in which the dither is added to the digital samples and subtracted from the DDS analog output signal [Twi94]. The requirement of dither subtraction at the DDS output makes the method complex and difficult to implement in practice. The novel spur reduction technique presented in this work uses high-pass filtered dither [Car87], [Ble87], which has most of its power in an unused spectral region between the band edge of the low-pass filter and the Nyquist frequency. After the DDS output has been passed through the low-pass filter, only a fraction of the dither power will remain. From this point of view, the low-pass filtering is a special implementation of the dither subtraction operation.

An error feedback (EF) technique is used to suppress low frequency quantization spurs [Lea91a], [Lea91b], [Laz94]. A novel tunable error feedback structures in the DDS is developed in Section 8.4. The drawback of conventional EF structures is that the output frequency is low with respect to the sampling frequency, because the transfer function of the EF has zero(s) at DC. In the proposed architecture, the sampling frequency needs only to be much greater than the bandwidth of the output signal, whereas the output frequency could be any frequency up to somewhat below the Nyquist rate. The coefficients of the EF are tuned according to the output frequency.

8.1 Nicholas Modified Accumulator

This method does not destroy the periodicity of the error sequences, but spreads the spur power into many spur peaks [Nic88]. If GCD (ΔP, 2^{j-k}) is equal to 2^{j-k-1}, the spur power is concentrated in one peak, see Figure 8-2. The worst case carrier-to-spur ratio is from (7.29)

$$\left(\frac{C}{S}\right) = (6.02k - 3.992)\text{ dBc}, \tag{8.1}$$

where k is the word length of the phase accumulator output used to address the LUT. If GCD (ΔP, 2^{j-k}) is equal to 1, the spur power is spread over many peaks in Figure 8-3. Then the carrier-to-spur ratio is approximately, from (7.30),

$$\left(\frac{C}{S}\right) = 6.02k\text{ dBc} \quad \text{when } (j - k) >> 1. \tag{8.2}$$

Comparing (8.1) and (8.2) shows that the worst-case spur can be reduced in magnitude by 3.922 dB by forcing GCD (ΔP, 2^{j-k}) to be unity, i.e. by forcing the phase increment word to be relatively prime to 2^{j-k}. This causes the phase accumulator output sequence to have a maximal numerical period for all values of ΔP, i.e. all possible values of the phase accumulator output sequence are generated before any values are repeated. In Figure 8-1, the hardware addition is to modify the existing j-bit phase accumulator structure to emulate the operation of a phase accumulator with a word length of j+1 bits under, the assumption that the least significant bit of the phase increment word is always one [Nic88]. It, too, has the effect of randomizing the errors introduced by the quantizied LUT samples, because, in a long output period, the error appears as "white noise" (7.39).

The disadvantage of the modification is that it introduces an offset of

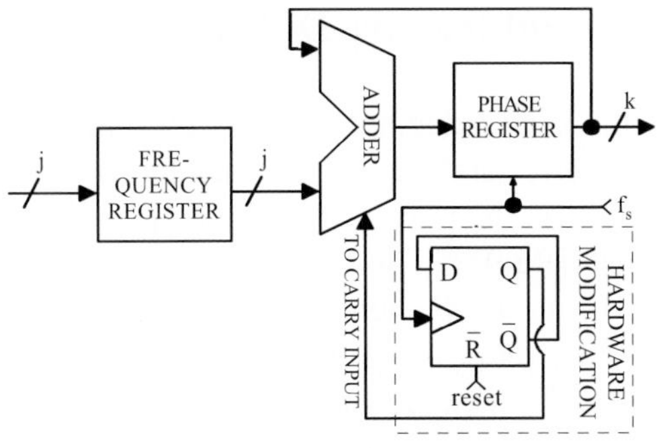

Figure 8-1. Hardware modification to force optional GCD $(\Delta P, 2^{j-k+1}) = 1$.

$$f_{offset} = \frac{f_{clk}}{2^{j+1}} \tag{8.3}$$

into the output frequency of the DDS. The offset will be small, if the clock frequency is low and the length of the phase accumulator is long. If there is no phase truncation error in the original samples (GCD $(\Delta P, 2^j) \geq 2^{j-k}$), then this method will make the situation worse for the phase error. If the carrier-to-spur ratio due to the amplitude and phase quantization is lower than $6.02k$ dBc (8.2) without the Nicholas modified accumulator, then the carrier-to-spur ratio can be improved up to maximum $6.02k$ dBc due to the spreading of the amplitude and phase quantization spurs. If the carrier-to-spur ratio due to the amplitude quantization (no phase truncation) is higher than $6.02k$ dBc without the Nicholas modified accumulator, then the degradation of the carrier-to-spur ratio occurs. It is good, therefore, that this spur reduction method is optional, depending on the phase increment word [Van98].

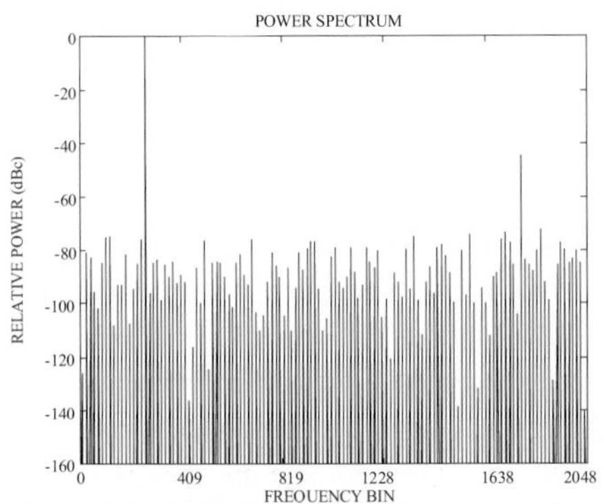

Figure 8-2. Spur due to the phase truncation, max. carrier-to-spur level 44.24 dBc (44.17 dBc (7.29) and (8.1)). There is a sea of amplitude spurs below the phase spur. The simulation parameters: j = 12, k = 8, m = 10, ΔP = 264.

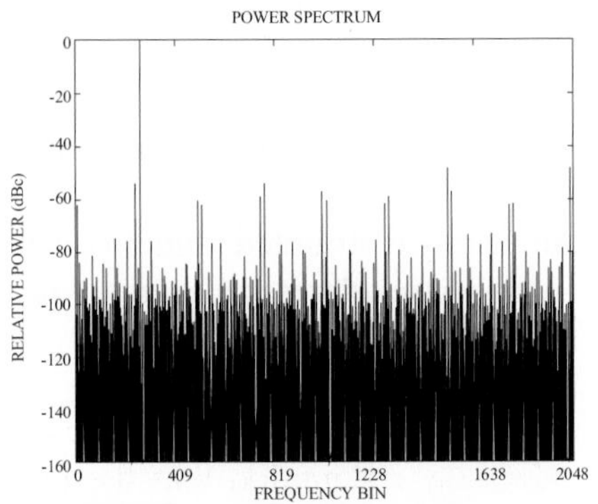

Figure 8-3. Spurs due to the phase truncation, max. carrier-to-spur level 48.08 dBc (48.16 dBc (7.30) and (8.2)). The simulation parameters same as Figure 8-2 but ΔP = 265.

8.2 Non-Subtractive Dither

In this section, methods of reducing the spurs by rendering certain statistical moments of the total error statistically independent of the signal are investigated [Fla95]. In essence, the power of the spurs is still there, but spreads out as a broadband noise [Rei93]. This broadband noise is more easily filtered out than the spurs. In the DDS there are different ways to dither: some designs have dithered the phase increment word [Whe83], the address of the sine wave table [Jas87], [Zim92] and the sine-wave amplitude [Rei91], [Ker90], [Fla95] with pseudo random numbers, in order to randomize the phase or amplitude quantization error.

The dither is summed with the phase increment word in the square wave output DDS [Whe83]. The technique could be applied to the sine output DDS (source 1 in Figure 8-4), too. It is important that the dither signal is canceled during the next sample, otherwise the dither will be accumulated in the phase accumulator and there will be frequency modulation. The circuit will be complex due to the previous dither sample canceling, therefore this method is beyond the scope of this work.

It is important that the period of the evenly distributed dither source (L) satisfies [Fla95]

$$\frac{\Delta^2}{6L} < P_{max}, \tag{8.4}$$

where $P_{\max}$ is the maximum acceptable spur power, and Δ is the step size for both the amplitude and phase quantization. In this work, first-order dither signals (evenly distributed) are considered. The use of higher-order dither accelerates spur reduction with the penalty of a more complex circuit and higher noise floor [Fla93], [Fla95].

8.2.1 Non-Subtractive Phase Dither

An evenly distributed random quantity $z_P(n)$ (source 2 in Figure 8-4) is added to the phase address prior to the phase truncation. The output sequence of the

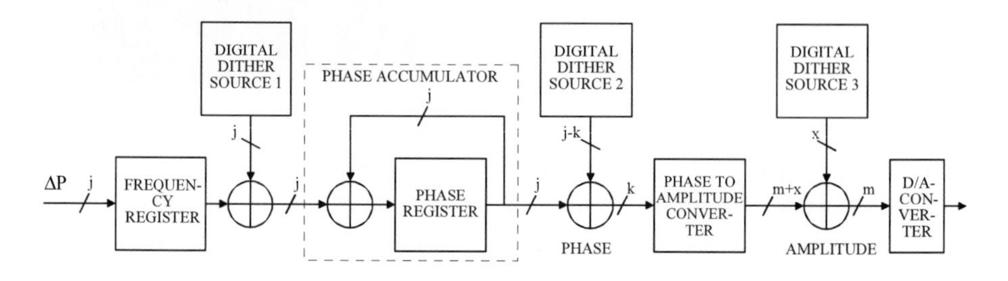

Figure 8-4. Different ways of dithering in the DDS.

DDS is given by

$$x(n) = \sin(\frac{2\pi}{2^j}(P(n) + \varepsilon(n))), \tag{8.5}$$

where $P(n)$ is a phase register value. The total phase truncation noise is

$$\varepsilon(\text{n}) = e_P(n) + z_P(n), \tag{8.6}$$

where the phase truncation error varies periodically as

$$e_p(n) = (P(n))\text{mod}2^{j-k}, \text{ when GCD}(\Delta P, 2^{j-k}) < 2^{j-k}, \tag{8.7}$$

and the period of the phase truncation error (M) is from (7.7).

Using small angle approximation

$$x(n) \approx \sin(\frac{2\pi}{2^j}P(n)) + \frac{2\pi}{2^j}\varepsilon(n)\cos(\frac{2\pi}{2^j}P(n)) + O((\max(\varepsilon(n)))^2), \tag{8.8}$$

where $\max(\varepsilon(n))$ is 2^{-k}. The number of bits, k, must be large enough to satisfy the small angle assumption, typically, $k \geq 4$. The total quantization noise will be examined by considering the first two terms above, and then the second-order, $O((\max(\varepsilon(n)))^2)$, effect.

8.2.2 First-Order Analysis

The total phase fluctuation noise will be proportional to $e_P(n)$ [Fla95], when the random value $z_P(n)$ is added to the phase address before truncation to k-bits, as in Figure 8-5. The evenly distributed random quantity $z_P(n)$ varies in the range $[0, 2^{j-k})$. If $z_P(n)$ is less than the quantity $(2^{j-k} - e_P(n))$, then $e_P(n) + z_P(n))$ will be truncated to (0). The total phase truncation noise will be

$$\varepsilon(\text{n}) = -e_P(n) \tag{8.9}$$

with probability

$$\frac{(2^{j-k} - e_P(n))}{2^{j-k}}, \tag{8.10}$$

because there are $(2^{j-k} - e_P(n))$ values of $z_P(n)$ less than $(2^{j-k} - e_P(n))$, and there are 2^{j-k} values of $z_P(n)$. If $z_P(n)$ is equal to, or greater than, the quantity $(2^{j-k} - e_P(n))$, then $(e_P(n) + z_P(n))$ will be truncated to (2^{j-k}). The total phase truncation noise will be

$$\varepsilon(\text{n}) = (2^{j-k} - e_P(n)) \tag{8.11}$$

with the probability

$$\frac{e_P(n)}{2^{j-k}}, \tag{8.12}$$

because there are $e_P(n)$ values of $z_P(n)$ that are equal to, or greater than, $(2^{j-k} - e_P(n))$.

At all sample times n, the first moment of the total phase truncation noise is zero

$$E\{\varepsilon(n)\} = -e_P(n)\frac{(2^{j-k} - e_p(n))}{2^{j-k}} + (2^{j-k} - e_p(n))\frac{e_P(n)}{2^{j-k}} = 0. \quad (8.13)$$

The second moment of the total phase truncation noise is

$$\begin{aligned} E\{\varepsilon^2(n)\} &= e_P^2(n)\frac{(2^{j-k} - e_p(n))}{2^{j-k}} + (2^{j-k} - e_p(n))^2\frac{e_P(n)}{2^{j-k}} \\ &= 2^{j-k}e_p(n) - e_P^2(n) \\ &= 2^{2(j-k)}\left(\frac{e_p(n)}{2^{j-k}} - \left(\frac{e_p(n)}{2^{j-k}}\right)^2\right). \end{aligned} \quad (8.14)$$

Two bounds are derived for the average value of the second moment (the power of the total truncation noise) based on the period of the error term (M). In the first case, GCD (ΔP, 2^{j-k}) is 2^{j-k-1} and M is 2 (7.7), and the average value of the sequence (8.14) reaches its minimum non-zero value. The phase truncation error sequence is 0, 2^{j-k-1}, 0, 2^{j-k-1}, 0, 2^{j-k-1} … from (8.7). Then the sequence (8.14) becomes

$$E\{\varepsilon^2\} = 0 + \frac{2^{2(j-k)}}{4} + 0 + \frac{2^{2(j-k)}}{4} + 0 + \frac{2^{2(j-k)}}{4}\ldots \quad (8.15)$$

The average value of this sequence is

$$\text{Avg}(E\{\varepsilon^2\}) = \frac{2^{2(j-k)}}{8}. \quad (8.16)$$

In the second case, GCD (ΔP, 2^{j-k}) is 1 and M is 2^{j-k} (7.7), and the average value of the sequence (8.14) reaches its maximum value. In this case the phase truncation error sequence takes on all possible error values ([0, 2^{j-k})) before any is repeated. Then the average value of the sequence (8.14) be-

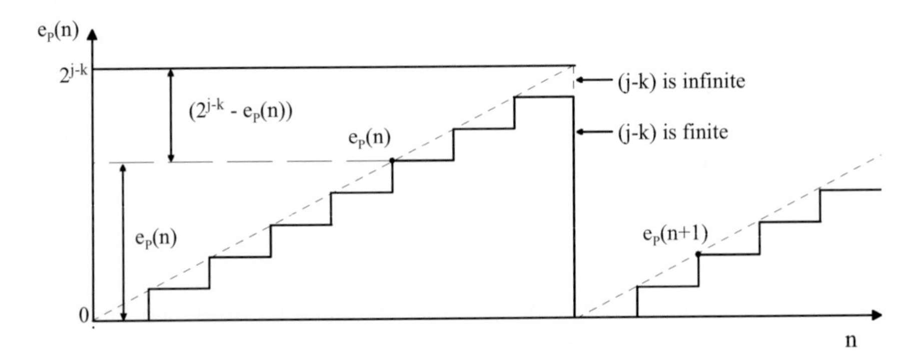

Figure 8-5. Phase truncation errors.

comes

$$\mathrm{Avg}(E\{\varepsilon^2\}) = \frac{2^{2(j-k)}}{6}, \text{ when } j >> k. \tag{8.17}$$

Information about the spurs and noise in the power spectrum of $x(n)$ is obtained from the autocorrelation function. The autocorrelation of $x(n)$ is [Fla95]

$$\begin{aligned} E\{x(n)\,x(n+m)\} &\approx \sin(\frac{2\pi}{2^j}P(n))\sin(\frac{2\pi}{2^j}P(n+m)) \\ &+\frac{4\pi^2}{2^{2j}}\cos(\frac{2\pi}{2^j}P(n))\cos(\frac{2\pi}{2^j}P(n+m))E\{\varepsilon(n)\,\varepsilon(n+m)\}+O(2^{-4k}). \end{aligned} \tag{8.18}$$

Spectral information is obtained by averaging over time [Lju87], resulting in [Fla95]

$$\bar{R}_{xx}[m] \approx \frac{1}{2}\left[1+\frac{4\pi^2}{2^{2j}}\bar{R}_{ee}[m]\right]\cos(\frac{2\pi}{2^j}P(m)), \tag{8.19}$$

where $\bar{R}_{ee}[m]$ = $\mathrm{Avg}_n(\mathrm{E}\{\varepsilon(n)\varepsilon(n+m)\})$, the time-averaged autocorrelation of the total quantization noise. It should be remembered that, for any fixed time n, the probability distribution of $\varepsilon(n)$, a function of $p(n)$, is determined entirely by the outcome of the dither signal $z(n)$. When $z(n)$ and $z(n+m)$ are independent random variables for non-zero lag m, $\varepsilon(n)$ and $\varepsilon(n+m)$ are also independent for $m \neq 0$, and hence $\varepsilon(n)$ is spectrally white. In this case, the autocorrelation becomes [Fla95]

$$\bar{R}_{xx}[m] \approx \frac{1}{2}\left[1+\frac{4\pi^2}{2^{2j}}\mathrm{Avg}(\varepsilon^2)\,\delta(m)\right]\cos(\frac{2\pi}{2^j}P(m)), \tag{8.20}$$

where $\delta(m)$ is the Kronecker delta function ($\delta(0) = 1$, $\delta(m) = 0$, $m \neq 0$).

The signal-to-noise ratio is derived from (8.20), when $m = 0$, as

$$\mathrm{SNR} \approx \frac{1}{\frac{4\pi^2}{2^{2j}}\mathrm{Avg}(E\{\varepsilon^2\})}. \tag{8.21}$$

The upper bound to the signal-to-noise ratio is from (8.16)

$$\mathrm{SNR} \approx 10\times\log_{10}\left(\frac{2}{\pi^2 2^{-2k}}\right) \approx (6.02k-6.93)\ \mathrm{dB}. \tag{8.22}$$

The lower bound to the signal-to-noise ratio is from (8.17)

$$\mathrm{SNR} \approx 10\times\log_{10}\left(\frac{6}{4\pi^2 2^{-2k}}\right) \approx (6.02k-8.18)\ \mathrm{dB}. \tag{8.23}$$

The sinusoid generated is a real signal, so its power is equally divided into negative and positive frequency components. The total noise power is divided to S spurs, where S is the number of samples and the period of the dither source is longer than S. Using these facts, the upper bound of the carrier-to-noise power spectral density is the same as in [Fla95]

$$\left(\frac{C}{N}\right) \approx (6.02k - 9.94 + 10\log_{10}(S))\ \text{dBc}. \tag{8.24}$$

The upper bound is achieved when GCD (ΔP, 2^{j-k}) is 2^{j-k-1}. The lower bound of the carrier-to-noise power spectral density is

$$\left(\frac{C}{N}\right) \approx (6.02k - 11.19 + 10\log_{10}(S))\ \text{dBc}. \tag{8.25}$$

The lower bound is achieved when $j >> k$ and GCD (ΔP, 2^{j-k}) is 1. The new bound (8.25) for the signal-to-noise spectral density is derived from these facts. A worst-case analysis of second-order effects was presented in [Fla95]. The phase dithering provides for acceleration beyond the normal 6 dB per bit spur reduction (7.29) to a 12 dB per bit spur reduction [Fla95]. Since the size of the LUT ($2^k \times m$) is exponentially related to the number of the phase bits, the technique results in a dramatic decrease in the LUT size. The expense of the phase dithering is the increased noise floor. However, the noise power is spread throughout the sampling bandwidth, so the carrier-to-noise spectral density could be raised by increasing the number of the samples in (8.24), (8.25). The phase dithering requires dither generation and an adder, which makes the circuit more complex. The overflows due to dithering cause no problems in the phase address, because the phase accumulator works according to the overflow principle.

The number of the samples is 4096 in all figures in Chapter 8. The carrier-to-noise power spectral density in Figure 8-6 is 74.35 dBc per FFT bin, in agreement with the lower bound 74.34 dBc (8.25). In Figure 8-7, the car-

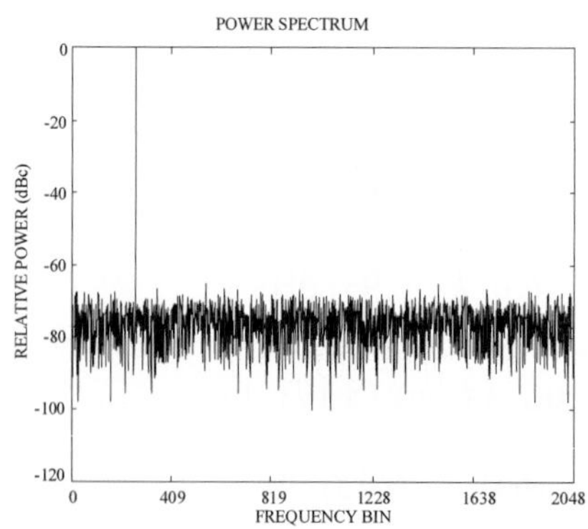

Figure 8-6. Dither is added into the phase address, when GCD (ΔP,2^{j-k}) = 1. Simulation parameters same as Figure 8-2.

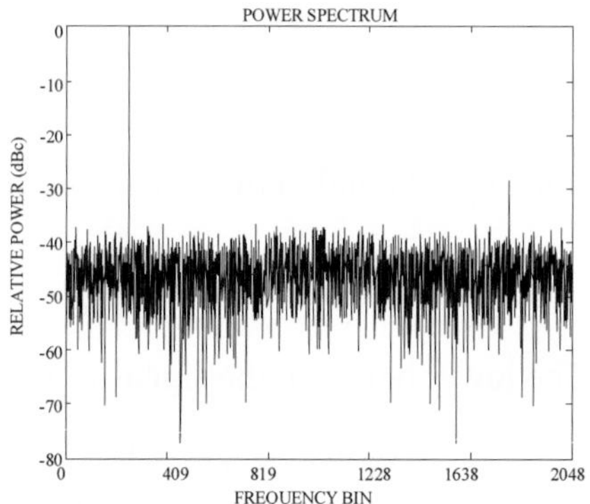

Figure 8-7. Dither is added into the phase address, when GCD (ΔP,2^{j-k}) = 2^{j-k-1} = 256. Simulation parameters: j = 12, k = 3, m = 10, ΔP = 256.

rier-to-spur level is 28.47 dBc (28.28 dBc [Fla95]), and the carrier-to-noise power spectral density is 44.20 dBc, in agreement with the upper bound 44.24 dBc (8.24).

8.2.3 Non-Subtractive Amplitude Dither

If a digital dither (from source 3 in Figure 8-4) is summed with the output of the phase to amplitude converter, then the output of the DDS can be expressed as

$$\sin(\frac{2\pi}{2^j}(\Delta P n - e_P(n))) + z_A(n) - e_A(n), \tag{8.26}$$

where $z_A(n)$ is the amplitude dither [Ker90], [Rei91], [Fla95]. The spurious performance of the D/A-converter input is the same as if the D/A-converter input were quantized to $(m + x)$ bits [Fla95], because the $z_A(n)$ randomizes a part of the quantization error (x bits) in Figure 8-4. If the $z_A(n)$ is wideband evenly distributed on $[-\Delta_A/2, \Delta_A/2)$, and independent of the $e_A(n)$, then the total amplitude noise power after dithering will be [Gra93]

$$E\{z_A^2\} + E\{e_A^2\} = \frac{\Delta_A^2}{12} + \frac{\Delta_A^2}{12}, \tag{8.27}$$

where $\Delta_A = 2^{-m}$, and $E\{e_A^2\}$ is from (7.39) or (7.41). The amplitude error power is doubled after dithering, but the error power is divided into all discrete frequency components. If the spur power is divided into the $Pe/2$ spurs (7.39), then, after dithering, the total noise power is divided into the Pe spurs and the carrier-to-spur power spectral density is not changed in the same measurement period (Pe). Then the carrier-to-noise power spectral density is the same as in (7.39)

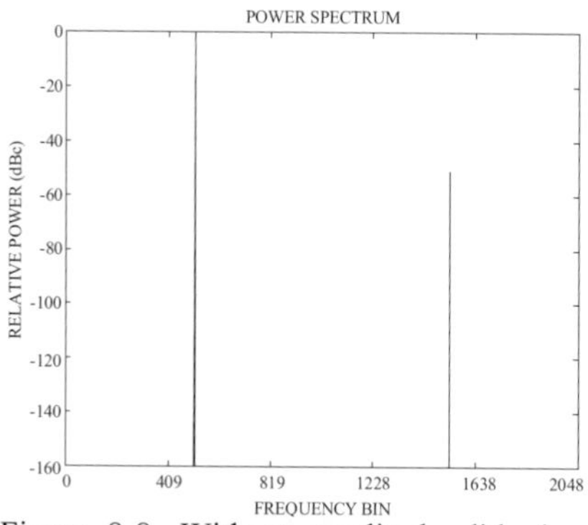

Figure 8-8. Without amplitude dithering, the carrier-to-spur level is 51.2 dBc. Simulation parameters: j, $k = 12$, $m = 8$, $x = 8$, $\Delta P = 512$.

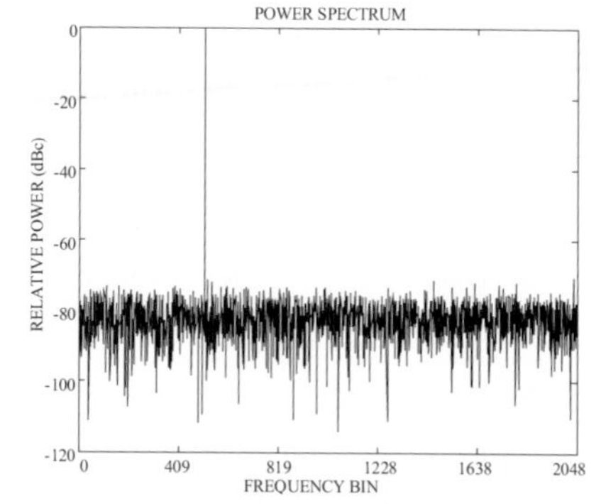

Figure 8-9. With amplitude dithering, the carrier-to-noise power spectral density is 80.1 dBc.

$$\left(\frac{C}{N}\right) = (1.76 + 6.02m + 10\times\log_{10}\left(\frac{Pe}{4}\right))\text{dBc}. \tag{8.28}$$

The penalty of amplitude dithering is a more complex circuit and a reduced dynamic range. In this method, the size of the LUT increases by $2^k \times x$, where k is the word length of the phase address and x is the word length of the amplitude error. The output of the LUT must be reduced (scaled) so that the original signal plus the dither will stay within the non-saturating region. The loss may be small, when the number of quantization levels is large.

Figure 8-8 shows the power spectrum of a sine wave without amplitude dithering. Figure 8-9 shows the power spectrum of a 16 bit sinusoid amplitude dithered with a random sequence, which is distributed evenly over $[-2^{-8}/2, 2^{-8}/2)$, prior to the truncation into 8 bits. The carrier-to-noise power spectral density is 80.1 dBc per FFT bin (80.02 dBc (8.28)) in Figure 8-9.

For example, QUALCOMM has used the non-subtractive amplitude dither in their device [Qua91a].

8.3 Subtractive Dither

Non-subtractive dither is used to reduce the undesired spurious components, but the penalty is that the broadband noise level is quite high after dithering. To alleviate the increase in noise, subtractive dither can be used, in which the dither is added to the digital samples and subtracted from the DDS analog output signal [Twi94]. The requirement of the dither subtraction at the DDS output makes the method complex and difficult to implement in practical applications. The technique presented in this work uses a high-pass filtered dither [Car87], [Ble87], which has most of its power in an unused spectral region between the band edge of the low-pass filter and the Nyquist frequency. After the DDS output has been passed through the low-pass filter, only a fraction of the dither power will remain [Ble87]. The low-pass filtering is a special implementation of the dither subtraction operation.

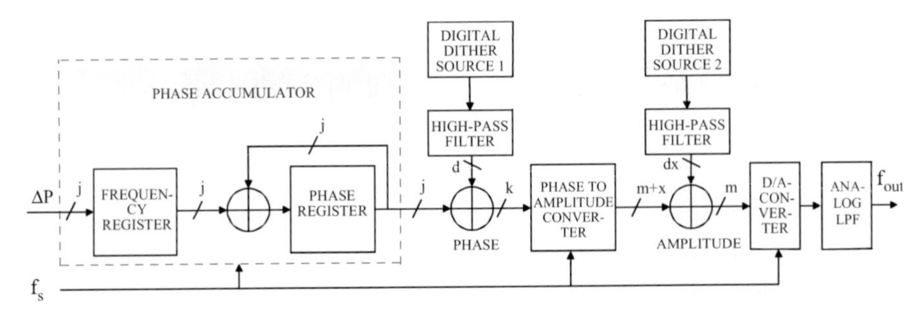

Figure 8-10. DDS with a high-pass filtered phase and amplitude dithering structures.

8.3.1 High-Pass Filtered Phase Dither

If a digital high-pass filtered dither signal $z_{HP}(n)$ (from source 1 in Figure 8-10) is added to the output of the phase accumulator, then the output of the DDS can be expressed as

$$\sin(\frac{2\pi}{2^j}(\Delta Pn - e_P(n) + z_{HP}(n))) - e_A(n). \tag{8.29}$$

If both the dither and the phase error are assumed to be small relative to the phase, then the DDS output signal (8.29) can be approximated by

$$\sin(2\pi\frac{f_{out}}{f_s}n) + \cos(2\pi\frac{f_{out}}{f_s}n)\frac{2\pi}{2^j}\left(z_{HP}(n) - e_P(n)\right) - e_A(n), \tag{8.30}$$

where f_{out} is the DDS output frequency and f_s is the DDS sampling frequency (4.1). The above phase dithering is in the form of an amplitude modulated sinusoid. The modulation translates the dither spectrum up and down in frequency by f_{out}, so that most of the dither power will be inside the DDS output bandwidth. So the high-pass filtered phase dither works only when the DDS output frequency is low with respect to the used clock frequency.

8.3.2 High-Pass Filtered Amplitude Dither

If a digital dither (from the source 2 in Figure 8-10) is summed with the output of the phase to amplitude converter, then the output of the DDS can be expressed as

$$\sin(\frac{2\pi}{2^j}(\Delta Pn - e_P(n))) + z_{HA}(n) - e_A(n), \tag{8.31}$$

where $z_{HA}(n)$ is the high-pass filtered amplitude dither, which has most of its

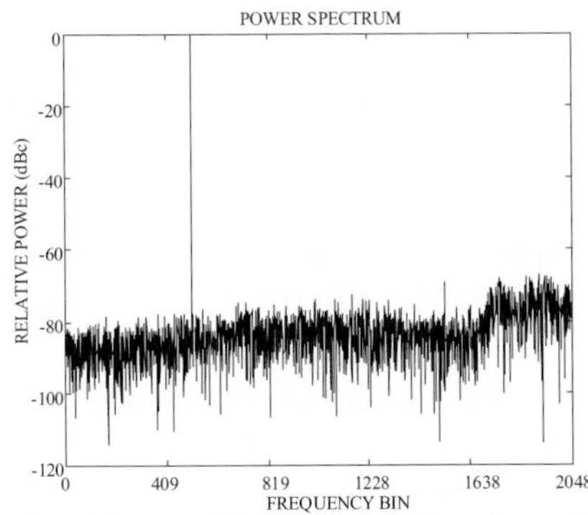

Figure 8-11. With high-pass filtered amplitude dithering, the carrier-to-spur level is increased to 69.25 dBc (see the level in Figure 8-8).

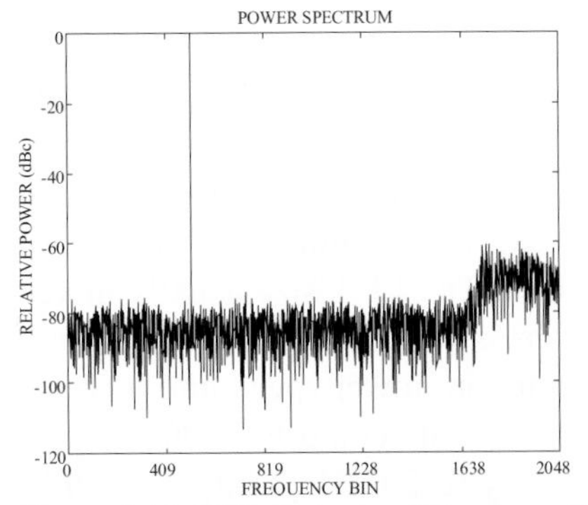

Figure 8-12. With high-pass filtered amplitude dithering, the carrier-to-noise power spectral density is 83.2 dBc (0 to 0.4 f_s).

power in an unused spectral region between the band edge of the low-pass filter and the Nyquist frequency. The benefits of the high-pass filtered amplitude dither are greater when it is used to randomize the D/A-converter non-linearities. The magnitude of the dither must be high in order to randomize the non-linearities of the D/A-converter [Wil91].

The high-pass filtered dither has poorer randomization properties than the wide band dither, which could be compensated by increasing the magnitude of the high-pass filtered dither [Ble87]. The spur reduction properties of the high-pass filtered amplitude dither are difficult to analyze theoretically, therefore only simulations are performed. The loss of the dynamic range is greater than in the case of the non-subtractive dither, because the magnitude of the high-pass filtered dither must be higher. However, the loss is small when the number of the quantization levels is large.

In this example, the digital high-pass filter is a 4th-order Chebyshev type I filter with the cut-off frequency of 0.42 f_s. Figure 8-8 shows the power spectrum of a sine wave without dithering. Figure 8-9 shows the power spectrum of a 16 bit sinusoid amplitude dithered with a random sequence that is distributed evenly over $[-2^{-8}/2, 2^{-8}/2)$, prior to truncation into 8 bits. Figure 8-11 shows the same example as Figure 8-9, but with a random sequence, which is distributed evenly over $[-2^{-7}/2, 2^{-7}/2)$. The processing is carried out by a digital high-pass filter, prior to dithering. In Figure 8-12, the amplitude range of the high-pass filtered dither is increased from over $[-2^{-7}/2, 2^{-7}/2)$ to over $[-2^{-6}/2, 2^{-6}/2)$ and so the spur reduction is accelerated. In Figure 8-12, the noise power spectral density is about 3 dB (half) lower in the DDS output bandwidth (0 to 0.4 f_s) than in Figure 8-9.

8.4 Tunable Error Feedback in DDS

The idea of the error feedback (EF) is to feed quantization errors back through a separate filter in order to correct the product at the following sampling occasions [Lea91a], [Lea91b], [Can92], [Har95]. The spectral purity of the conventional direct digital synthesizer (DDS) is also determined by the resolution of the phase to amplitude converter. Therefore, it is desirable to increase the resolution of the phase to amplitude converter (conventionally LUT). The LUT size usually increases exponentially in powers of 2 with the address width. Unfortunately, larger LUT storage means higher power consumption, lower speed and increased costs. The number of the needed phase address bits could be reduced by using the phase EF. In DDSs (output frequency (≥ 20MHz), output (≥ 10-bit)), most spurs are generated less by digital errors (quantization errors) than by dynamic non-linearities in the D/A-converter analog output response, because the development of D/A converters is not keeping up with the capabilities of digital signal processing with faster technologies [Van01]. The needed wordlength of the D/A converter could be reduced by using the amplitude EF. With the amplitude EF, lower accuracy D/A-converters with a better inner spurious performance could be used.

The phase value is generated using the modulo 2^j overflowing property of a j-bit phase accumulator as shown in Figure 8-13 and Figure 8-14. The rate of the overflows is the output frequency (4.1). The notches of the phase and amplitude EF are tuned according to the DDS output frequency (4.1) in order to reduce phase and amplitude quantization spurs at in-band in Figure 8-13 and Figure 8-14. The analog filter removes the out of band quantization noise along with any out-of-band noise and distortion introduced by the D/A converter. The passband of the analog filter is tuned according to the DDS output frequency. The quadrature DDS requires a complex amplitude EF filter and analog filter as shown in Figure 8-14. If only the real analog output is needed, then an analog half-complex filter can be used.

8.4.1 Phase EF

The EF has been placed between the phase accumulator and the phase-to-amplitude converter, as shown in Figure 8-13 [Lea91b], [Van97]. The phase error word (j-k least significant bits of the phase word) are fed to the phase EF filter ($H(z)$). The filter output is added to the phase value, so the transfer function between the phase input $P(z)$ and the phase output $Y(z)$ shown in Figure 8-15 can be characterized by

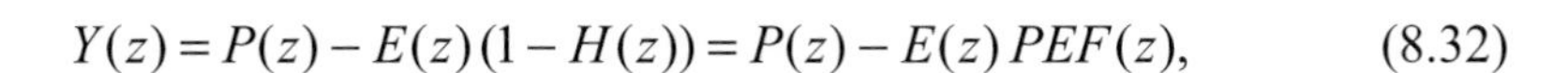

$$Y(z) = P(z) - E(z)(1 - H(z)) = P(z) - E(z)\,PEF(z), \tag{8.32}$$

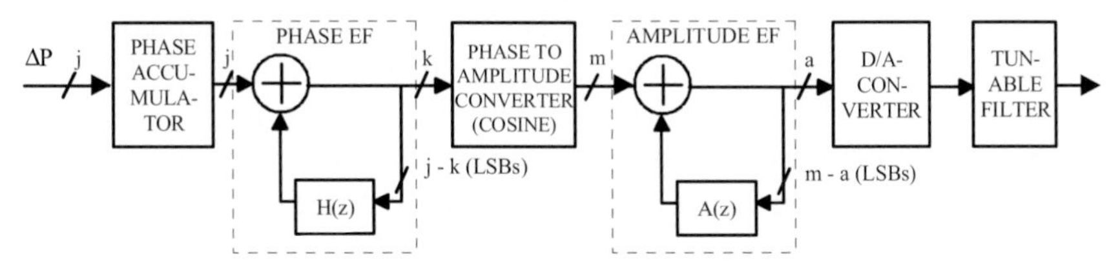

Figure 8-13. Direct digital synthesizer with phase and amplitude EF.

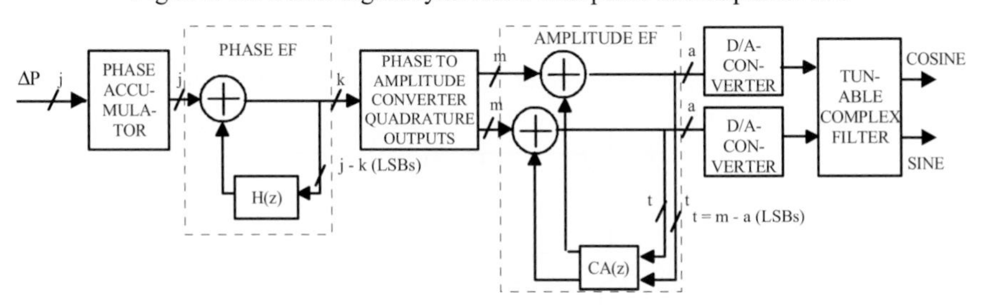

Figure 8-14. Quadrature direct digital synthesizer with phase and amplitude EF.

where $P(z)$ and $E(z)$ are the z transforms of the phase input and phase error signal and $PEF(z) = 1 - H(z)$ is the phase EF transfer function. If the phase EF filter output is fed back to the phase accumulator input along with the phase increment word, as shown in Figure 8-16, then the phase EF transfer function is given by

$$PEF(z) = \frac{1 - z^{-1} - z^{-1}G(z)}{1 - z^{-1}} = 1 - \frac{z^{-1}G(z)}{1 - z^{-1}}, \tag{8.33}$$

where the phase accumulator transfer function is $z^{-1}/(1-z^{-1})$. The phase accumulator with phase EF in Figure 8-16 introduces an extra pole at dc and zero to the phase EF transfer function (8.33). The digital filter transfer function in Figure 8-16 is obtained from (8.33)

$$G(z) = z\ (1 - z^{-1})(1 - PEF(z)). \tag{8.34}$$

8.4.1.1 Phase EF for cosine DDS

It is possible to derive the following approximation for the DDS output signal (real) by using the first two terms in the Taylor series around $\omega_{out}n$:

$$y(n) = A\cos(\omega_{out}n - npef(n)) \approx A\cos(\omega_{out}n) + A\sin(\omega_{out}n)npef(n), \tag{8.35}$$

where the shaped phase truncation error is $npef(n) \ll 1$ and with no amplitude quantization. The Fourier transform of (8.35) is

$$Y(\omega) \approx \frac{Af_s}{2}\sum_{k=-\infty}^{\infty}\begin{bmatrix}(\delta(\omega-\omega_{out}+2\pi k)+\delta(\omega+\omega_{out}+2\pi k))+\\ j(NPEF(\omega-\omega_{out}+2\pi k)-NPEF(\omega+\omega_{out}+2\pi k))\end{bmatrix}, \tag{8.36}$$

where $NPEF(\omega)$ is the Fourier transform of the shaped phase truncation error. If the $NPEF(\omega)$ is zero at $\omega = 0$, the expression $NPEF(\omega + \omega_{out}) - NPEF(\omega - \omega_{out})$ is not zero at the frequencies plus and minus ω_{out}, as shown in Figure 8-17(a). The cosine output DDS will contain the subtraction of the phase errors at $NPEF(0)$ and $NPEF(2\omega_{out})$ at the output frequency ω_{out}. In the previous work [Lea91b], [Van97], the $NPEF(\omega)$ is zero at dc and nonzero elsewhere, thus the system is only effective at low ω_{out} or near $\omega_s/2$ where $NPEF(2\omega_{out})$ is still somewhat notched out. If the $NPEF(\omega)$ is zero at $\omega = 0$ and $\omega = 2\omega_{out}$, the expression $NPEF(\omega + \omega_{out}) - NPEF(\omega - \omega_{out})$ is zero at the frequencies plus and minus ω_{out}, as shown in Figure 8-17(b). Therefore, the

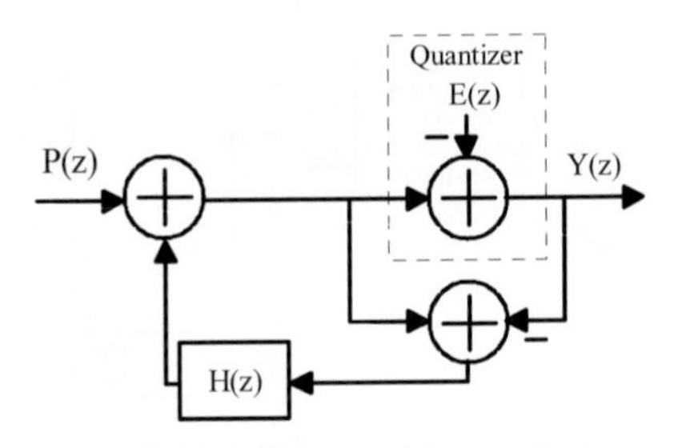

Figure 8-15. Phase EF noise model in Figure 8-13.

phase EF transfer function in the cosine output DDS should have J zeros at dc and K zeros at 2 ω_{out}

$$PEF(z) = (1 - z^{-1})^J \; (1 + b\,z^{-1} + z^{-2})^K = 1 - H(z), \tag{8.37}$$

where b = -2 cos2ω_{out}. If we want the same amount of attenuation at dc and 2ω_{out}, then J must be equal to K. The spectral density of the shaped noise in (8.36) using the phase EF transfer function in (8.37) is given for $z = e^{j\omega/fs}$

$$\left|NPEF(f)\right| = \left|2^{J+2K} \sin(\frac{\pi f}{f_s})^J \; (\sin\frac{\pi(2 f_{out} + f)}{f_s} \sin\frac{\pi(2 f_{out} - f)}{f_s})^K \; EP(f)\right|, \tag{8.38}$$

where

$$EP(f) = \frac{\sqrt{EP}}{f_s / 2}, \tag{8.39}$$

where EP is the sum of the powers of the spurs due to phase truncation (from (7.17))

$$EP = 1 - \left|A(0, L, M, N)\right|^2 \approx (\pi^2 / 2^{2k} - \pi^2 / Pe^2)/3, \tag{8.40}$$

where Pe is the period of the DDS output from (4.4), and k is the word length of the phase accumulator output used to address the phase to amplitude converter. The error feedback destroys the periodicity of the phase error signal (see Section 7.1), and all of its power spreads into the frequency band $0 \le f \le f_s/2$ in (8.39). This is a valid assumption for the phase error when the EF filter coefficient b in (8.37) is "approximately" irrational (a rational number with a large denominator) and/or the order of the EF structure (J, K in (8.37)) is high. In this case, the phase error signal has a very long period and the error appears as “white noise”.

The transfer function of the phase EF (FIR (finite impulse response)) has zero at dc and at 2 ω_{out} in Figure 8-18. The transfer function is characterized by

$$\begin{aligned} PEF(z) &= (1 - z^{-1})(1 + b\,z^{-1} + z^{-2}) \\ &= 1 + (b-1)\,z^{-1} - (b-1)\,z^{-2} - z^{-3} = 1 - H(z). \end{aligned} \tag{8.41}$$

The phase EF filter coefficients are tuned according to the DDS output frequency.

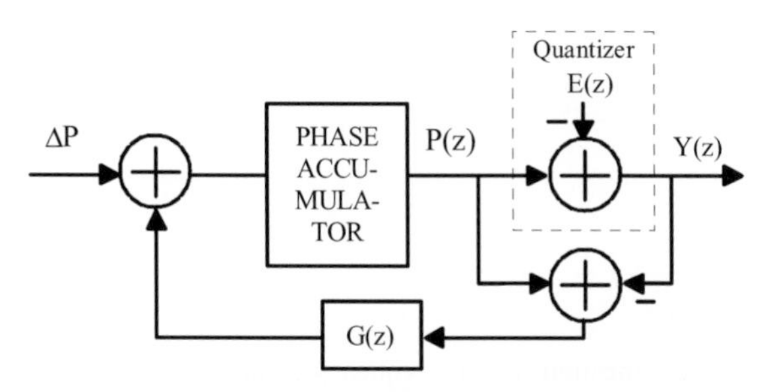

Figure 8-16. Phase accumulator with phase EF.

If the phase EF is fed back to the phase accumulator input, as shown in Figure 8-15, then the phase EF filter is given by (8.34)

$$\begin{aligned} G(z) &= z\,(1-z^{-1})\,(1-PEF(z)) \\ &= (1-b)+2\,(b-1)\,z^{-1}+(2-b)\,z^{-2}-z^{-3}. \end{aligned} \tag{8.42}$$

Comparing (8.41) and (8.42), using the structure of Figure 8-15 does not decrease the order of the phase EF filter (FIR), only the coefficient values of the digital filter are changed, because the extra zero at dc is cancelled by the extra pole in (8.33).

8.4.1.2 Phase EF for Quadrature DDS

For the quadrature DDS, it is possible to derive the following approximation for the output signal by using the first two terms in the Taylor series around $\omega_{out}n$:

$$\begin{aligned} c(n) &= A\cos(\omega_{out}n-cnpef(n))+j\,A\sin(\omega_{out}n-cnpef(n)) \\ &\approx Ae^{j\omega_{out}n}-j\,Ae^{j\omega_{out}n}\;cnpef(n), \end{aligned} \tag{8.43}$$

where the shaped phase truncation error is $cnpef(n) \ll 1$ and the Fourier transform of (8.43) is

$$C(\omega) \approx Af_s\sum_{k=-\infty}^{\infty}\left[\delta(\omega-\omega_{out}+2\pi k)-j\,CNPEF(\omega-\omega_{out}+2\pi k)\right]. \tag{8.44}$$

If the $CNPEF(\omega)$ is zero at $\omega = 0$, the expression $CNPEF(\omega - \omega_{out})$ is zero at the output frequency of ω_{out}, as shown in Figure 8-17(a). The spectrum of the shape phase error is modulated to the quadrature DDS output fre-

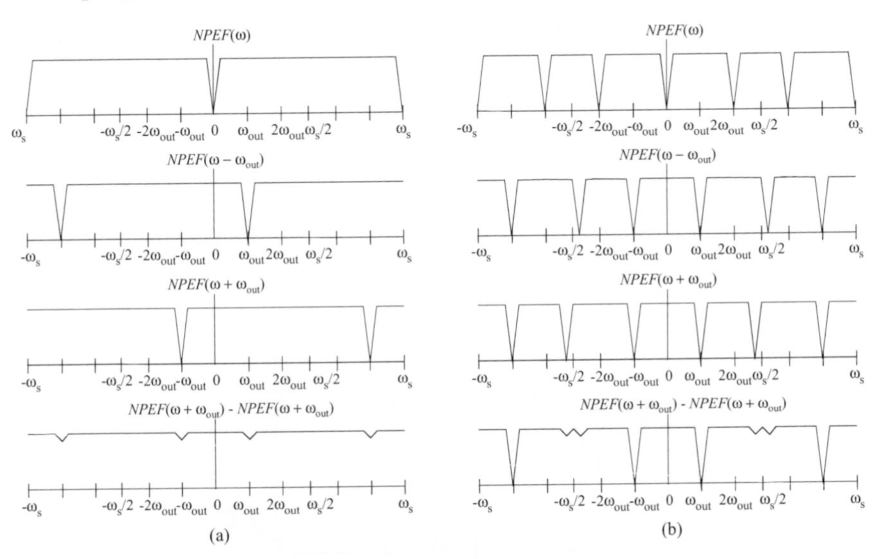

Figure 8-17. Phase truncation error in frequency domain: (a) $NPEF(\omega)$ is zero at 0. (b) $NPEF(\omega)$ is zero at 0 and $2\omega_{out}$.

quency ω_{out}. Therefore, the phase EF transfer function coefficients for quadrature DDS has zeros at dc

$$PEF(z) = (1 - z^{-1})^J = 1 - H(z), \tag{8.45}$$

where J is the number of zeros at dc. The spectral density of the noise in (8.43) is given for $z = e^{j\omega/fs}$

$$|CNPEF(f)| = |EP(f)\,PEF(f)| = \left|EP(f)\,2^J\,(\sin\frac{\pi f}{f_s})^J\right|, \tag{8.46}$$

where

$$EP(f) = \frac{\sqrt{EP}}{f_s}, \tag{8.47}$$

and EP is defined by (8.40). It is assumed that the error feedback destroys the periodic of the phase error signal, and that all of its power spreads into frequency band $0 \le f < f_s$.

The multiplication process is greatly simplified by replacing the fixed phase EF filter coefficients with canonic signed digit (CSD) numbers (see section 11.6). In Figure 8-19, the phase EF transfer function has two zeros at dc.

If the second phase EF with two zeros at dc is fed back to the phase accumulator input, as shown in, Figure 8-15, then the phase EF filter is given by (8.34)

$$G(z) = z\ (1 - z^{-1})\,(1 - PEF(z)) = -2 + 3z^{-1} - z^{-2}. \tag{8.48}$$

So using the structure of Figure 8-15 does not decrease the order of the phase EF filter (FIR), only the coefficient values of the digital filter are changed.

The phase error feedback methods presented here are based on approximations (8.35) and (8.43). These two methods work only if $npef(n)$ or $cnpef(n) \ll 1$, which makes the effect of the higher order terms in the Taylor series not significant. If the approximations in (8.35) and (8.43) are not valid, the higher order terms spread the shaped noise and the notch around the output frequency will disappear. There are two factors that affect the level of $npef(n)$ and $cnpef(n)$, the length of the phase-to-amplitude converter address (k), and the gain of the EF filter. The gain of the EF filter depends on the notch frequency, the order of the filter and the type of the filter.

8.4.2 Amplitude EF

The EF filter has been placed after the phase-to-amplitude converter in Figure 8-13 and Figure 8-14. The amplitude error word (m-a least significant bits of the amplitude word) is fed to the amplitude EF filter ($A(z)$). It is possible to derive the following equation from (8.35) and (8.43) for the cosine and quadrature DDS output

$$y(n) \approx A\cos(\omega_{out} n) + A\sin(\omega_{out} n)\, npef(n) - naef(n) - ae(n), \qquad (8.49)$$

$$c(n) \approx Ae^{j\omega_{out} n} - jAe^{j\omega_{out} n}\, cnpef(n) - cnaef(n) - cae(n), \qquad (8.50)$$

where *cnaef*(*n*)/*naef*(*n*) is the shaped amplitude error and *ea*(*n*)/*cae*(*n*) is quantization noise due to finite wordlength in the phase to amplitude converter output. The amplitude of the phase to amplitude converter output (A) must be reduced sufficiently to ensure the output plus amplitude EF output stays within the no-overload region. This loss may be acceptable (and small) when the number of the quantization levels is large. The overflows due to the phase EF cause no problems in the phase address, because the phase accumulator works according to the overflow principle. The second order real and complex amplitude EF filters are used in Figure 8-18 and Figure 8-19, respectively. The phase to amplitude converters output values were scaled to prevent overflows.

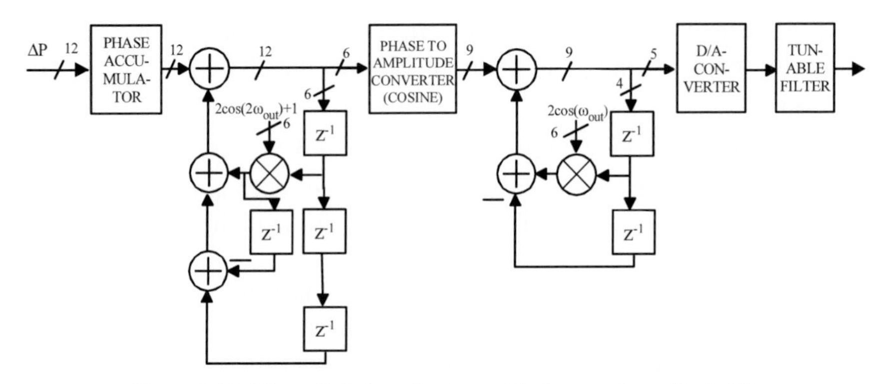

Figure 8-18. Direct digital synthesizer with phase and amplitude EF.

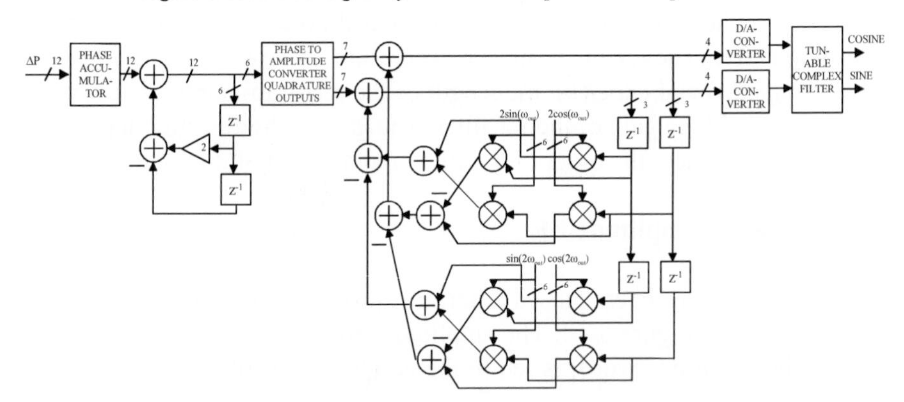

Figure 8-19. Quadrature direct digital synthesizer with phase and amplitude EF.

8.4.2.1 Amplitude EF for Cosine DDS

In the cosine DDS (Figure 8-13), the amplitude EF transfer function is

$$AEF(z) = (1 + b z^{-1} + z^{-2})^P = 1 - A(z), \tag{8.51}$$

where $A(z)$ is the amplitude error feedback filter in Figure 8-13 and $b = -2 \cos\omega_{out}$. The amplitude EF filter coefficients are chosen according to the output frequency of the DDS. The spectral density of this noise is in (8.49) and is given for $z = e^{j\omega/fs}$

$$\begin{aligned} |NAEF(f)| &= (\left|\frac{(\Delta_a^2 - \Delta_m^2)}{12} AEF(f)\right| + E\{ea^2\})\frac{1}{f_s} \\ &= (\left|\frac{2^{-2a} - 2^{-2m}}{12} 4^P (\sin\frac{\pi(f_{out}+f)}{f_s}\sin\frac{\pi(f_{out}-f)}{f_s})^P\right| + \frac{2^{-2m}}{12})\frac{1}{f_s}, \end{aligned} \tag{8.52}$$

where $AEF(f)$ is from (8.51), $(\Delta_a^2 - \Delta_m^2)/12$ is the noise variance of the amplitude error word [Con99], $E\{ea^2\}$ is the phase to amplitude converter output noise power from (7.37) (residual noise power), m is the phase to amplitude converter output wordlength and a is the D/A converter wordlength. The amplitude EF destroys the symmetry property of the error sequences (see Section 7.2). Therefore amplitude quantization power spreads into frequency band $0 \le f < f_s$. The cosine generated is a real signal, so its power is equally divided into negative and positive frequency components. Using these facts, the power spectral density of the cosine output in (8.49) is given for $z = e^{j\omega/fs}$

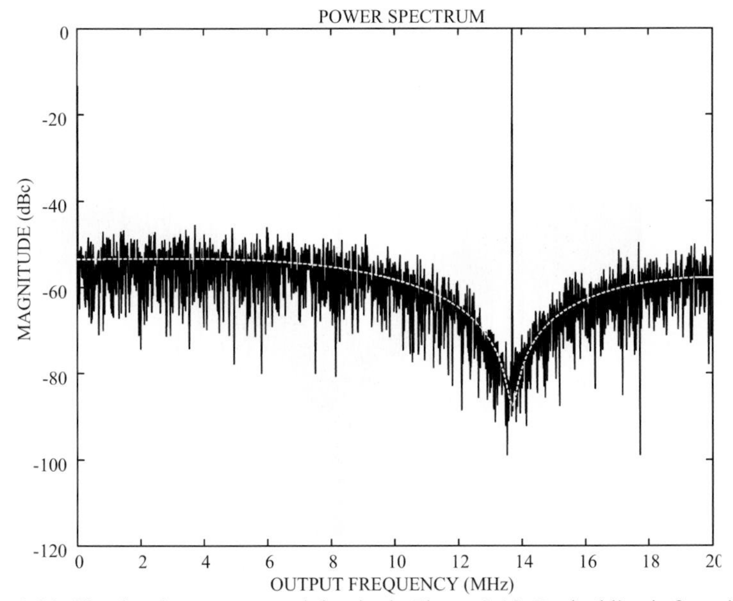

Figure 8-20. Simulated power spectral density in Figure 8-18. Dashed line is from (8.53).

$$|P(f)|^2 \approx \frac{\frac{A^2/2}{2}}{\frac{A^2/2}{2}(|NPEF(f+f_{\mathrm{out}})|^2+|NPEF(f-f_{\mathrm{out}})|^2)+|NAEF(f)|^2}, \quad (8.53)$$

where $A^2/2$ is the cosine wave power, $NPEF(f)$ is from (8.38) and $NAEF(f)$ is from (8.52).

A computer program (in Matlab) has been created to simulate the direct digital synthesizer in Figure 8-18. The simulated and theoretical from (8.53) signal to noise power spectrum densities are shown in Figure 8-20. The EF structures spread quantization noise to all the frequency bins as shown in Figure 8-20, because the coefficients of the EF structures in Figure 8-18 are "approximately" irrational numbers at the notch frequency ($f_{notch}/f_s = 0.3438$).

8.4.2.2 Amplitude EF for Quadrature DDS

In the quadrature DDS (Figure 8-14), the complex amplitude EF transfer function is

$$CAEF(z) = (1-e^{j\omega_{out}}z^{-1})^P = 1-CA(z), \quad (8.54)$$

and $CA(z)$ is the amplitude error feedback filter in Figure 8-14. The spectral density of this amplitude quantization noise in (8.50) is given for $z = e^{j\omega/fs}$

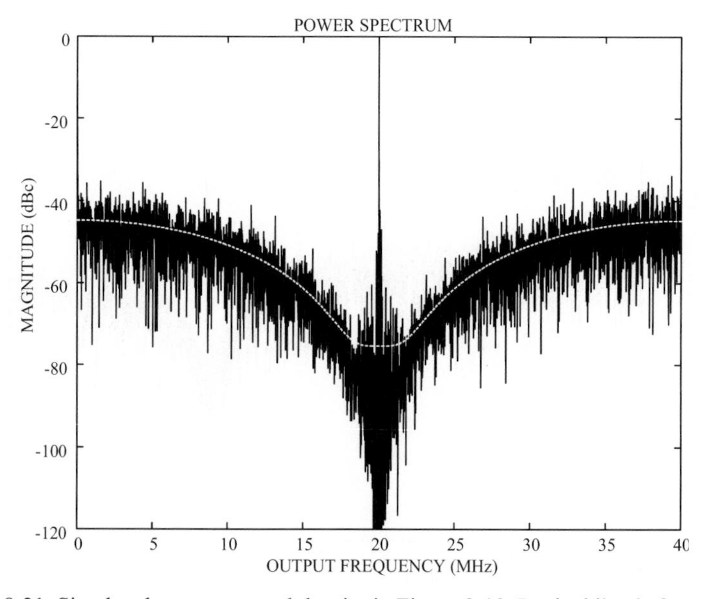

Figure 8-21. Simulated power spectral density in Figure 8-19. Dashed line is from (8.56).

$$|CNAEF(f)| = (\frac{(2^{-2a} - 2^{-2m})}{6}|CAEF(f)| + E\{cea^2\})\frac{1}{f_s}$$
$$= (\frac{(2^{-2a} - 2^{-2m})}{6} 2^P (\sin\frac{\pi(f - f_{out})}{f_s})^P + \frac{2^{-2m}}{6})\frac{1}{f_s}, \tag{8.55}$$

where $E\{cea^2\}$is from (7.38). The EF destroys the symmetry property of the amplitude error sequences. Therefore, amplitude quantization power spreads into frequency band $0 \leq f < f_s$. The power spectral density of the quadrature

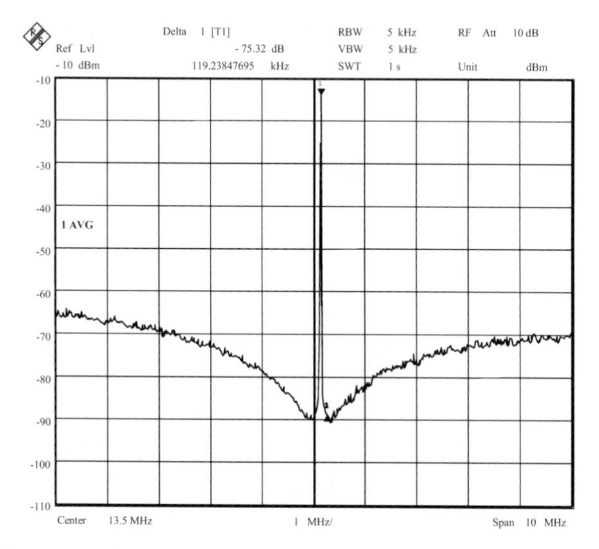

Figure 8-22. Spectrum of 13.7 MHz signal at 5-bit D/A converter output, where sampling frequency is 40 MHz.

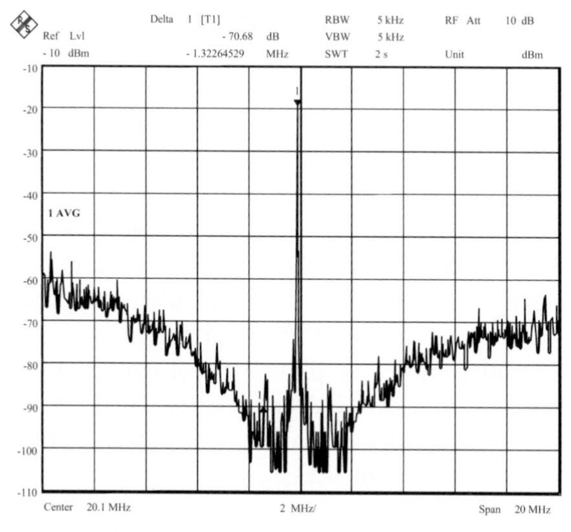

Figure 8-23. Spectrum of 20 MHz signal at 4-bit D/A converter output, where sampling frequency is 40 MHz.

output is from (8.50) and is given for $z = e^{j\omega/fs}$

$$|P(f)|^2 \approx \frac{A^2}{A^2\,|NPEF(f-f_{\mathrm{out}})|^2 + |CNAEF(f)|^2}, \tag{8.56}$$

where A^2 is the quadrature wave power, *NPEF(f)* is from (8.46) and *CNAEF(f)* is from (8.55).

A computer program (in Matlab) language was written to simulate the direct digital synthesizer in Figure 8-19. The simulated and theoretical (from 8.56) signal to noise power spectrum densities are shown in Figure 8-21. The spectrum has spurious signals in Figure 8-21 because of the phase EF filter coefficient is integer (2), but the EF still spreads part of the spur power to broadband noise.

8.5 Implementations

The cosine and quadrature DDS with phase and amplitude EF were implemented with the Altera FLEX 10KA-1 series devices [Fle98]. The cosine DDS (in Figure 8-18) with real phase and amplitude EF filters requires 282 (5% of the total) logic elements (LEs) in the EPF10K100A device. The quadrature DDS (in Figure 8-19) with real phase and complex amplitude EF filters requires 586 (11% of the total) LEs in the EPF10K100A device. The maximum operating frequency of the cosine and quadrature realization is 42 and 43 MHz, respectively. The D/A converters in Figure 8-13 and Figure 8-14 were implemented using the MSB bits of the 14-bit D/A converter [Kos01].

8.6 Measurement Results

To evaluate the cosine and quadrature DDS with phase and amplitude EF, a test board was built and a computer program was developed to control the measurement. The phase increment and the EF filters tuning words are loaded into the test board via the parallel port of a personal computer. A spectrum plot of 13.7 MHz output from the cosine DDS with real tunable phase and amplitude EF (Figure 8-13) is illustrated in Figure 8-22. The power spectral density agrees with the simulation (and theoretical) results from Figure 8-20. A spectrum plot of 20 MHz output from the quadrature DDS with phase and complex amplitude EF (Figure 8-14) is illustrated in Figure 8-23. The power spectral density agrees with the simulation (and theoretical) results from Figure 8-21. The spur power is in tones around the carrier in Figure 8-21 and Figure 8-23.

8.7 Conclusions

The reason why the dither techniques have not been applied very often to reduce the spurs due to the finite word length of the digital part of the DDS is because the effect of the D/A-converter non-linearities nullifies the contribution. The benefits of the high-pass filtered amplitude dither would be greater when it is used to randomize the D/A-converter non-linearities because the magnitude of the dither must be high in order to randomize the non-linearities of the D/A-converter. The cosine and quadrature DDS with tunable phase and amplitude EF were designed and implemented. The drawback of the conventional phase EF is that it reduces the phase spurs only at DDS output frequencies which are near dc or half of the sampling frequency. New phase EF architectures with an arbitrary center frequency for cosine and quadrature output DDS are introduced. The EF structures can be used in conjunction with different phase to amplitude converter compression techniques.

REFERENCES

[Ble87] B. A. Blesser, and B. N. Locanthi, "The Application of Narrow-Band Dither Operating at the Nyquist Frequency in Digital Systems to Provide Improved Signal-to-Noise Ratio over Conventional Dithering," J. Audio Eng. Soc., Vol. 35, pp. 446-454, June 1987.

[Can92] J. C. Candy, and G. C. Temes, "Oversampling Delta-Sigma Data Converters," IEEE Press, New York, 1992

[Car87] L. R. Carley, "An Oversampling Analog-to-Digital Converter Topology for High-Resolution Signal Acquisition Systems," IEEE Trans. Circuits and Syst., CAS-34, pp. 83-90, Jan. 1987.

[Con99] G. A. Constantinides, P. Y. K. Cheung, and W. Luk, "Truncation Noise in Fixed-Point SFGs [digital filters]," Electron. Lett., Vol. 35, No. 23, pp. 2012 -2014, Nov. 1999.

[Fla93] M. J. Flanagan, and G. A. Zimmerman, "Spur-Reduced Digital Sinusoid Generation Using Higher-Order Phase Dithering," Asilomar Conf. on Signals, Syst. and Comput., Nov. 1993, pp. 826-830.

[Fla95] M. J. Flanagan, and G. A. Zimmerman, "Spur-Reduced Digital Sinusoid Synthesis," IEEE Trans. Commun., Vol. COM-43, pp. 2254-2262, July 1995.

[Fle98] FLEX 10K Embedded Programmable Logic Family Data Sheet, Altera Corp., San Jose, CA, Oct. 1998.

[Gra93] R. M. Gray, and T. G. Stockholm, "Dithered Quantizers," IEEE Trans. Inform. Theory., Vol. 39, pp. 805-812, May 1993.

[Har95] F. Harris and B. McKnight, "Error Feedback Loop Linearizes Direct Digital Synthesizers," in Proc. IEEE 29th Asilomar Conf. on Signals, Systems, and Computers, Pacific Grove, USA, 30 Oct. - 2 Nov. 1995, pp. 98-102.

[Jas87] S. C. Jasper, "Frequency Resolution in a Digital Oscillator," U. S. Pat. 4,652,832, Mar. 24, 1987.

[Ker90] R. J. Kerr, and L. A. Weaver, "Pseudorandom Dither for Frequency Synthesis Noise," U. S. Pat. 4,901,265, Feb. 13, 1990.

[Kos01]M. Kosunen, J. Vankka, M. Waltari, and K. Halonen, "A Multicarrier QAM Modulator for WCDMA Base Station with on-chip D/A converter," Proceedings of CICC 2001 Conference, May 6-9 2001, San Diego, USA, pp. 301-304.

[Laz94] G. Lazzari, F. Maloberti, G. Oliveri, and G. Torelli, "Sinewave Modulation for Data Communication by Direct Digital Synthesis and Sigma Delta Techniques," European Trans. Telecommun. and Related Technologies, Vol. 5, pp. 689-695, Nov.-Dec. 1994.

[Lea91a] P. O' Leary, M. Pauritsch, F. Maloberti, and G. Raschetti, "An Oversampling-Based DTMF Generator," IEEE Trans. Commun., Vol. COM-39, pp. 1189-1191, Aug. 1991.

[Lea91b] P. O' Leary, and F. Maloberti, "A Direct Digital Synthesizer with Improved Spectral Performance," IEEE Trans. Commun., Vol. COM-39, pp. 1046-1048, July 1991.

[Lju87] L. Ljung, "System Identification: Theory for the User," Englewood Cliffs, NJ: Prentice-Hall, 1987.

[Nic88] H. T. Nicholas, H. Samueli, and B. Kim, "The Optimization of Direct Digital Frequency Synthesizer in the Presence of Finite Word Length Effects Performance," in Proc. 42nd Annu. Frequency Contr. Symp., June 1988, pp. 357-363.

[Qua91a] Qualcomm Q2334, Technical Data Sheet, June 1991.

[Rei91] V. S. Reinhardt, and et al., "Randomized Digital/Analog Converter Direct Digital Synthesiser," U. S. Pat. 5,014,231, May 7, 1991.

[Rei93] V. S. Reinhardt, "Spur Reduction Techniques in Direct Digital Synthesizers," in Proc. IEEE Int. Frequency Cont. Symp., June 1993, pp. 230-241.

[Twi94]E. R. Twitchell, and D. B. Talbot, "Apparatus for Reducing Spurious Frequency Components in the Output Signal of a Direct Digital Synthesizer," U. S. Pat. 5,291,428, Mar. 1, 1994.

[Van01] J. Vankka, and K. Halonen, "Direct Digital Synthesizers: Theory, Design and Applications," Kluwer Academic Publishers, 2001.

[Van97] J. Vankka, "A Direct Digital Synthesizer with a Tunable Error Feedback Structure," IEEE Transactions on Communications, Vol. 45, pp. 416-420, April 1997.

[Van98] J. Vankka, M. Waltari, M. Kosunen, and K. Halonen, "Direct Digital Syntesizer with on-Chip D/A-converter," IEEE Journal of Solid-State Circuits, Vol. 33, No. 2, pp. 218-227, Feb. 1998.

[Whe83] C. E. Wheatley, III, "Digital Frequency Synthesiser with Random Jittering for Reducing Discrete Spectral Spurs," U. S. Pat. 4,410,954, Oct. 18, 1983.
[Wil91] M. P. Wilson, and T. C. Tozer, "Spurious Reduction Techniques for Direct Digital Synthesis," IEE Coll. Digest 1991/172 on Direct Digital Frequency Synthesis, Nov. 1991, pp. 4/1-4/5.
[Zim92] G. A. Zimmerman, and M. J. Flanagan, "Spur Reduced Numerically-Controlled Oscillator for Digital Receivers," Asilomar Conf. on Signals, Syst. and Comput., Dec. 1992, pp. 517-520.

Chapter 9

9. BLOCKS OF DIRECT DIGITAL SYNTHESIZERS

The direct digital synthesizer (DDS) is shown in a simplified form in Figure 4-1. In this chapter, the blocks of the DDS: phase accumulator, phase to amplitude converter and filter are investigated. The D/A converter was described in Section 10. The methods of accelerating the phase accumulator are described in detail. Different sine memory compression and algorithmic techniques and their trade-offs are investigated.

9.1 Phase Accumulator

In practice, the phase accumulator circuit cannot complete the multi-bit addition in a short single clock period because of the delay caused by the carry bits rippling through the adder. In order to provide the operation at higher clock frequencies, one solution is a pipelined accumulator [Cho88], [Ekr88], [Gie89], [Lia97], as shown in Figure 9-1. To reduce the number of the gate delays per clock period, a kernel 4-bit adder is used in Figure 9-1, and the carry is latched between successive adder stages. In this way, the length of the accumulator does not reduce the maximum operating speed. To maintain the valid accumulator data during the phase increment word transition, the new phase increment value is moved into the pipeline through the delay circuit. All the bits of the input phase increment word must be delay equalized. The phase increment word delay equalization circuitry is thus very large. For example, in Figure 9-1, a 32-bit accumulator with 4-bit pipelined segments requires 144 D-flip-flops (DFFs) for input delay equalization alone. These D-flip-flop circuits would impact the loading of the clock network. To reduce the number of pipeline stages, a carry increment adder (CIA) [Kos99] and a conditional sum adder in [Tan95a] are used. To reduce the cycle time and size of pipeline stages further, the outputs of the adder and the D-flip-flops could be combined to form "logic-flip-flop" (L-FF) pipeline stages

[Yua89], [Rog96], [Kos99]; thereby their individual delays are shared, resulting in a shorter cycle time and smaller area.

Pre-skewing latches with pipeline control are used to eliminate the large number of D-flip-flops required by the input delay equalization registers [Che92], [Lu93]. The cost of this simplified implementation is that the frequency can be updated only at f_s/PS, where *PS* is the number of the pipelined stages.

The phase increment inputs to the phase accumulator are normally generated by a circuitry that runs from a clock that is much lower in frequency than, and often asynchronous to, the DDS clock. To allow this asynchronous loading of the phase increment word, double buffering is used at the input of the phase accumulator.

The output delay circuitry is identical to the input delay equalization circuitry, inverted so that the low-order bits receive a maximum delay while the most significant bits receive the minimum delay. In Figure 9-1 the data from the most significant 12 bits of the phase accumulator are delayed in pipelined registers to reach the phase to amplitude converter with full synchronization. A hardware simplification is provided by eliminating the de-skewing registers for the least significant *j-k* bits of the phase accumulator output. This is possible because only the *k* most significant phase bits are used to calculate the sine function. The only output bits that have to be delay equalized are those that form the address of the phase to amplitude converter. The processing delay is from the time a new value is loaded into the phase register to the time when the frequency of the output signal actually changes, and the pipeline latency associated with frequency switching is 9 clock pulses, see Figure 9-1.

In [Tho92], a progression-of-states technique, rather than pipelining, was incorporated as shown in Figure 9-2. The outputs of the phase accumulator are shown in (9.1)-(9.4) when the phase increment word is held constant for

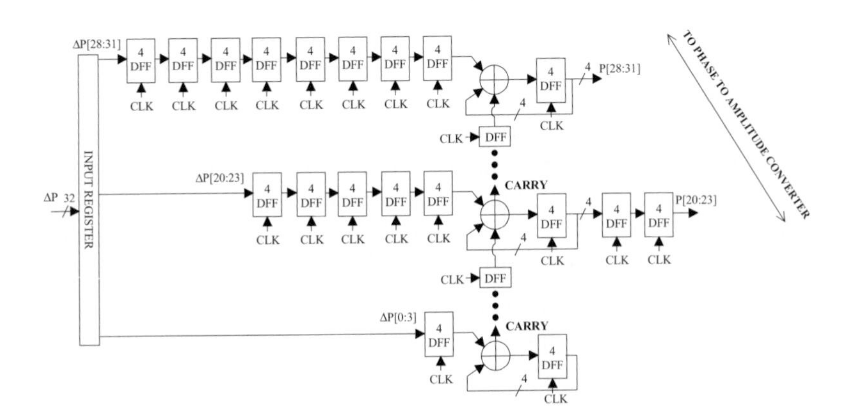

Figure 9-1. Pipelined 32-bit phase accumulator.

four clock cycles. Holding the phase increment limits the phase increment update rate. The outputs of the phase accumulator are

$$P(n+1) = P(n) + \Delta P = P(n) + \Delta P, \tag{9.1}$$

$$P(n+2) = P(n+1) + \Delta P = P(n) + 2\Delta P, \tag{9.2}$$

$$P(n+3) = P(n+2) + \Delta P = P(n) + 3\Delta P, \tag{9.3}$$

$$P(n+4) = P(n+3) + \Delta P = P(n) + 4\Delta P, \tag{9.4}$$

where ΔP is the phase increment word and $P(n)$ is the output of the phase accumulator at nth sampling instant. To generate $P(n+2)$ and $P(n+4)$ in (9.2) and (9.4), ΔP is shifted up 1bit and 2bits before they are added respectively. To generate $P(n+3)$ in (9.3), $P(n+2)$ and ΔP are used instead of $3\Delta P$. The progression-of-states technique in Figure 9-2 demands D-flip-flop circuits and four adders so it has the area and power overheads. To increase the operation speed of the phase accumulator using the progression-of-states technique, the pipelining technique is used in [Yan02].

The use of parallel phase accumulators to attain a high throughput has been utilized in [Has84], [Gol90], [Tan95b]. The structure in Figure 9-3 uses a phase accumulator and four adders [Tan95b]. Four adders are required to make four sequential phase outputs $P(n+1)$, $P(n+2)$, $P(n+3)$, $P(n+4)$. The four phase outputs are generated by adding 0, ΔP, $2\Delta P$, $3\Delta P$ phase offsets to the phase accumulator output. To generate every fourth sample at the phase accumulator output before adders in Figure 9-3, the phase increment word must be multiplied by four ($4\Delta P$).

The phase accumulator could be accelerated by introducing a Residue Number System (RNS) representation into the computation, and eliminating the carry propagation from each addition [Chr95]. The conversion and the re-conversion to/from the RNS representation reduces the gain in the computation speed.

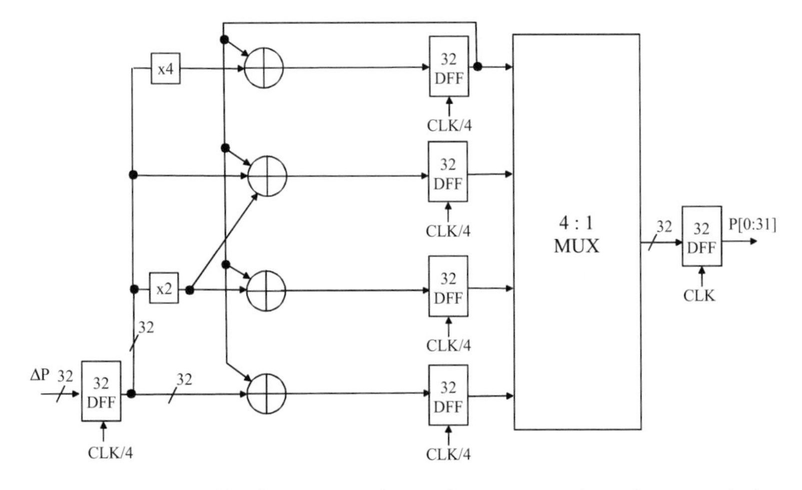

Figure 9-2. 32-bit phase accumulator using a progression-of-states technique.

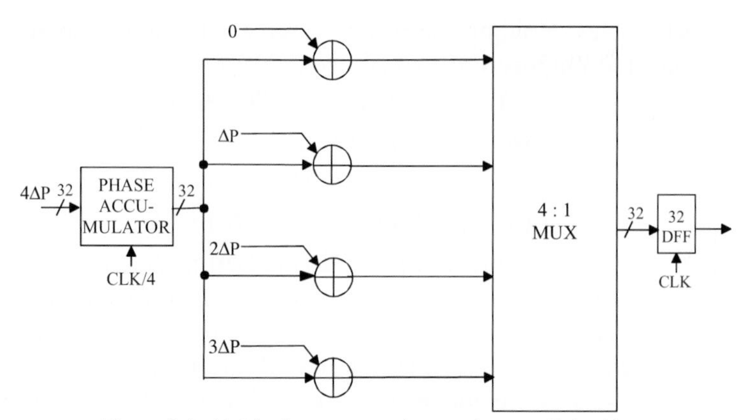

Figure 9-3. 32-bit phase accumulator using parallel adders.

The frequency resolution is from (4.2), when the modulus of the phase accumulator is 2^j. In this case, the implementation of the modulo operation is straight forward. Few techniques have been devised to use a different modulus [Jac73], [Gol88], [McC91b], [Gol96], [Har91], [Nos01a], [Uus01]. The penalty of those designs is a more complicated phase address decoding [Gol96]. The benefit is a more exact frequency resolution (the divider is not restricted to a power of two in (4.2), when the clock frequency is fixed [Gol96]. For example, 10 MHz is the industry standard for electronic instrumentation requiring accurate frequency synthesis [McC91b]. To achieve a one hertz resolution in these devices, the phase accumulator modulus must be set equal to 10^6 (decimal) [Jac73], [Gol88]. The modulus of the phase accumulator is variable in [McC91b], [Ell96], [Nos01a]. This allows a flexible output tuning resolution.

For a phase accumulator with phase truncated output, the only useful information produced by the adder below the truncation point is the carry-out signal. Therefore, the lower part of the phase accumulator can be replaced

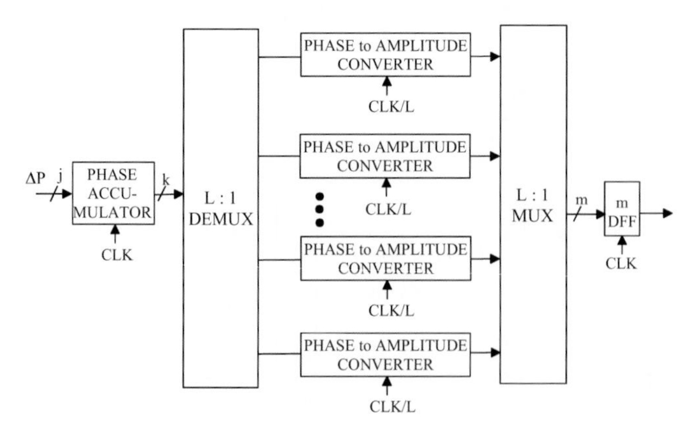

Figure 9-4. A DDS using parallel phase to amplitude converters.

with a hardware block producing an identical (or approximate) carry-out signal. This was the idea used in [Ert96], [Mer02].

9.2 Phase to Amplitude Converter

The spectral purity of the conventional direct digital synthesizer (DDS) is also determined by the resolution of the phase to amplitude converter. Therefore, it is desirable to increase the resolution of the phase to amplitude converter. Unfortunately, a larger resolution means higher power consumption, lower speed and greatly increased costs.

The use of parallel phase to amplitude converters to attain a high throughput has been utilized in [Has84], [Gol90], [Tan95b]. The DDS in Figure 9-4 utilizes the parallelism for the phase to amplitude converter to achieve high throughput. The DDS attains L times the maximum speed of a single phase to amplitude converter by paralleling L phase to amplitude converters. A L to 1 DEMUX transfers phases of the phase accumulator to phase to amplitude converters with f_{clk}/L. A L to 1 MUX alternatively selects the outputs of the phase to amplitude converters with f_{clk}.

The most elementary technique of compression is to store only π/2 rad of sine information, and to generate the LUT samples for the full range of 2π by exploiting the quarter-wave symmetry of the sine function. After that, methods of compressing the quarter-wave memory include: trigonometric identities, Nicholas method, the polynomial approximations or the CORDIC algorithm. A different approach to the phase-to-sine-amplitude mapping is

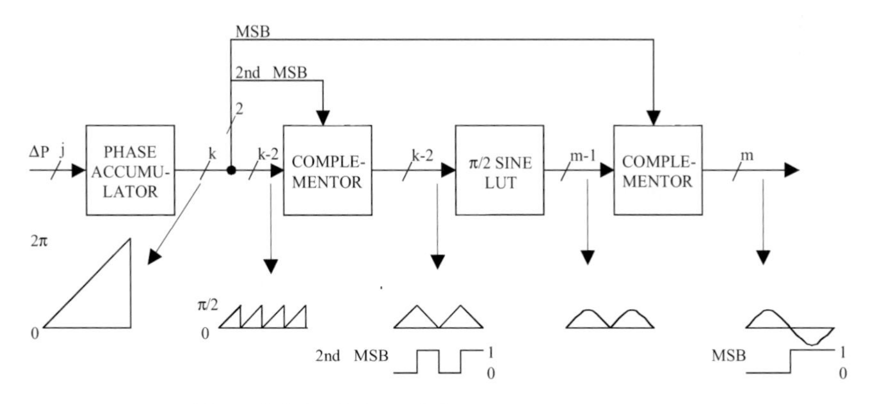

Figure 9-5. Logic to exploit quarter-wave symmetry.

Table 9-1. Coding table to obtain full sine wave, where θ is from 0 to π/2.

Quadrant (q)	MSB	2nd MSB	sin(θ+ qπ/2)
0	0	0	sin(θ)
1	0	1	sin(π/2-θ)
2	1	0	-sin(θ)
3	1	1	-sin(π/2-θ)

the CORDIC algorithm, which uses an iterative computation method. The costs of the different methods are an increased circuit complexity and distortions that will be generated when the methods of memory compression are employed. Because the possible number of generated frequencies is large, it is impossible to simulate all of them to find the worst-case situation. If the least significant bit of the phase accumulator input is forced to one, then only one simulation is needed to determine the worst-case carrier-to-spur level (see Section 7.3). In this chapter, 14-bit phase to 12-bit amplitude mapping is investigated. The results are only valid for these requirements. Some examples of commercial circuits using the above methods are also presented.

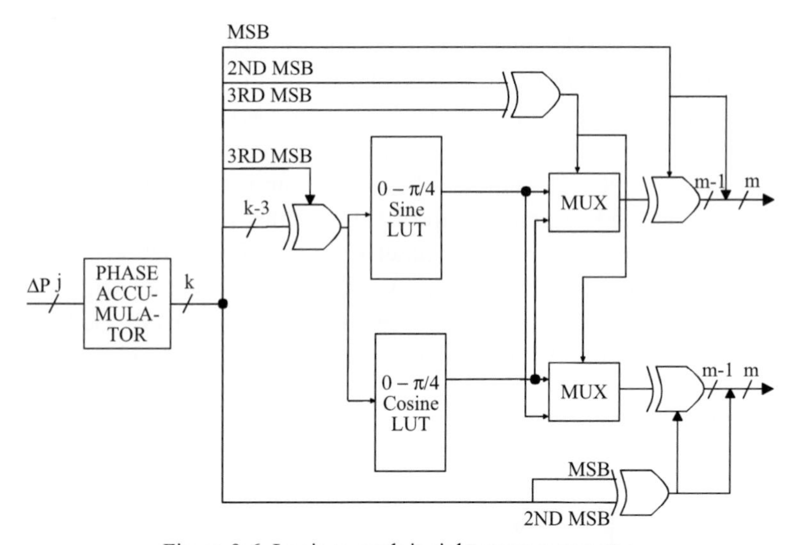

Figure 9-6. Logic to exploit eight-wave symmetry.

Table 9-2. Coding table to obtain full sine and cosine waves, where θ is from 0 to π/4.

Octant (o)	MSB	2nd MSB	3rd MSB	$\sin(\theta+ o\pi/4)$	$\cos(\theta+ o\pi/4)$
0	0	0	0	$\sin(\theta)$	$\cos(\theta)$
1	0	0	1	$\cos(\pi/4-\theta)$	$\sin(\pi/4-\theta)$
2	0	1	0	$\cos(\theta)$	$-\sin(\theta)$
3	0	1	1	$\sin(\pi/4-\theta)$	$-\cos(\pi/4-\theta)$
4	1	0	0	$-\sin(\theta)$	$-\cos(\theta)$
5	1	0	1	$-\cos(\pi/4-\theta)$	$-\sin(\pi/4-\theta)$
6	1	1	0	$-\cos(\theta)$	$\sin(\theta)$
7	1	1	1	$-\sin(\pi/4-\theta)$	$\cos(\pi/4-\theta)$

9.2.1 Non-Linear D/A Converter

A non-linear D/A-converter is used in the place of the sine look-up table (LUT) for the phase-to-sine amplitude conversion and linear D/A converter [Bje91], [Mor99], [Jia02], [Sha01], [Ait02], [Moh02]. A technique was proposed in [Jia02] to split the nonlinear D/A converter into a coarse D/A converter and a fine D/A converter. Gutierrez-Aitken et al. [Ait02] have reported a DDS with a 9.2 GHz clock rate in an advanced InP technology. It segments the phase angle into two sections, and uses the first two terms of the Taylor series for sine (see Section 9.2.3.6.3). Mohieldin et al. [Moh02] have recently described a nonlinear D/A converter architecture that uses a piecewise linear approximation of the sine function. The amplitude modulation is performed with analogue components in [Ait02], [Moh02]. The drawback of the non-linear D/A-converter is that the digital amplitude modulation cannot be incorporated into the DDS (see Figure 4-3).

9.2.2 Exploitation of Sine Function Symmetry

A well-known technique is to store only $\pi/2$ rad of sine information, and to generate the sine look-up table samples for the full range of 2π by exploiting the quarter-wave symmetry of the sine function. The decrease in the look-up table capacity is paid for by the additional logic necessary to generate the complements of the accumulator and the look-up table output.

The sine wave is separated into $\pi/2$ intervals in Table 9-1, and each interval is called a quadrant (q). The two most significant phase bits represent the quadrant number. The full sine wave can be represented by the first quadrant sine, using the symmetric and antisymmetric properties of sine, as shown in Table 9-1 [Tie71], [McC84]. Therefore, the two most significant phase bits are used to decode the quadrant, while the remaining k-2 bits are used to address a one-quadrant sine look-up table in Figure 9-5. As shown in Figure 9-5, the sampled waveform at the output of the look-up table is a full, rectified version of the desired sine wave.

Figure 9-6 shows the architecture for both sine and cosine functions (quadrature output). The architecture is a simple extension of Figure 9-5 with no increase in lookup table size, because one could take advantage of an eight wave symmetry of a sine and cosine waveform [McC84], [Tan95a]. The sine and cosine waves are separated into $\pi/4$ intervals in Table 9-2; each interval is called an octant (o). The three most significant phase bits represent the octant number. The sine and cosine curves for all octants can be represented by one octant of sine and one octant of cosine, using the symmetric and antisymmetric properties of sine and cosine. Hence, only the sine and cosine samples from 0 to $\pi/4$ in Figure 9-6 need to be stored, where the cosine wave from 0 to $\pi/4$ is the same as the sine from $\pi/2$ to $\pi/4$. The Exclusive Ors (XORs), negators and multiplexers (MUXs) in Figure 9-6 are used to obtain full sine and cosine waves by changing polarity and/or exchanging the first octant sine and cosine values according to the scheme defined by Table 9-2 [McC84].

In most practical DDS digital implementations, numbers are represented in a 2's complement format. Therefore 2's complementing must be used to invert the phase and multiply the output of the look-up table by -1. However, it can be shown that if a 1/2 LSB offset is introduced into a number that is to be complemented, then a 1's complementor may be used in place of the 2's complementor without introducing error [Nic88], [Rub89]. This provides savings in hardware since a 1's complementor may be implemented as a set of simple exclusive-or gates. This 1/2 LSB offset is provided by choosing look-up table samples such that there is a 1/2 LSB offset in both the phase and amplitude of the samples [Nic88], [Rub89], as shown in Figure 9-7 and Figure 9-8. In Figure 9-7, the phase offset must be used to reduce the address

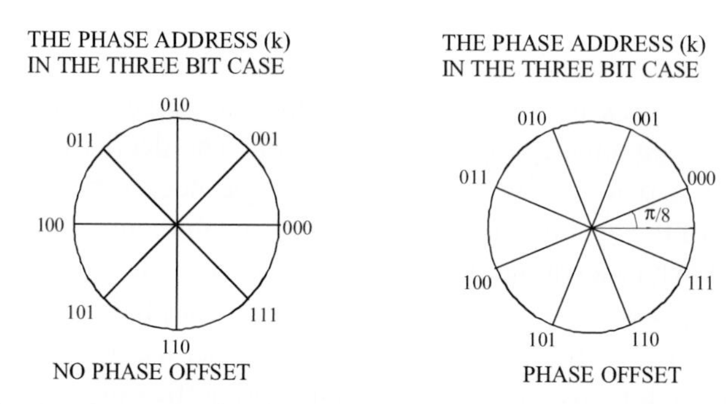

Figure 9-7. ½ LSB phase offset is introduced in all phase addresses. In this case ½ LSB correspond $\pi/16$. The 1/2 LSB phase offset is added to all the sine look-up table samples. In this figure it is shown, that 1's complementor maps the phase values to the first quadrant without error.

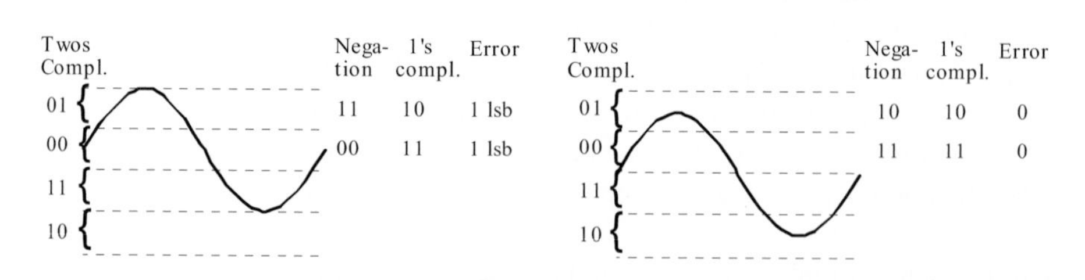

Figure 9-8. -1/2 LSB offset is introduced into the amplitude that is to be complemented; then the negation can be carried out with the 1's complementor in Figure 9-5. The -1/2 LSB offset introduces DC offset in the sine wave, which can be removed by adding +1/2 LSB offset to the D/A-converter output.

bits by two. If there is no phase offset, 0 and $\pi/2$ have the same phase address, and one more address bit is needed to distinguish these two values.

Garvey has shown if the phase to amplitude converter exploits the symmetry ($\sin(\theta) = -s(\theta+\pi)$), then distortion is minimized [Gar90]. Furthermore, if the phase to amplitude converter is designed to preserve the symmetry, then the spurs corresponding to all even DFT frequency bins will be zero for any input phase increment word (ΔP) [Tie71].

9.2.3 Compression of Quarter-Wave Sine Function

In this section, the quarter-wave memory compression is investigated. The width of the sine look-up table is reduced before taking advantage of the quadrant symmetry of the sine function (see Figure 9-5). The compression techniques are the trigonometric approximations, the so-called Nicholas architecture, the polynomial approximations and the CORDIC algorithm. For each method, the total compression ratio, the size of memory, the worst-case spur level and additional circuits are presented in Table 9-1. The amplitude values of the quarter-wave compression could be scaled to provide an improved performance in the presence of amplitude quantization [Nic88]. The optimization of the value scaling constant provides only a negligible improvement in the amplitude quantization spur level, so it is beyond the scope of this work.

9.2.3.1 Difference Algorithm

The difference algorithm approximates the sine function over the first quarter period by another function $D(P)$ that can be easily implemented. In this algorithm, a look-up table is used to store the error (see Figure 9-9)

$$f(P) = \sin(\frac{\pi P}{2}) - D(P), \quad 0 \le P < 1, \tag{9.5}$$

which results in less memory wordlength requirements [McC84], [Nic88], [Lia97], [Yam98], [Sod00], [Pii01], [Lan01], [Liu01]. The simplest form is the sine-phase difference method [Nic88] that uses a straight-line approxi-

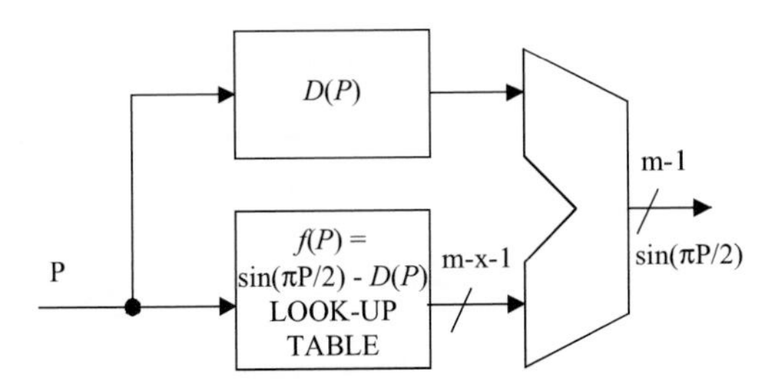

Figure 9-9. Difference algorithm.

mation for the sine function. Compression of the storage required for the quarter-wave sine function is obtained by storing the function

$$f(P) = \sin(\frac{\pi P}{2}) - P \tag{9.6}$$

instead of $\sin(\pi P/2)$ in the look-up table. Because (see Figure 9-10)

$$\max\left[\sin(\frac{\pi P}{2}) - P\right] \approx 0.21 \max\left[\sin(\frac{\pi P}{2})\right], \tag{9.7}$$

2 bits of amplitude in the storage of the sine function ($x = 2$ in Figure 9-9) are saved [Nic88]. The penalty for this storage reduction is the introduction of an extra adder at the output of the look-up table to perform the operation

$$\left[\sin(\frac{\pi P}{2}) - P\right] + P. \tag{9.8}$$

Another similar work of this category is a double trigonometric approximation, devised by Yamagishi et al. [Yam98], where the stored function is defined as

$$f(P) = \begin{cases} \sin(\frac{\pi P}{2}) - 1.25P & 0 \leq P < 0.5 \\ \sin(\frac{\pi P}{2}) - 0.75P - 0.25 & 0.5 \leq P < 1. \end{cases} \tag{9.9}$$

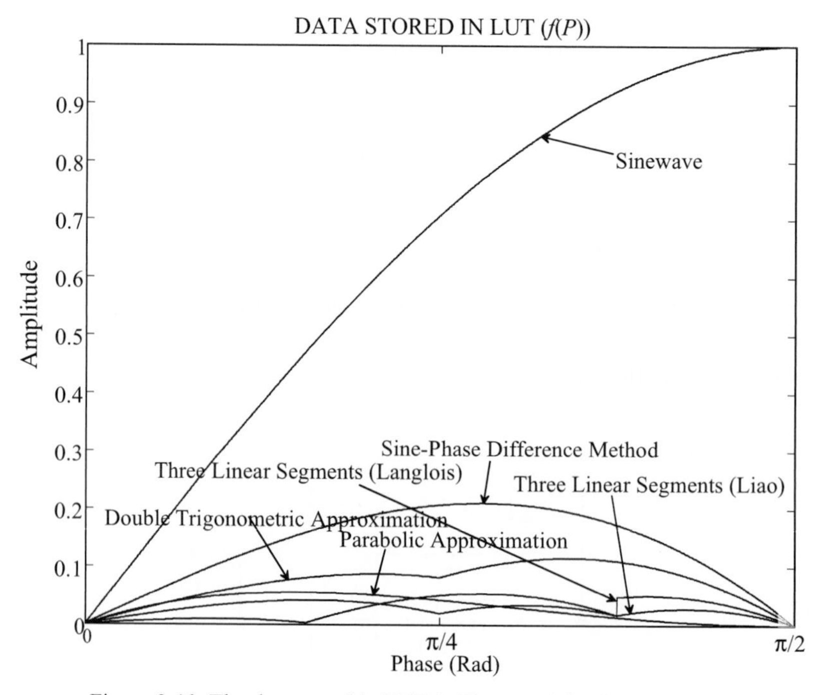

Figure 9-10. The data stored in LUT in Figure 9-9 for different methods.

This leads to a memory wordlength reduction of 3 bits ($\max(f(P)) \approx 0.117$ in Figure 9-10), but the trade-off is a complementor and two adders [Yam98].

Parabolic approximation has been introduced [Sod00]. The stored function is defined as

$$f(P) = \sin(\frac{\pi P}{2}) - P(2 - P) \qquad 0 \leq P < 2. \tag{9.10}$$

This results in saving four bits of memory wordlength, but it requires an adder, two's complementor and a multiplier.

The sinewave amplitude is approximated with three lines in [Lia97]. The storing function is [Lia97]

$$f(P) = \begin{cases} \sin(\frac{\pi P}{2}) - 1.375P & 0 \leq P < 0.5 \\ \sin(\frac{\pi P}{2}) - 0.875P - 0.25 & 0.5 \leq P < 0.75 \\ \sin(\frac{\pi P}{2}) - 0.375P - 0.625 & 0.75 \leq P < 1. \end{cases} \tag{9.11}$$

The word length of the sine LUT could be shortened by 4 bits ($\max(f(P)) \approx 0.043$ in Figure 9-10). The trade-off is three adders, two MUXs and control logic.

Langlois and Al-Khalili have proposed a method that involves splitting first quadrant angles into three sections, and to use a linear approximation in each case [Lan01]. The stored function is defined as

$$f(P) = \begin{cases} \sin(\frac{\pi P}{2}) - 1.5P & 0 \leq P < 0.3125 \\ \sin(\frac{\pi P}{2}) - P - 0.1563 & 0.3125 \leq P < 0.75 \\ \sin(\frac{\pi P}{2}) - 0.5P - 0.5 & 0.75 \leq P < 1. \end{cases} \tag{9.12}$$

It also results in saving four bits of memory wordlength ($\max(f(P)) \approx 0.055$ in Figure 9-10) and requires two adders, two MUXs and control logic.

Figure 9-10 compares the function stored in the LUT (Figure 9-9) for the sinewave (reference), the sine-phase difference method [McC84], [Nic88], the double trigonometric approximation [Yam98], the parabolic approximation [Sod00], the three piecewise linear segments approximation [Lia97] and the three piecewise linear segments approximation [Lan01].

9.2.3.2 Splitting into Coarse and Fine LUTs

The following procedure and technique for LUT compression was presented in [Gol96]. The block diagram of the architecture is shown in Figure 9-11.

The phase address "P" represents the input phase [0, $\pi/2$] scaled to a binary fraction in the interval [0, 1]. The phase address of the quarter of the sine wave is decomposed to $P = a + b$, with the word lengths of the variables being $a \rightarrow A$ and $b \rightarrow B$. Then, the function $\sin(\pi a/2)$ is stored in a coarse lookup table and the function $g(a + b)$ selected so that

$$\sin(\frac{\pi}{2}(a+b)) = \sin(\frac{\pi}{2}a) + g(a+b). \quad (9.13)$$

More specifically, the fine LUT are the differences between the true values of the sine waveform and the coarse LUT. Summing up the outputs of the coarse and fine LUT, the derived sine waveform is exactly the same as this one obtained from a single LUT of equivalent size. Clearly, this method does not decrease the total number of memory locations, but because the fine LUT output wordlength is smaller than m-1, it reduces the total number of bits. It is worth noticing that this algorithm is an error-free algorithm.

9.2.3.3 Angle Decomposition

The phase address of the quarter of the sine wave is decomposed to $P = a + b$, with the word lengths of the variables being $a \rightarrow A$ and $b \rightarrow B$. The following trigonometric approximation is made [Tie71], [Gor75], [Cur01b]

$$\sin(\frac{\pi}{2}(a+b)) \approx \sin(\frac{\pi}{2}a) + \cos(\frac{\pi}{2}a)\sin(\frac{\pi}{2}b), \quad (9.14)$$

which is valid if B is sufficiently small that $\cos(\pi b/2) \approx 1$. For this purpose, $\sin(\pi a/2)$, $\cos(\pi a/2)$ and $\sin(\pi b/2)$ are stored in lookup tables. As distinct from the Sunderland algorithm (see Section 9.2.3.4), the second term of the approximation (9.14) is computed rather than stored, as shown in Figure 9-12. Due to the symmetry of the sine and cosine about $\pi/4$, the size of the lookup tables containing $\sin(\pi a/2)$ and $\cos(\pi a/2)$ can be halved. Two 2:1 MUXs, which are controlled by the MSB bit of a, are needed to obtain the full quarter period sine wave, as shown in Figure 9-12.

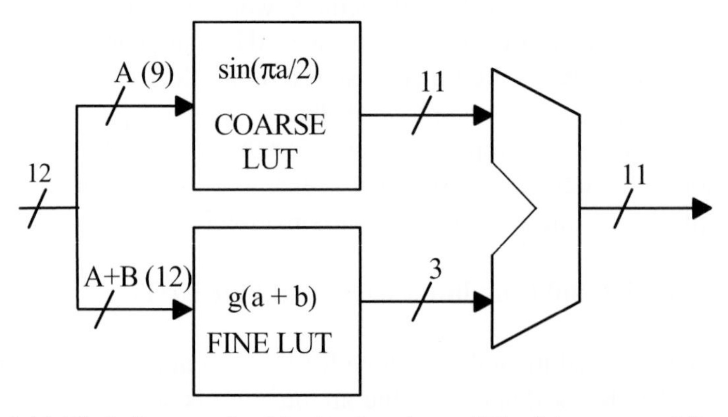

Figure 9-11. Block diagram of architecture based on splitting into coarse and fine LUTs.

The drawback of this method is the introduction of a multiplier. Alternatively, the product of cos($\pi a/2$) sin($\pi b/2$) can be stored to the LUT, but it increases the LUT size [Gor75]. The hardware could be further reduced by replacing the standard multiplier with a truncated multiplier [Sch99], [Jou99], [Cur01a].

In (9.14), the term sin($\pi a/2$) provides low resolution phase samples, and the term cos($\pi a/2$) sin($\pi b/2$) gives additional phase resolution by interpolating between the low resolution phase samples. The hardware could be reduced by performing the interpolation about a point in the middle of the interpolation range (see Figure 9-14). In order to be able to perform interpolation around a point in the middle of the interpolation range, an offset $\pi/2^{A+2}$ has to be added to the coarse phase values (a). The term sin($\pi a/2$) is looked up and cos($\pi a/2$) sin($\pi b/2$) added or subtracted depending on the MSB bit of the fine phase (b). Therefore, we can halve the memory size by storing sin($\pi b/2$) and using the MSB of b to control the sign of the lookup table's output [Gor75], [Cur01b]. However, a small degradation in spectral purity is expected. This method is beyond the scope of this book.

The phase address "P'" represents the input phase [0, π/4] scaled to a binary fraction in the interval [0, 1]. The phase address for the quadrature wave is decomposed to $P' = a' + b' + c'$. The trigonometric approximation for quadrature wave (Figure 9-6) is given by [Gor75], [Cur01b]

$$\sin(\frac{\pi}{4}(a'+b')) \approx \sin(\frac{\pi}{4}a') + \cos(\frac{\pi}{4}a')\sin(\frac{\pi}{4}b') \qquad (9.15)$$

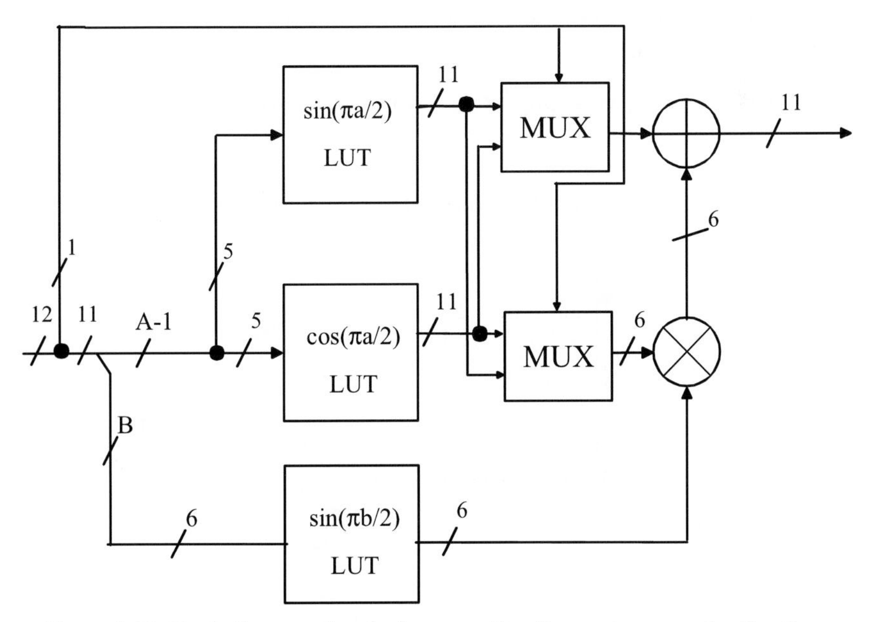

Figure 9-12. Block diagram of angle decomposition for quarter-wave sine function compression.

$$\cos(\frac{\pi}{4}(a'+b')) \approx \cos(\frac{\pi}{4}a') - \sin(\frac{\pi}{4}a')\sin(\frac{\pi}{4}b'). \qquad (9.16)$$

The sin($\pi b'/4$) term appears in both (9.15) and (9.16), which allows memory to be shared between sine and cosine blocks [Gor75], [Chu98], [Cur01b].

9.2.3.4 Modified Sunderland Architecture

The original Sunderland technique is based on simple trigonometric identities [Sun84]. There are modifications to the original Sunderland paper. After this paper was published, a method of performing the two's complement negation function with only an exclusive-or, which does not introduce errors when reconstructing a sine wave, was published [Nic88], [Rub89]. This method works by introducing the 1/2-LSB offsets into the phase and amplitude of the sine LUT samples, as described in Section 9.2.2.

The phase address of the quarter of the sine wave is decomposed to $P = a + b + c$, with the word lengths of the variables being $a \rightarrow A$, $b \rightarrow B$, and $c \rightarrow C$. In Figure 9-13, the twelve phase bits are divided into three 4-bit fractions such that $a < 1$, $b < (2^{-4})$, $c < (2^{-8})$. The desired sine function is given by

$$\begin{aligned}\sin(\frac{\pi}{2}(a+b+c)) &= \sin(\frac{\pi}{2}(a+b))\cos(\frac{\pi}{2}c) \\ &\quad + \cos(\frac{\pi}{2}(a+b))\sin(\frac{\pi}{2}c).\end{aligned} \qquad (9.17)$$

Given the relative sizes of a, b, and c, this expression can be approximated by

$$\sin(\frac{\pi}{2}(a+b+c)) \approx \sin(\frac{\pi}{2}(a+b)) + \cos(\frac{\pi}{2}(a+\bar{b}))\sin(\frac{\pi}{2}c). \qquad (9.18)$$

The approximation is improved by adding the average value of b to a in the second term. In Figure 9-13, the coarse LUT provides low resolution phase

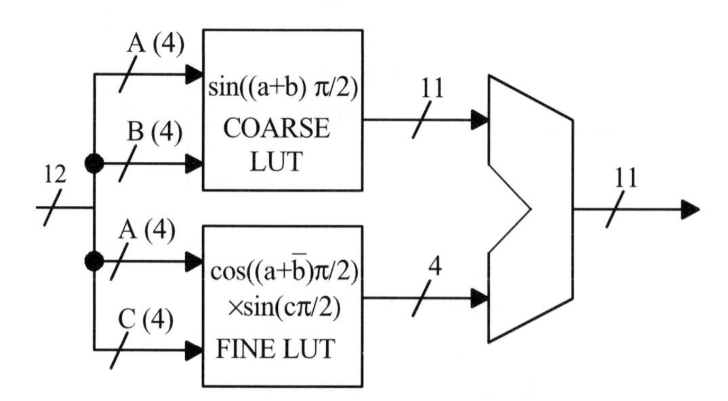

Figure 9-13. Block diagram of the modified Sunderland architecture for quarter-wave sine function compression.

samples, and the fine LUT gives additional phase resolution by interpolating between the low resolution phase samples.

The trigonometric approximation for quadrature wave (Figure 9-6) is given by

$$\sin(\frac{\pi}{4}(a'+b'+c')) \approx \sin(\frac{\pi}{4}(a'+b')) + \cos(\frac{\pi}{4}(a'+\bar{b}')) \sin(\frac{\pi}{4}c') \quad (9.19)$$

$$\cos(\frac{\pi}{4}(a'+b'+c') \approx \cos(\frac{\pi}{4}(a'+b')) - \sin(\frac{\pi}{4}(a'+\bar{b}')) \sin(\frac{\pi}{4}c'). \quad (9.20)$$

The phase address for the quadrature wave is decomposed to $P' = a' + b' + c'$, with the word lengths of the variables being $a' \rightarrow 3$, $b' \rightarrow 4$ and $c' \rightarrow 4$. The $\sin(\pi c'/4)$ term appears in both (9.19) and (9.20).

The architecture in [Yan01] is a modified version of the Sunderland architecture. In [Yan01], the size of the coarse and fine lookup tables is reduced by using the splitting technique (see Section 9.2.3.2).

9.2.3.5 Nicholas Architecture

An alternative methodology for choosing the samples to be stored in the LUTs is based on numerical optimization [McC84], [Nic88]. The phase address of the quarter of the sine wave is defined as $P = a + b + c$, where the word length of the variable a is A, the word length of b is B, and of c is C. The variables a, b form the coarse LUT address, and the variables a, c form the fine LUT address. In Figure 9-14, the coarse LUT samples are represented by the dot along the solid line, and the fine LUT samples are chosen to be the difference between the value of the "error bars" directly below and above that point on the solid line. In Figure 9-14 the function is divided into 4 regions, corresponding to a = 00, 01, 10, and 11. Within each region, only

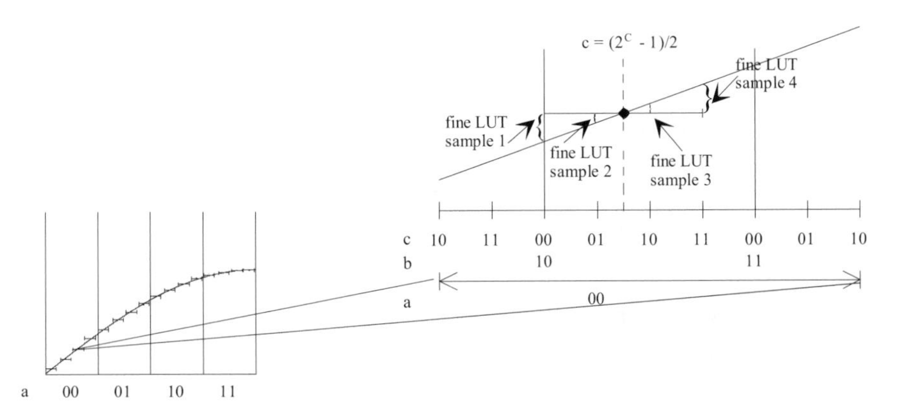

Figure 9-14. Fine LUT samples are used to interpolate a higher phase resolution function from the coarse samples, and the symmetry in the fine LUT samples around the $c=(2^C-1)/2$ point. Here $C = 2$.

one interpolation value may be used between the error bars and the solid line for the same c values. The interpolation value used for each value of c is chosen to minimize either the mean square or the maximum absolute error of the interpolation within the region [Nic88]. Further storage compression is provided by exploiting the symmetry in the fine LUT correction factors, Figure 9-14. If the coarse LUT samples are chosen in the middle of the interpolation region, then the fine LUT samples will be approximately symmetric around the $c = (2^C - 1)/2$ point (C is the word length of the variable c). Thus, by using an adder/subtractor instead of an adder to sum the coarse and fine LUT values, the size of the fine LUT may be halved. Since the fine LUT is generally not in the critical speed path, the effective resolution of the fine LUT may be doubled, rather than halving the LUT. It allows the segmentation of the compression algorithm to be changed, effectively adding an extra bit of phase resolution to the look-up table, which thereby reduces the magnitude of the worst-case spur due to phase accumulator truncation.

The simulations showed that the mean square criterion gives a better total spur level than the maximum absolute error criterion in this segmentation. The architecture for sine wave generation employing this look-up table compression technique is shown in Figure 9-15. The amplitude values of the coarse and fine LUTs could be scaled to provide an improved performance in the presence of amplitude quantization [Nic88]. The optimization of the value scaling constant provides only a negligible improvement in the amplitude quantization spur level, so it is beyond the scope of this book.

In a modified version of the above architecture the symmetry in the fine LUT samples is not utilized [Tan95a], so the extra bit of the phase resolution to the LUT address is not achieved. Therefore, the modified Nicholas architecture uses a 14-to-12-bit instead of 15-to-12-bit phase to amplitude mapping in this case. Some hardware is saved, because an adder instead of an adder/subtractor is used to sum the coarse and fine LUT values, so the adder/subtractor control logic is not needed. The difference between the modi-

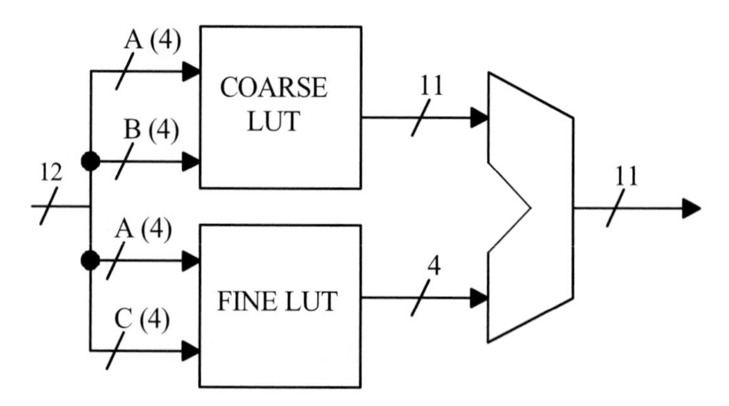

Figure 9-15. Nicholas' architecture.

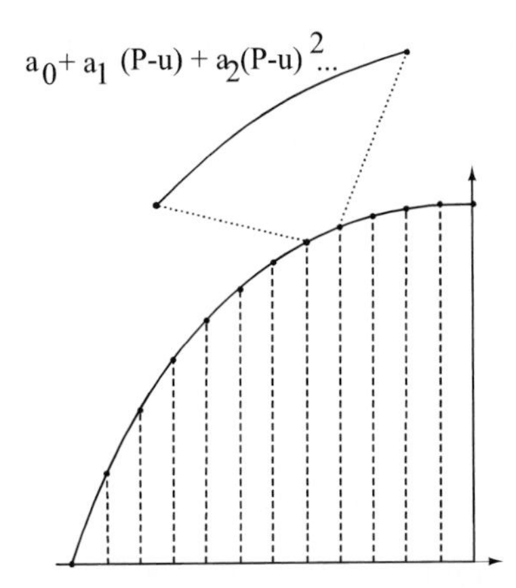

Figure 9-16. Piecewise polynomial interpolation.

fied Nicholas architecture and the modified Sunderland architecture is that the samples stored in the sine LUT are chosen using the numerical optimization in the modified Nicholas architecture.

The IC realization of the Nicholas architecture is presented in [Nic91], where a CMOS chip has the maximum clock frequency of 150 MHz. Analog Devices has also used this sine memory compression method in their CMOS device, which has the output word length of 12 bits and 100 MHz clock frequency [Ana94]. The IC realization of the modified Nicholas architecture is presented in [Tan95a], where the CMOS quadrature digital synthesizer operates at a 200 MHz clock frequency. The modified Nicholas architecture has also been used in [Tan95b], where a CMOS chip has four parallel LUT to achieve four times the throughput of a single DDS. The chip that uses only one LUT has the clock frequency of 200 MHz [Tan95a]. Using the parallel architecture with four LUT, the chip attains the speed of 800 MHz [Tan95b]. The silicon compiler for the Nicholas architecture was presented in [Lau92].

9.2.3.6 Polynomial Approximations

Instead of using a single polynomial approximation, it is possible to subdi-

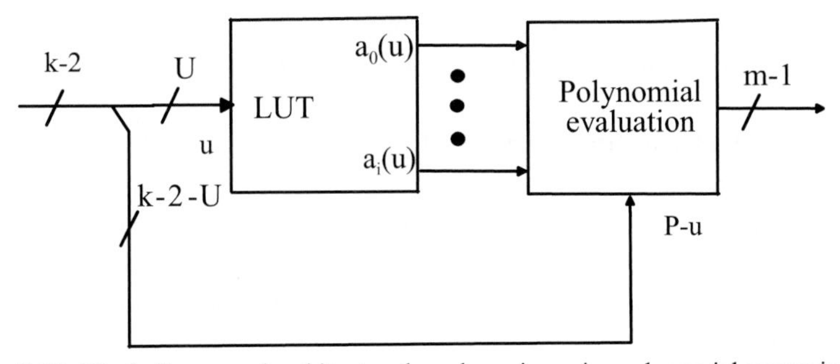

Figure 9-17. Block diagram of architecture based on piecewise polynomial approximation.

vide the interval in which functions have to be evaluated ([0, π/4] for quadrature DDS (see Figure 9-6) and [0, π/2] for single-phase DDS (see Figure 9-5) in shorter sub-intervals. Polynomials of a given degree are used in each subinterval to approximate trigonometric functions, resulting in piecewise polynomial interpolation architectures (see Figure 9-16). The phase address of the quarter of the sine wave "*P*" is divided into the upper phase address "*u*" and the lower phase address "*P-u*" with the word lengths of the variables being $u \rightarrow U$ and $P \rightarrow k - 2$. Thus, the piecewise polynomial approximation can be defined as:

$$\sin(\frac{\pi}{2}P) = a_0(u) + a_1(u)\,(P-u) + a_2(u)\,(P-u)^2 ..., \tag{9.21}$$

where $a_i(u)$ is the polynomial coefficient as a function of the approximation interval. To implement (9.21) we have to store the polynomial coefficients, $a_i(u)$, into a lookup table and perform the arithmetic operations required to evaluate the polynomial. The block diagram of the piecewise polynomial approximation based architecture is shown in Figure 9-17. It is important to note that for this architecture there is a trade-off between the number of approximation intervals and the polynomial order for a targeted spectral purity. From the point of view of hardware implementation, this means that by increasing the lookup table locations it is possible to decrease the hardware used for the polynomial evaluation. Moreover, if the polynomial order is fixed, the increase of the lookup table locations allows the reduction of the multiplier's wordlength used for polynomial evaluation. The architectures presented in [Fre89], [Wea90a], [Liu01], [Bel00], [Pal00], [Fan01], [Car02], [Elt02], [Sai02], [Str02], [Lan02a], [Lan02b], [Lan03a], [Lan03b], [Pal03], [Van04] are particular instances of this generic architecture. The main difference between them comes from the method used for computing the polynomial coefficients, the polynomial order, and the technique utilized for implementing the polynomial evaluation block. This differentiates them in terms of spectral purity, operating speed, area, and power consumption.

An algorithm is presented in [Sch92] where combinatorial binary logic is used to approximate the sine function. The quarter wave sine and cosine samples have been calculated by Chebyshev approximation [Whi93], [McI94], [Pal00], [Car02], Legendre approximation [Car02] and the Taylor series [Wea90a], [Edm98], [Bel00], [Sai02], [Pal03].

9.2.3.6.1 Piecewise Linear Interpolation

The case of piecewise linear interpolation is investigated in [Fre89], [Liu01], [Bel00], [Sai02], [Str02], [Lan02a], [Lan02b], [Lan03a], [Lan03b], [Str03a]. All of these architectures approximate the sine function using a set of first-order polynomials

$$\sin(\frac{\pi}{2}P) \approx a_0(u) + a_1(u)(P-u). \tag{9.22}$$

The block diagram of the architecture based on linear interpolation is shown in Figure 9-18. The polynomial coefficients are computed using various techniques, such as optimized computer procedure [Fre89], [Ken93], genetic algorithms [Lan02b], exhaustive search [Lan03b], Chebyshev expansion [Str02], Taylor series [Bel00]. Although these designs share the same basic architecture, they provide different results because the spectral purity and the hardware cost are dependent on the polynomial coefficients. It can be observed in Figure 9-18 that in order to reduce the lookup table address wordlength, i.e., the number of memory locations, we have to increase the wordlength of the multiplier. However, the number of memory locations, i.e., the number of polynomials, limits the spectral purity of the generated wave. It was shown in [Lan03b] that, for a given number of polynomials, e.g., 2^U, the spurious free dynamic range is limited by:

$$\text{SFDR} \leq (24+12U)\,\text{dBc}. \tag{9.23}$$

Therefore, this architecture implies a tradeoff between the lookup table size and the multiplier wordlength for a targeted spectral purity.

A subdivision using segments of unequal lengths is considered only in [Liu01]; other papers resort to the simpler and more effective case in which segments are equal to each other and the segment number is a power of two. Freeman proposed a polynomial approximation decomposing the first quadrant of the sine function into 16 linear segments that are equal in length [Fre89]. The approximation is

$$\sin(\frac{\pi}{2}P) \approx \sin(\frac{\pi}{2}u) + \cos(\frac{\pi}{2}u)\frac{\pi}{2}(P-u) + e(P), \tag{9.24}$$

where u is an upper phase address identifying one of 16 segments, P-u is the fine angle inside such a segment, and $e(u,P\text{-}u)$ are correction values that reduce the Maximum Amplitude Error (MAE) between an ideal sine amplitude and the approximation. A paper by Kent and Sheng [Ken93] describes a DDS implementation based on this method.

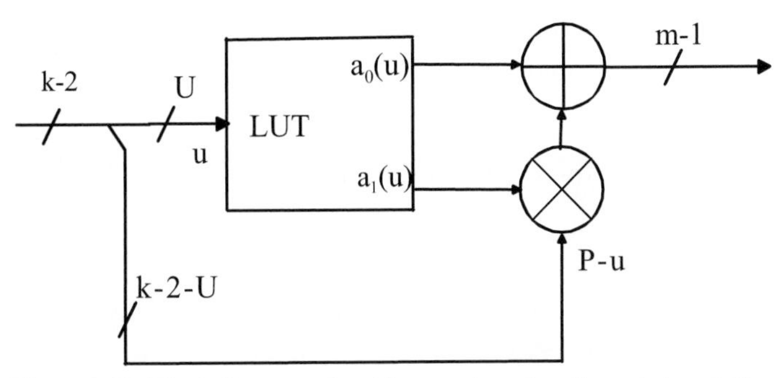

Figure 9-18. Block diagram of architecture based on linear interpolation.

A different approach is used in [Lan02b], [Lan03b] in which the second coefficients of (9.22), a_1, are selected so that their values can be represented with a limited number of signed digits. The selection is made by performing an exhaustive search starting from the standard linear interpolation solution [Lan03b]. This method reduces the number of the multiplication's partial products procedure, which might result in increased multiplication speed. However, the disadvantage is that the partial products can be also negative and have to be selectively shifted. In order to diminish this drawback, the negation operation is performed using one's complementers [Lan02b]. The hardware cost is reduced further by truncating the partial products prior to addition. Moreover, the lookup table and the multiplier are replaced by multibit multiplexers.

The technique called the dual-slope approach uses a set of first order polynomials to approximate the sine function [Str03a]. However, in [Str03a], each of the approximation intervals is divided into two equal sub-intervals having their own polynomial. These two polynomials have identical a_0 coefficients, but different a_1 coefficients. As a result, the lookup table storing the coefficients a_0 has half of the locations of the lookup table containing the coefficients a_1.

9.2.3.6.2 High Order Piecewise Interpolation

By using a high order polynomial approximation, it is possible to generate high spectral purity waves or to significantly reduce or even eliminate the lookup tables. Several of the architectures proposed in the literature based on high order polynomial approximation [Car02], [Pal00], [Sod00], [Fan01], [Sod01], [Elt02], [Pal03], [Van04] will be discussed in the following.

Piecewise parabolic interpolation is proposed in [Pal00], [Fan01], [Elt02], [Pal03], [Van04]. Fanucci, Roncella and Saletti [Fan01] used four piecewise-continuous 2nd degree polynomials to approximate the first quadrant of the sine function. The authors state that, in order to maximize the SFDR, they selected polynomial coefficients that reduced the MAE. Eltawil and Daneshrad [Elt02] used the same architecture as Fanucci et al. [Fan01] for a second degree piecewise-continuous parabolic interpolation, necessitating two multipliers and two adders. The polynomial coefficients are calculated according to a Farrow structure realizing interpolation through digital filtering. An algorithm presented in [Van04] approximates the first quadrant of the sine function with sixteen equal length second degree polynomial segments (see Section 14.3). The coefficients are chosen using least-squares polynomial fitting.

The architectures from [Car02], [Pal00], [Pal03] use only two polynomials for approximating the sine/cosine function over the first quadrant. Consequently, the lookup table is completely eliminated, as the polynomial coefficients can be hardwired. Unfortunately, the elimination of the lookup tables increases the polynomial order and/or decreases the SFDR.

De Caro, Napoli and Strollo [Car02] introduced two designs implementing the quadrature architecture shown in Figure 9-6. The designs approximate the first octant of the sine and cosine functions with single 2nd and 3rd degree polynomials, respectively. In each case, the selection of the polynomial coefficients was made to optimize spectral purity. Several selection processes were considered by the authors, including the Taylor series, Chebyshev polynomials and Legendre polynomials. The minimax polynomial can be calculated by using numerical iterative algorithms, like the one originating from Remez [Pre92]. The Chebyshev approximation, however, is in practice so close to the minimax polynomial that additional numerical effort to improve approximation is hardly required [Car02]. The polynomial that minimizes the square error can easily be obtained using Legendre polynomials [Car02]. However, it was found in [Car02] that the best SFDR was achieved using non-linear Nelder-Mead simplex optimization [Nel65].

The method given in [Sod00] is based on quadratic polynomial interpolation, but it has the major drawback of its low SFDR (28.4dBc). This problem has been overcome in the next paper [Sod01] by using another quadratic interpolator for the error. The addition of the new interpolator makes the entire system implement a fourth degree interpolation. The obtained SFDR by this method is 62.8dBc.

Polynomials of degree $n > 3$ can be evaluated in fewer than n multiplications. For example,

$$P(x) = a_0 + a_1 x + a_2 x^2 + a_3 x^3 + a_4 x^4, \tag{9.25}$$

where $a_4 > 0$, can be evaluated with 3 multiplications and 5 additions as follows [Pre92]

$$P(x) = \left[(Ax+B)^2 + Ax + C\right]\left[(Ax+B)^2 + D\right] + E, \tag{9.26}$$

where A, B, C, D and E are precomputed

$$\begin{aligned} A &= (a_4)^{1/4} \\ B &= \frac{a_3 - A^3}{4A^3} \\ C &= \frac{a_2}{A^2} - 2B - 6B^2 - D \\ D &= 3B^2 + 8B^3 + \frac{a_1 A - 2a_2 B}{A^2} \\ E &= a_0 - B^4 - B^2(C+D) - CD. \end{aligned} \tag{9.27}$$

Using Horner rule [Fan01], [Elt02] the equation (9.25) is computed as:

$$P(x) = a_0 + x(a_1 + x(a_2 + x(a_3 + xa_4))). \quad (9.28)$$

This requires 4 multiplications and 4 additions. However, it is less parallel than (9.25) and the operation speed might be slower.

9.2.3.6.3 Taylor Series Approximation

The phase address "P" is divided into the upper phase address "u" and the lower phase address "P-u" [Wea90a]. The Taylor series is performed around the upper phase address (u) for sine

$$\sin(\frac{\pi}{2}P) = \sin(\frac{\pi}{2}u) + k_1\,(P-u)\cos(\frac{\pi}{2}u) - \frac{k_2\,(P-u)^2\sin(\frac{\pi}{2}u)}{2} + R_3, \quad (9.29)$$

where k_n represents a constant used to adjust the units of each series term. The adjustment in units is required because the phase values have angular units. Therefore it is necessary to have a conversion factor k_n, which includes a multiple of $\pi/2$ to compensate for the phase units. The remainder is for sine

$$R_n = \frac{d^n(\sin(\frac{\pi}{2}r))}{dr}\frac{(P-u)^n}{n!}. \quad \text{where } r\,\varepsilon\,[u,P], \quad (9.30)$$

Since sine and cosine both have upper limits of 1, the following upper estimate defines the accuracy:

$$|R_n| = \left|\frac{k_n\,(P-u)^n}{n!}\right| \le \left|\frac{k_n\,|P-u|^n_{\max}}{n!}\right|. \quad (9.31)$$

The Taylor series (9.29) are approximated in Figure 9-19 by taking two terms. The estimated accuracy is 7.5×10^{-5} (9.31), while additional terms can be employed, their contribution to the accuracy is very small and, therefore, of little weight in this application. Other inaccuracies present in the operation of current DDS designs override the finer accuracy provided by successive series terms. The seven most significant bits of the input phase are selected as the upper phase address "u", which is transferred simultaneously to a sine LUT and a cosine LUT as address signals, as shown in Figure 9-19. The output of the sine LUT is the first term of the Taylor series and is transferred to a first adder, where it will be summed with the remaining terms involved. The output of the cosine LUT is configured to incorporate the predetermined unit conversion value k_1. The cosine LUT output is the first derivative of the sine. The least significant bits (P-u) are multiplied by the output of the cosine LUT to produce the second term.

Further storage compression is provided by exploiting the symmetry in the cosine LUT samples in Figure 9-19 (similar idea to that in Figure 9-14) [Sai02]. This results in a reduction in coefficient storage by approximately half, at the expense of a slight increase in multiplier complexity [Sai02]. In this book, the symmetry is not utilized in the Taylor series approximation and Nicholas architecture.

QUALCOMM has used the Taylor series approximations in their device, which has the output word length of 12 bits and 50 MHz clock frequency [Qua91a].

9.2.3.6.4 Chebyshev Approximation

The Chebyshev approximation is a method for approximating any arbitrary function $f(x)$ in the interval $x \in [-1, +1]$. The Chebyshev approximation is very close to the minimax polynomial, which has the smallest maximum deviation from the true function [Pre92]. The $(N - 1)$th order Chebyshev approximation for $f(x)$ is given

$$f(x) \approx \sum_{i=0}^{N-1} c_i(u) T_i(x), \tag{9.32}$$

where c_i, $i = 0, \ldots, N - 1$ and $T_i(x)$ are defined by

$$\begin{aligned} c_0 &= \frac{1}{\pi} \int_{-1}^{+1} \frac{T_0(x) f(x)}{\sqrt{1-x^2}} dx \\ c_i &= \frac{2}{\pi} \int_{-1}^{+1} \frac{T_i(x) f(x)}{\sqrt{1-x^2}} dx, \quad i \geq 1 \\ T_i(x) &= \cos(i \arccos(x)). \end{aligned} \tag{9.33}$$

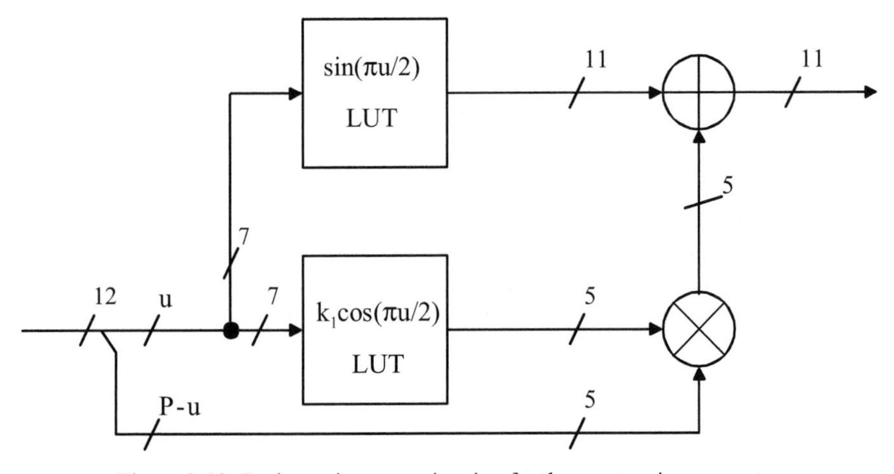

Figure 9-19. Taylor series approximation for the quarter sine converter.

By using trigonometric identities, it is possible to yield explicit expressions for $T_N(x)$

$$\begin{aligned} T_0(x) &= 1 \\ T_1(x) &= x \\ T_{i+1}(x) &= 2xT_i(x) - T_{i-1}(x), \quad i \geq 1. \end{aligned} \tag{9.34}$$

The phase address of the quarter of the sine wave "P" is divided into the upper phase address "u" and the lower phase address "P-u" with the word lengths of the variables being $u \rightarrow U$ and $P \rightarrow k$ - 2. The phase address "P" represents the input phase [0, π/2] scaled to a binary fraction in the interval [0, 1]. The range [0, 1] is subdivided into 2^U sub-intervals. The U most significant bits of "P" encode the segment starting point u and are used as address to look-up tables that store polynomials coefficients. The remaining bits of "P" represent the offset "P-u". In order to obtain the Chebyshev expansion of sin(πP/2) function in [u, (u + 2^{-U})] interval, we firstly construct the function $f(x)$, which, for x ϵ [-1, +1], assumes the same values that the function sin(πP/2) assumes for P ϵ [u, (u + 2^{-U})].

$$f(x) = \sin\left(\frac{\pi}{2}\left(\frac{x}{2^{U+1}} + \frac{1}{2^{U+1}} + u\right)\right), \tag{9.35}$$

where

$$x = 2^{U+1}(P-u) - 1, \quad x \, \varepsilon \, [-1, +1], \quad (P-u) \, \varepsilon \, [0, 2^{-U}]. \tag{9.36}$$

By using (9.32), (9.33), and (9.34), it is possible to derive a piecewise linear approximation for sine

$$\begin{aligned} \sin\left(\frac{\pi}{2}P\right) &\approx c_0(u) + c_1(u)T_1(2^{U+1}(P-u) - 1) \\ &= (c_0(u) - c_1(u)) + 2^{U+1}c_1(u)(P-u). \end{aligned} \tag{9.37}$$

For sinusoidal signals, the integrals (9.33) have closed form solutions as [Str03b]

$$\begin{aligned} c_0(u) &= J_0\left(\frac{\pi}{2^{U+2}}\right)\sin\left(\frac{\pi}{2}\left(u + \frac{1}{2^{U+1}}\right)\right), \\ c_i(u) &= 2J_i\left(\frac{\pi}{2^{U+2}}\right)\sin\left(\frac{\pi}{2}\left(u + \frac{1}{2^{U+1}} + \frac{i\pi}{2}\right)\right), \quad i \geq 1, \end{aligned} \tag{9.38}$$

where J_i is the Bessel function of the first kind with index i.

The truncation error in (9.32) can be approximated as the first term dropped from the series: $c_N\ T_N(x)$. Furthermore, since $T_i(x)$ is bounded between -1 and +1 (see (9.33)), the absolute value of the error can be upper bounded by

$$e_{\max} \approx |c_N|. \tag{9.39}$$

Considering equations (9.38) and (9.39) we can evaluate the maximum error introduced by the first order Chebyshev approximation over all the $[u, (u + 2^{-U})]$ intervals

$$e_{\max Cheb} \approx 2J_2\left(\frac{\pi}{2^{U+2}}\right) \approx \frac{1}{16}\left(\frac{\pi}{2^{U+1}}\right)^2 \text{ for } U >> 1. \tag{9.40}$$

This can be compared with

$$e_{\max Tay} \approx \frac{1}{2}\left(\frac{\pi}{2^{U+1}}\right)^2 \text{ for } U >> 1 \tag{9.41}$$

that is the worst case error obtained choosing the coefficients according to Taylor approximation (9.31). Note that an eight-fold error reduction is obtained by using Chebyshev (9.40) instead of Taylor approximation (9.41). Equations (9.40) and (9.41) also show that using Chebyshev approximation the length of each sub-interval can be halved, with respect to Taylor approximation, while still having a better approximation. Table 9-1 shows how much memory and how many additional circuits are needed in each memory compression and algorithmic technique to meet the spectral requirement for the worst case spur level, which is about -85 dBc, due to the sine memory compression. Table 9-1 shows that, for piecewise-linear case, the Chebyshev polynomial approximation gives better compression ratio (at the expense of an increase in adder and multiplier complexity) when compared with the Taylor series approximation.

In [Str02] piecewise-linear Chebyshev expansion is employed. In [Pal00] a fourth-order Chebyshev approximation is employed for both sine and cosine functions, while second-order and third-order polynomial approximations are investigated in [Car02].

9.2.3.6.5 Legendre Approximation

The polynomial that minimizes the square error for: $u < P < (u + 2^{-U})$ can easily be obtained using Legendre polynomials [Car02]. The Legendre polynomials are defined in a very elegant and compact way by the Rodrigues formulas:

$$P_i(x) = \frac{1}{2^i\, i!}\frac{d^i}{dx^i}(x^2 - 1)^i. \tag{9.42}$$

The following relations hold:

$$P_0(x) = 1, \quad P_1(x) = x, \quad P_{i+1}(x) = \frac{2i+1}{i+1}xP_i(x) - \frac{i}{i+1}P_{i-1}(x). \tag{9.43}$$

The least mean square polynomial approximation of degree N - 1 is given by:

$$f(x) \approx \sum_{i=0}^{N-1} c_i(u) P_i(x), \tag{9.44}$$

where

$$c_i(u) = \frac{(2i+1)}{2} \int_{-1}^{1} f(x) P_i(x)\, dx, \tag{9.45}$$

where $f(x)\Big|_{x=2^{U+1}(P-u)-1} = \sin(\frac{\pi}{2}(\frac{x}{2^{U+1}} + \frac{1}{2^{U+1}} + u)), \quad x\,\varepsilon\,[-1,+1].$

By solving the integral one obtains:

$$c_0(u) = \frac{2^{U+1}}{\pi}(\cos(u\pi/2) - \cos((u+2^{-U})\pi/2))$$

$$c_1(u) = 6(\frac{2^U}{\pi})^2(4(\sin((u+2^{-U})\pi/2) - \sin(u\pi/2)) \tag{9.46}$$

$$-\pi(\cos((u+2^{-U})\pi/2) + \cos(u\pi/2))/2^U).$$

By using (9.44), (9.45), and (9.46), it is possible to derive piecewise linear approximation for sine

$$\sin(\frac{\pi}{2}P) \approx c_0(u) + c_1(u) P_1(2^{U+1}(P-u)-1)$$
$$= (c_0(u) - c_1(u)) + 2^{U+1} c_1(u)(P-u). \tag{9.47}$$

The approximation error can be estimated as the first dropped term in the sum (9.44). Table 9-1 shows that, for piecewise-linear case, the Chebyshev and Legendre polynomial approximations give very similar results.

9.2.3.7 Using CORDIC Algorithm as a Sine Wave Generator

The CORDIC algorithm performs vector coordinate rotations by using simple iterative shifts and add/subtract operations, which are easy to implement in hardware [Vol59]. The details of the CORDIC algorithm are presented in Chapter 6. If the initial values are constant, (I_0, Q_0) and P_0 is formed using

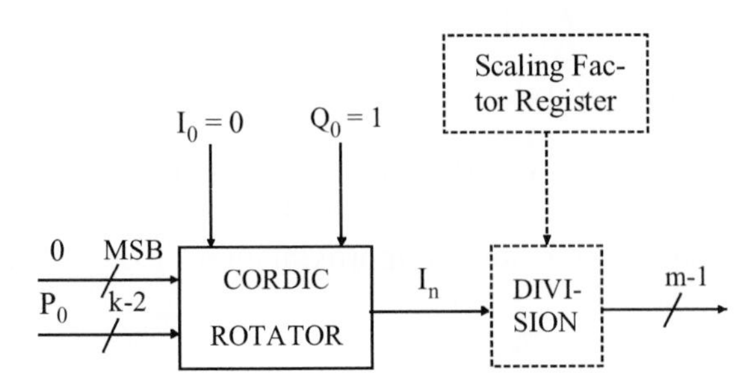

Figure 9-20. CORDIC rotator for a quarter sine converter.

the remaining k-2 bits of the phase register value from the phase accumulator, then the result will be from (6.8)

$$I_n = G_n \sqrt{I_0^2 + Q_0^2} \cos(A - \arctan(\frac{Q_0}{I_0}))$$

$$Q_n = G_n \sqrt{I_0^2 + Q_0^2} \sin(A - \arctan(\frac{Q_0}{I_0}))$$

$$G_n = \prod_{i=0}^{n-1} \sqrt{1 + 2^{-2i}} \tag{9.48}$$

$$A = P_0 - P_n,$$

where P_n is the angle approximation error.

The initial values (I_0, Q_0) can be chosen so that there is no need for a scaling operation after the CORDIC iterations. If the initial values are chosen to be $I_0 = 0$, $Q_0 = 1/G_n$, then there is no need for a scaling operation after the CORDIC iterations. The architecture for quarter sine wave generation employing this technique is shown in Figure 9-20. It consists of a CORDIC module with a convergence range of [-π/2, π/2]. The input phase to the CORDIC processor takes values in the [0, π/2) range (the MSB bit of the phase input to the CORDIC rotator is set to zero in Figure 9-20). The input phase has to be flipped, i.e. one's complemented, introducing distortion to the output waveform. This problem is solved in LUT-based methods by introducing ½ LSB offset to the phase address (see Figure 9-7); however, this solution is impracticable in CORDIC-based methods. In [Tor01], the input phase to the CORDIC processor takes values in the [-π/2, π/2) range. The input phase is not flipped in [Tor01], which reduces distortion.

The QDDS is implemented by the partioned hybrid CORDIC algorithm (see Section 6.4.2) [Mad99], [Jan02], [Cur03]. The QDDS architectures in [Mad99], [Jan02] take advantage of the eighth wave symmetry of a sine and cosine waveform [McC84], [Tan95a]. The hybrid CORDIC generates sine and cosine samples from 0 to π/4. The partioned hybrid CORDIC has coarse and fine rotation as shown in Figure 6-3. The coarse rotation is performed by the look-up table in [Mad99], [Jan02], followed by the CORDIC iterations, where the rotation directions are obtained from (6.25).

The hardware costs of the CORDIC and LUT based phase to amplitude converters were estimated in FPGA, which shows that the CORDIC based architecture becomes better than the LUT based architecture when the required accuracy is 9 bits or more [Par00]. The CORDIC algorithm is also effective for solutions where quadrature mixing is performed (see Section 16.5). The conventional quadrature mixing requires four multipliers, two adders and sine/cosine memories (see Figure 4-4). It replaces sine/cosine LUTs, four multipliers and two adders.

For example, the GEC-Plessey I/Q splitter has a 20 MHz clock frequency with a 16-bit phase and amplitude accuracy [GEC93], and the Raytheon

Table 9-1. Memory compression and algorithmic techniques with the worst-case spur level due to the sine memory compression specified to be about –85 dBc.

Method	Needed LUT	Total compression ratio	Additional Circuits (not include quarter logic)	Worst case Spur (below carrier)	Comments
Uncompressed memory	$2^{14} \times 12$ bits	1 : 1	-	-97.23 dBc	Reference
Angle decomposition	$2^5 \times 11$ bits $2^5 \times 11$ bits $2^6 \times 6$ bits	180.7 : 1	11-bit Adder 6×6 bit Multiplier Two MUXs	-84.05 dBc	Need multiplier and two MUXs
Mod. Sunderland architecture	$2^8 \times 11$ bits $2^8 \times 4$ bits	51.2 : 1	11-bit Adder	-86.91 dBc	Simple
Mod. Nicholas architecture	$2^8 \times 11$ bits $2^8 \times 4$ bits	51.2 : 1	11-bit Adder	-86.81 dBc	Simple
Taylor series approximation with two terms	$2^7 \times 11$ bits $2^7 \times 5$ bits	96 : 1	11-bit Adder 5×5 bit Multiplier	-85.88 dBc	Need multiplier
Chebyshev approximation with two terms	$2^6 \times 12$ bits $2^6 \times 6$ bits	171 : 1	12-bit Adder 6×6 bit Multiplier	-89.50 dBc	Need multiplier
Legendre approximation with two terms	$2^6 \times 12$ bits $2^6 \times 6$ bits	171 : 1	12-bit Adder 6×6 bit Multiplier	-88.55 dBc	Need multiplier
CORDIC algorithm	–	–	14 pipelined stages, 18-bit inner word length	-84.25 dBc	Much computation

Semiconductor's DDS has a 25 MHz clock frequency with a 16-bit phase and amplitude accuracy [Ray94].

9.2.4 Simulation

A computer program (in Matlab) has been created to simulate the direct digital synthesizer in Figure 4-1. The memory compression and algorithmic techniques have been analyzed with no phase truncation (the phase accumulator length = the phase address length), and the spectrum is calculated prior to the D/A-conversion. The number of points in the DDS output spectrum depends on ΔP (phase increment word) via the greatest common divisor of ΔP and 2^j (GCD(ΔP,2^j)) (4.4). Any phase accumulator output vector can be formed from a permutation of another output vector regardless of the initial phase accumulator contents, when GCD(ΔP,2^j) = 1 for all values of ΔP (see Section 7.3). A permutation of the samples in the time domain results in an identical permutation of the discrete Fourier transform (DFT) samples in the frequency domain (see Section 7.3). This means that the spurious spectrum due to all system non-linearities can be generated from a permutation of another spectrum, when GCD(ΔP,2^j) = 1 for all ΔP, because each spectrum will differ only in the position of the spurs and not in the magnitudes. When the least significant bit of the phase accumulator input is forced to one, it causes all of the phase accumulator output sequences to belong to the number theoretic class GCD(ΔP,2^j) = 1, regardless of the value of ΔP. Only one simulation needs to be performed to determine the value of the worst-case spurious response due to system non-linearities. The number of samples has been chosen (an integer number of cycles in the time record) so that problems of leakage in the fast Fourier transform (FFT) analysis can be avoided and unwindowed data can be used. The FFT was performed over the output period (4.4). The size of the FFT was 16384 points.

9.2.5 Summary of Memory Compression and Algorithmic Techniques

Table 9-1 comprises the summary of memory compression and algorithmic techniques. Table 9-1 shows how much memory and how many additional circuits are needed in each memory compression and algorithmic technique to meet the spectral requirement for the worst case spur level, which is about -85 dBc, due to the sine memory compression. In the DDS, most spurs are normally not generated by digital errors, but rather by the analog errors in the D/A-converter. The spur level (-85 dBc) from the sine memory compression is not significant in DDS applications because it will stay below the spur level of a high speed 12-bit D/A-converter [Sch03]. The memory compression and algorithmic techniques in Table 9-1 are comparable with almost the same worst case spur. There is a trade-off between the LUT sizes (compression ratio) and additional circuits. In Table 9-1 the modified Nicholas architecture [Tan95a] is used; therefore the compression ratio and the worst-case spur level are different from that in the Nicholas architecture [Nic88]. The difference between the modified Nicholas architecture and the modified Sunderland architecture is that the samples stored in the sine LUT are chosen according to the numerical optimization in the modified Nicholas architecture. In the 14-to-12-bit phase-to-amplitude mapping the numerical optimization gives no benefit, because the modified Sunderland architecture and the modified Nicholas architecture give almost the same spur levels.

9.3 Filter

There are many classes of filters existing in the literature. However, for most applications, the field can be narrowed down to three basic filter families. Each is optimized for a particular characteristic in either the time or frequency domain. The three filter types are the Chebyshev, Gaussian, and Legendre families of responses [Zve67]. Filter applications that require fairly sharp frequency response characteristics are best served by the Chebyshev family of responses. However, it is assumed that ringing and overshoot in the time domain do not present a problem in such applications.

The Chebyshev family can be subdivided into four types of responses, each with its own special characteristics. The four types are the Butterworth response, the Chebyshev response, the inverse Chebyshev response, and elliptical response.

The Butterworth response is completely monotonic. The attenuation increases continuously as the frequency increases, i.e. there are no ripples in the attenuation curve. Of the Chebyshev family of filters, the passband of the Butterworth response is the flattest. Its cut-off frequency is identified by the 3dB attenuation point. Attenuation continues to increase with frequency, but the rate of attenuation after cut-off is rather slow.

The Chebyshev response is characterized by attenuation ripples in the passband followed by monotonically increasing attenuation in the stopband. It has a much sharper passband to stopband transition than the Butterworth response. However, the cost for the faster stopband roll-off is ripples in the passband. The steepness of the stopband roll-off is directly proportional to the magnitude of the passband ripples; the larger the ripples, the steeper the roll-off.

The inverse Chebyshev response is characterized by monotonically increasing attenuation in the passband with ripples in the stopband. Similar to the Chebyshev response, larger stopband ripples yields a steeper passband to stopband transition.

The elliptical response offers the steepest passband to stopband transition of any of the filter types. The penalty, of course, is attenuation ripples, in this case both in the passband and stopband.

REFERENCES

[Ait02] A. Gutierrez-Aitken et al., "Ultrahigh-speed Direct Digital Synthesizer Using InP DHBT Technology," IEEE Journal of Solid State Circuits, Vol. 37, No. 9, pp. 1115-1119, Sept. 2002.

[Ana94] Analog Devices AD 9955 data sheet, Rev. 0, 1994, and AD9712A data sheet, Rev. 0, 1994.

[Bel00] A. Bellaouar, and et al., "Low-Power Direct Digital Frequency Synthesis for Wireless Communications," IEEE J. Solid-State Circuits, Vol. 35, No. 3, pp. 385-390, Mar. 2000.

[Bje91] B. E. Bjerede, "Suppression of Spurious Frequency Components in Direct Digital Frequency Synthesizer," U. S. Patent 5 073 869, Dec. 17, 1991.

[Car02] D. D. Caro, E. Napoli, and A. G. M. Strollo, "Direct Digital Frequency Synthesizers Using High-order Polynomial Approximation," Proc. IEEE 2002 ISSCC, pp. 134-135, Feb. 2002.

[Che92]B. W. Cheney, D. C. Larson and A. M. Frisch, "Delay Equalization Emulation for High Speed Phase Modulated Direct Digital Synthesis," U. S. Patent 5,140,540, Aug. 18, 1992.

[Cho88] J. Chow, F. F. Lee, P. M. Lau, C. G. Ekroot, and J. E. Hornung, "1.25 GHz 26-bit Pipelined Digital Accumulator," 1988 GaAs IC Symp. Technical Digest, Nov. 1988, pp. 131-134.

[Chr95] W. A. Chren, Jr., "RNS-Based Enhancements for Direct Digital Frequency Synthesis," IEEE Trans. Circuits Syst. - Analog and Digital Signal Processing, Vol. CAS-42, No. 8, pp. 516-524, Aug. 1995.

[Chu98] S. W. Chu, and et al., "Digital Phase to Digital Sine and Cosine Amplitude Translator," U. S. Patent 5,774,082, June 30, 1998.

[Cur01a] F. Curticăpean, and J. Niittylahti, "A Hardware Efficient Direct Digital Frequency Synthesizer," in Proc. IEEE Int. Conference on Electronics Circuits and Systems, Malta, Sept. 2001, pp. 51-54.

[Cur01b] F. Curticăpean, K. I. Palomäki, and J. Niittylahti, "Direct Digital Frequency Synthesiser with High Memory Compression Ratio," Electronic Letters, Vol. 37, No. 21, pp. 1275-1277, Oct. 2001.

[Cur03] F. Curticăpean, K. I. Palomäki, and J. Niittylahti, "Quadrature Direct Digital Frequency Synthesizer Using an Angle Rotation Algorithm," in

Proc. IEEE Int. Symposium on Circuits and Systems, Thailand, May 2003, pp. 81-84.
[Edm98] A. Edman, A. Björklid, and I. Söderquist, "A 0.8 μm CMOS 350 MHz Quadrature Direct Digital Frequency Synthesizer with Integrated D/A Converters," Symposium on VLSI Circuits 1998 Digest of Technical Papers, pp. 54-55.
[Ekr88] C. G. Ekroot, and S. I. Long, "A GaAs 4-b Adder-Accumulator Circuit for Direct Digital Synthesis," IEEE J. Solid-State Circuits, Vol. 23, pp. 573-580, April 1988.
[Ell96] P. M. Elliott, "Numerically Controlled Oscillator for Generating a Digitally Represented Sine Wave Output Signal," U. S. Patent 5,521,534, May 28, 1996.
[Elt02] A. M. Eltawil, and B. Daneshrad, "Piece-Wise Parabolic Interpolation for Direct Digital Frequency Synthesis," Proceedings of the IEEE Custom Integrated Circuits Conference, Orlando, May 2002, pp. 401-404.
[Ert96] R. Ertl, and J. Baier, "Increasing the Frequency Resolution of NCO-Systems Using a Circuit Based on a Digital Adder," IEEE Trans. Circuits and Systems II, Vol. 43, No. 3, pp. 266-269, Mar. 1996.
[Fan01] L. Fanucci, R. Roncella, and R. Saletti, "A Sine Wave Digital Synthesizer Based on a Quadratic Approximation," Proceedings of the IEEE International Frequency Control Symposium, 2001, pp. 806-810.
[Fre89] R. A. Freeman, "Digital Sine Conversion Circuit for Use in Direct Digital Synthesizers," U. S. Patent 4,809,205, Feb. 28, 1989.
[Gar90] J. F. Garvey, and D. Babitch, "An Exact Spectral Analysis of a Number Controlled Oscillator Based Synthesizer," in Proc. 44th Annu. Freq. Contr. Symp., May 1990, pp. 511-521.
[GEC93] GEC-Plessey Semiconductors, Data Sheet PDSP16350 I/Q Splitter/NCO, Dec. 1993.
[Gie89] B. Giebel, J. Lutz, and P. L. O'Leary, "Digitally Controlled Oscillator," IEEE J. Solid-State Circuits, Vol. 24, pp. 640-645, June 1989.
[Gol88] B. G. Goldberg, "Digital Frequency Synthesizer," U. S. Patent 4,752,902, June 21, 1988.
[Gol90] B. G. Goldberg, "Digital Frequency Synthesizer Having Multiple Processing Paths," U. S. Patent 4,958,310, Sep. 18, 1990.
[Gol96] B. G. Goldberg, "Digital Techniques in Frequency Synthesis," McGraw-Hill, 1996.
[Gor75] J. Gorski-Popiel, "Frequency-Synthesis: Techniques and Applications," IEEE Press, New York, USA, 1975.
[Har91] M. V. Harris, "A J-Band Spread-Spectrum Synthesiser Using a Combination of DDS and Phaselock Techniques," IEE Coll. Digest 1991/172 on Direct Digital Frequency Synthesis, Nov. 1991, pp. 8/1-10.
[Has84] R. Hassun, and A. W. Kovalick, "Waveform Synthesis Using Multiplexed Parallel Synthesizers," U. S. Patent 4,454,486, June 12, 1984.

[Jac73] L. B. Jackson, "Digital Frequency Synthesizer," U. S. Patent 3,735,269, May 22, 1973.
[Jan02] I. Janiszewski, B. Hoppe, and H. Meuth, "Numerically Controlled Oscillators with Hybrid Function Generators," IEEE Transactions on Ultrasonics, Ferroelectrics and Frequency Control, Vol. 49, pp. 995-1004, July 2002.
[Jia02] J. Jiang, and E. K. F. Lee, "A Low-Power Segmented Nonlinear DAC-based Direct Digital Frequency Synthesizer," IEEE J. Solid-State Circuits, Vol. 37, No. 10, pp. 1326-1330, Oct. 2002.
[Jou99] J. M. Jou, S. R. Kuang, and R. D. Chen, "Design of Low-Error Fixed-Width Multipliers for DSP Applications," IEEE Trans. on Circuits and Systems, Vol. 46, pp. 836-842, June 1999.
[Ken93] G. W. Kent, and N.-H. Sheng, "A High Purity, High Speed Direct Digital Synthesizer," Proceedings of the IEEE International Frequency Control Symposium, 1993, pp. 207-211.
[Kos99] M. Kosunen, J. Vankka, M. Waltari, L. Sumanen, K. Koli, and K. Halonen, "A CMOS Quadrature Baseband Frequency Synthesizer/Modulator," Analog Integrated Circuits and Signal Processing, Vol. 18, No. 1, pp. 55-67, Jan. 1999.
[Lan01] J. M. P. Langlois, and D. Al-Khalili, "ROM Size Reduction with Low Processing Cost for Direct Digital Frequency Synthesis," Proceedings of the IEEE Pacific Rim Conference on Communications, Computers and Signal Processing, Victoria, BC, Aug. 2001, pp. 287-290.
[Lan02a] J. M. P. Langlois, and D. Al Khalili, "Hardware Optimized Direct Digital Frequency Synthesizer Architecture with 60 dBc Spectral Purity," Proc. of IEEE Symp. on Circuits and Systems (ISCAS 2002), Vol. 5, May 2002, pp. 361–364.
[Lan02b] J. M. P. Langlois, and D. Al Khalili, "A New Approach to the Design of Low-Power Direct Digital Frequency Synthesizers," in Proc. of IEEE Int.Frequency Control Symp., May 2002, pp. 654-661.
[Lan03a] J. M. P. Langlois, and D. Al Khalili, "Piecewise Continuous Linear Interpolation of the Sine Function for Direct Digital Frequency Synthesis," Proc. of 2003 IEEE Radio Freq. Integr. Cyrcuits Symp., 2003, pp. 579–582.
[Lan03b] J. M. P. Langlois, and D. Al Khalili, "Novel Approach to the Design of Direct Digital Frequency Synthesizers Based on Linear Interpolation," IEEE Trans. on Circuits and System II: Analog and Digital Signal Processing, Vol. 50, No. 9, pp. 567-578, Sept. 2003.
[Lau92] L. K. Lau, R. Jain, H. Samueli, H. T. Nicholas III, and E. G. Cohen, "DDFSGEN: A Silicon Compiler for Direct Digital Frequency Synthesizers," Journal of VLSI Signal Processing, Vol. 4, pp. 213-226, April 1992.

[Lia97] S. Liao, and L-G. Chen, "A Low-Power Low-Voltage Direct Digital Frequency Synthesizer," in Proc. Int. Symp. on VLSI Technology, Systems, and Applications, Hsinchu, Taiwan, June 1997, pp. 265-269.
[Liu01] S. I. Liu, T. B. Yu, and H. W. Tsao, "Pipeline Direct Digital Frequency Synthesiser Using Decomposition Method," IEE Proceedings Circuits, Devices and Systems, Vol. 148, No. 3, pp. 141-144, June 2001.
[Lu93] F. Lu, H. Samueli, J. Yuan, and C. Svensson, "A 700-MHz 24-b Pipelined Accumulator in 1.2 μm CMOS for Application as a Numerically Controlled Oscillator," IEEE J. Solid-State Circuits, Vol. 26, pp. 878-885, Aug. 1993.
[Mad99] A. Madisetti, A. Kwentus, and A. N. Wilson, Jr., "A 100-MHz, 16-b, Direct Digital Frequency Synthesizer with a 100-dBc Spurious-Free Dynamic Range," IEEE J. of Solid State Circuits, Vol. 34, No. 8, pp. 1034-1044, Jan. 1999.
[McC84] R. D. McCallister, and D. Shearer, III, "Numerically Controlled Oscillator Using Quadrant Replication and Function Decomposition," U. S. Patent 4,486,846, Dec. 4, 1984.
[McC91b] E. W. McCune, Jr., "Variable Modulus Digital Synthesizer," U. S. Patent 5,053,982, Oct. 1, 1991.
[McI94] J. A. McIntosh, and E. E. Swartzlander, "High-speed Cosine Generator," Asilomar Conf. on Signals, Syst. and Comput., 1994, pp. 273-277.
[Mer02] E. Merlo, K. H. Baek, and M. J. Choe, "Split Accumulator with Phase Modulation for High Speed Low Power Direct Digital Synthesizers," in Proc. 15th Annual IEEE International ASIC/SOC Conference, New-York, USA, Sept. 25-28 2002, pp. 97-101.
[Moh02] A. N. Mohieldin, A. A. Emira, and E. Sánchez-Sinencio, "A 100-MHz 8-mW ROM-less Quadrature Direct Digital Frequency Synthesizer," IEEE Journal of Solid State Circuits, Vol. 37, No. 10, pp. 1235-1243, Oct. 2002.
[Mor99] S. Mortezapour, and E. K. F. Lee, "Design of Low-Power ROM-Less Direct Digital Frequency Synthesizer Using Nonlinear Digital-to Analog Converter," IEEE J. Solid-State Circuits, Vol. 34, No. 10, pp. 1350-1359, Oct. 1999.
[Nel65] J. A. Nelder, and R. Mead, "A Simplex Method for Function Minimization," Computer Journal, Vol. 7, 1965, pp. 308-313.
[Nic88] H. T. Nicholas, H. Samueli, and B. Kim, "The Optimization of Direct Digital Frequency Synthesizer in the Presence of Finite Word Length Effects Performance," in Proc. 42nd Annu. Frequency Contr. Symp., June 1988, pp. 357-363.
[Nic91] H. T. Nicholas, and H. Samueli, "A 150-MHz Direct Digital Frequency Synthesiser in 1.25-μm CMOS with -90-dBc Spurious Performance," IEEE J. Solid-State Circuits, Vol. 26, pp. 1959-1969, Dec. 1991.

[Nos01a] H. Nosaka, Y. Yamaguchi, and M. Muraguchi, "A Non-Binary Direct Digital Synthesizer with an Extended Phase Accumulator," IEEE Transactions on Ultrasonics, Ferroelectrics and Frequency Control, Vol. 48, pp. 293-298, Jan. 2001.

[Pal00] K. I. Palomaki, and J. Niittylahti, "Direct Digital Frequency Synthesizer Architecture Based on Chebyshev Approximation," in Proc. 34th Asilomar Conf. on Signals, Systems and Computers, 29 Oct.-1 Nov. 2000, Vol. 2, pp. 1639-1643.

[Pal03] K. I. Palomäki, and J. Niittylahti, "A Low-Power, Memoryless Direct Digital Frequency Synthesizer Architecture," Proceedings of the IEEE International Symposium on Circuits and Systems, May 2003, pp. II-77-80.

[Par00] M. Park, K. Kim, and J. A. Lee, "CORDIC-Based Direct Digital Synthesizer: Comparison with a ROM-Based Architecture in FPGA Implementation," IEICE Trans. Fundam., Vol. E83-A, No. 6 June 2000.

[Pii01] O. Piirainen, "Method of Generating Signal Amplitude Responsive to Desired Function, and Converter," U. S. Patent 6,173,301, Jan. 9, 2001.

[Pre92] W. H. Press, S. A. Teukolsky, W. T. Vetterling, and B. P. Flannery, "Numerical Recipes in C, The Art of Scientific Computing," Second Edition, Press Syndicate of the University of Cambridge, New York, NY, USA, 1992.

[Qua91a] Qualcomm Q2334, Technical Data Sheet, June 1991.

[Ray94] Raytheon Semiconductor Data Book, Data Sheet TMC2340, 1994.

[Rog96] R. Rogenmoser, and Q. Huang, "A 800-MHz 1-µm CMOS Pipelined 8-b Adder Using True Single-Phase Clocked Logic-Flip-Flops," IEEE J. of Solid State Circuits, Vol. 31, No. 3, pp. 401-409, Mar. 1996.

[Rub89] P. W. Ruben, E. F. Heimbecher, II, and D. L. Dilley, "Reduced Size Phase-to-Amplitude Converter in a Numerically Controlled Oscillator," U. S. Patent 4,855,946, Aug. 8, 1989.

[Sai02] M. M. El Said, and M. I. Elmasry, "An Improved ROM Compression Technique for Direct Digital Frequency Synthesizers," Proceedings of the IEEE International Symposium on Circuits and Systems, Phoenix AZ, May 2002, pp. 437-440.

[Sch03] W. Schofield, D. Mercer, and L. St. Onge, "A 16b 400MS/s DAC with <-80dBc IMD to 300 MHz and <-160dBm/Hz Noise Power Spectral Density," ISSCC Digest of Technical Papers, February 2003, San Francisco, USA, pp. 126-127.

[Sch92] E. M. Schwarz, and M. J. Flynn, "Approximating the Sine Function with Combinational Logic," Asilomar Conf. on Signals, Syst. and Comput., Oct. 1992, pp. I-386-390.

[Sch99] M. J. Schulte, J. E. Stine, and J. G. Jansen, "Reduced Power Dissipation Through Truncated Multiplication," Proc. IEEE Workshop on Low-Power Design, 1999, pp. 61-69.

[Sha01] S. S. Shah, and S. Collins, "A 200 MHz Analogue-ROM Based Direct Digital Frequency Synthesizer with Amplitude Modulation," Proceedings of the IEEE International Symposium on VLSI Technology, Systems and Applications, Hsinchu, Taiwan, April 2001, pp. 53-56.
[Sod00] A. M. Sodagar, and G. R. Lahiji, "Mapping from Phase to Sine-Amplitude in Direct Digital Frequency Synthesizers Using Parabolic Approximation," IEEE Trans. Circuits Syst. II, Vol. 47, pp. 1452–1457, Dec. 2000.
[Sod01] A. M. Sodagar, and G. R. Lahiji, "A Pipeline ROM-Less Architecture for Sine-Output Direct Digital Frequency Synthesizers Using the Second-Order Parabolic Approximation," IEEE Transactions on Circuit and Systems II, Vol. 48, No. 9, pp. 850-857, Sep. 2001.
[Str02] A. G. M. Strollo, E. Napoli, and D. De Caro, "Direct Digital Frequency Synthesizers Using First-Order Polynomial Chebyshev Approximation," in Proc. of 28th European Solid-State Circuits Conf. (ESSCIRC 2002), Sept. 2002, pp. 527-530.
[Str03a] A. Strollo, D. De Caro, E. Napoli, and N. Petra, "Direct Digital Frequency Synthesis with Dual-Slope Approach," in Proc. 29th European Solid-State Circuits Conference, Estoril, Portugal, Sept. 16-18 2003, pp. 397-400.
[Str03b] A. G. M. Strollo, and D. De Caro, "Direct Digital Frequency Synthesizers Exploiting Piecewise Linear Chebyshev Approximation," Microelectronics Journal, Vol. 34, pp.1099-1106, Nov. 2003.
[Sun84] D. A. Sunderland, R. A. Strauch, S.S. Wharfield, H. T. Peterson, and C. R. Cole, "CMOS/SOS Frequency Synthesizer LSI Circuit for Spread Spectrum Communications," IEEE J. of Solid State Circuits, Vol. SC-19, pp. 497-505, Aug. 1984.
[Tan95a] L. K. Tan, and H. Samueli, "A 200 MHz Quadrature Digital Synthesizer/Mixer in 0.8 μm CMOS," IEEE J. of Solid State Circuits, Vol. 30, No. 3, pp. 193-200, Mar. 1995.
[Tan95b] L. K. Tan, E. W. Roth, G. E. Yee, and H. Samueli, "A 800 MHz Quadrature Digital Synthesizer with ECL-Compatible Output Drivers in 0.8 μm CMOS," IEEE J. of Solid State Circuits, Vol. 30, No. 12, pp. 1463-1473, Dec. 1995.
[Tho92] M. Thompson, "Low-Latency, High-Speed Numerically Controlled Oscillator Using Progression-of-States Technique," IEEE J. Solid-State Circuits, Vol. 27, pp. 113-117, Jan. 1992.
[Tie71] J. Tierney, C. Rader, and B. Gold, "A Digital Frequency Synthesizer," IEEE Trans. Audio and Electroacoust., Vol. AU-19, pp. 48-57, Mar. 1971.
[Tor01] F. Cardells-Tormo, and J. Valls-Coquillat, "Optimisation of Direct Digital Frequency Synthesisers Based on CORDIC," Electronic Letters, Vol. 37, No. 21, pp. 1278 -1280, Oct. 2001.

[Uus01]R. Uusikartano, and J. Niittylahti, "A Periodical Frequency Synthesizer for a 2.4-GHz Fast Frequency Hopping Transceiver," IEEE Trans. on Circuits and Systems Part II, Vol. 48, No. 10, pp. 912-918, Oct. 2001.
[Van04] J. Vankka, J. Lindeberg, and K. Halonen, "A Direct Digital Synthesizer with Tunable Delta Sigma Modulator," Analog and Integrated Circuit and Signal Processing, Vol. 38, No.1, pp. 7-15, Jan. 2004.
[Vol59] J. E. Volder, "The CORDIC Trigonometric Computing Technique," IRE Trans. on Electron. Comput., Vol. C-8, pp. 330–334, Sept. 1959.
[Wea90a] L. A. Weaver, and R. J. Kerr, "High Resolution Phase To Sine Amplitude Conversion," U. S. Patent 4,905,177, Feb. 27, 1990.
[Whi93] S. A. White, "A Simple High-Performance Sine/Cosine Reference Generator," Asilomar Conf. on Signals, Syst. and Comput., Nov. 1993, pp. 612-616.
[Yam98] A. Yamagishi, M. Ishikawa, and T. Tsukahara, "A 2-V 2-GHz Low-Power Direct Digital Frequency Synthesizer Chip Set for Wireless Communication," IEEE J. Solid-State Circuits, Vol. 33, No. 2, pp. 210-217, Feb. 1998.
[Yan01] B.-D. Yang, and et al., "A Direct Digital Frequency Using A New ROM Compression Method," in Proc. ESSCIRC'01, Sept. 2001, pp. 288-291.
[Yan02] B.-D. Yang, L.-S. Kim, and H.-K. Yu, "A High Speed Direct Digital Frequency Synthesizer Using a Low Power Pipelined Parallel Accumulator," Proceedings of the IEEE International Symposium on Circuits and Systems, May 2002, pp. IV-373-376.
[Yua89] J. Yuan, and C. Svensson, "High-Speed CMOS Circuit Technique," IEEE J. of Solid State Circuits, Vol. 24, No. 1, pp. 62-70, Feb. 1989.
[Zve67]A. Zvever, "Handbook of Filter Synthesis," John Wiley & Sons, New York, 1967.

Chapter 10

10. CURRENT STEERING D/A CONVERTERS

For high-speed and high-resolution applications (>50 MHz, >10 bits), the current source switching architecture is preferred, since it can drive a low impedance load directly without the need for a voltage buffer. The core of a current-steering D/A converter consists of an array of current sources that are switched to the output according to the digital input code. The current switches of the D/A converter could be implemented by CMOS differential pairs (see Figure 10-1), which are driven by the differential switching control signal V_{ctrl} and $\overline{V}_{ctrl}$. The dominant parasitic capacitance (C_0) is shown in Figure 10-1. The current is almost completely in one complementary output, when the voltage difference between the two switching transistors is [All87]

$$\left|V_{ctrl} - \overline{V_{ctrl}}\right| > \sqrt{\frac{2I_{SS}L_{SW}}{\mu C_{ox}W_{SW}}}, \tag{10.1}$$

where L_{sw} is effective channel length, W_{sw} is effective channel width, μ is the mobility of electronics and C_{ox} is the gate capacitance per unit area. Design for minimum voltage difference is not recommended because it is difficult to ensure that the minimum voltage is achieved in practice. Consequently, a safety margin is introduced. The current sources are implemented by current mirrors, for which the operation point is set by an internal or external current reference.

10.1 D/A Converter Specifications

Figure 10-2 illustrates an offset, gain error, differential and integral non-linearity (DNL and INL) as approximations to this transfer function. The output offset is usually defined as a constant DC offset in the transfer curve. The gain defines the full-scale output of the converter in relation to its reference circuit. The DNL is typically measured in the LSBs as the worst-case deviation from an average LSB step between adjacent code transitions. Formally, the DNL (in LSBs) at digital code k ($0 \le k \le 2^N$-1) is given by

$$DNL(k) = (I_{out}(k) - I_{out}(k-1)) - \frac{I_{out(\max)} - I_{out(\min)}}{2^N - 1}, \qquad (10.2)$$

where N is the number of bits. The DNL can have negative or positive values. Other DNL specifications have been presented in [Bur01], [Gus00]. The INL is defined as the deviation of the analog output (in LSBs) at that code from the ideal transfer response, which is a straight line from the output at 0 LSB to the output at 2^N-1 LSB. Formally, the INL (in LSBs) at digital code k ($0 \le k \le 2^N$-1) is given by

$$INL(k) = \sum_{i=0}^{k} DNL(i) = I_{out}(k) - \left(\frac{I_{out(\max)} - I_{out(\min)}}{2^N - 1} k + I_{out(\min)}\right). \qquad (10.3)$$

A monotonic D/A converter requires that each voltage in the transfer curve is larger than the previous voltage for a rising ramp input or less than the previous voltage for a falling ramp input. Guaranteeing that the D/A converter is monotonic requires

$$|DNL| < 1\,LSB \qquad (10.4)$$

$$|INL| < \frac{1}{2} LSB. \qquad (10.5)$$

However, some D/A converter structures are guaranteed to be always monotonic (see Section 10.7).

10.2 Static Non-Linearities

The INL is mainly determined by the matching behavior of the current

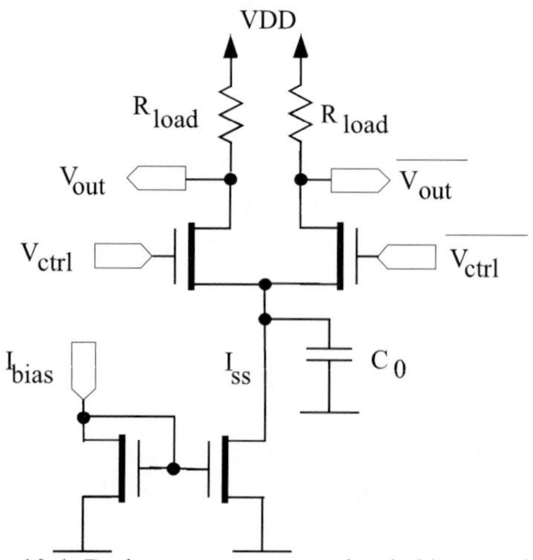

Figure 10-1. Basic current source and switching transistors.

sources and the finite output impedance of the current source. Inaccuracies in the matching can be caused by random variations and systematic influences. The random error of the current source is determined by the matching properties, which determine the dimensions of the unit current source. The systematic influences are caused by the process gradient errors, finite output impedance of current sources, temperature gradients and voltage drop in the power supply lines.

10.2.1 Random Errors

For a D/A converter to be fully functional, the INL error has to be smaller than 1/2 LSB (least significant bit). A direct relationship exists between the INL error and the matching properties of the used technology expressed by the relative unit current source standard deviation $\sigma(I_{LSB})/I_{LSB}$. This parameter determines the dimensions of the current sources of the D/A converter [Pel89]

$$W_{\mathrm{LSB}}L_{\mathrm{LSB}} = \frac{A_{\beta}^2}{\dfrac{2\sigma^2(I_{LSB})}{I_{LSB}^2}} + \frac{4\,A_{VT}^2}{\dfrac{2\sigma^2(I_{LSB})}{I_{LSB}^2}(V_{GS}-V_T)^2}, \tag{10.6}$$

where A_β and A_{VT} are technological parameters, $(V_{GS} - V_T)$ the gate overdrive voltage of the current source and $\sigma(I_{LSB})/I_{LSB}$ the unit current source relative standard deviation. By increasing $(V_{GS} - V_T)$, the minimum area requirement is decreased. For very large values, however, the mismatch is mainly determined by the A_β term and barely decreases with the $(V_{GS} - V_T)$. Consequently, a convenient criterion to determine the gate override voltage of the current source transistor is to make the two mismatches contributing in (10.6) about equal. Assuming that each unit current source has a value that follows a

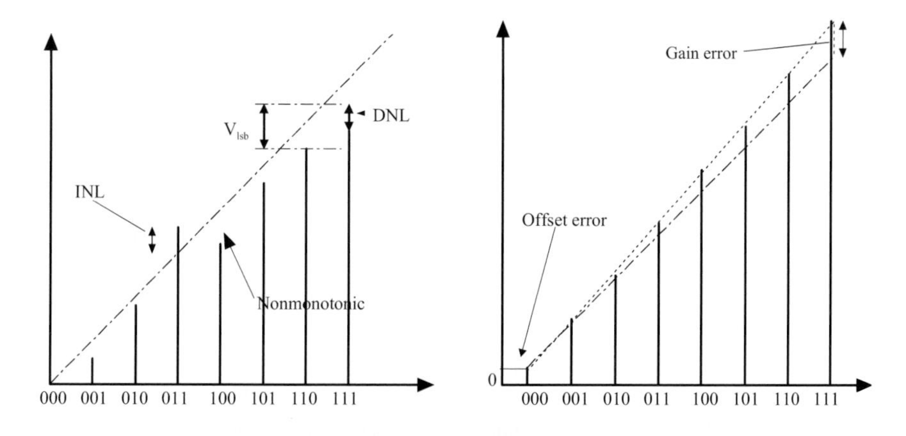

Figure 10-2. Static errors.

normal distribution, the required accuracy is given by [Bos01a]

$$\frac{\sigma(I_{LSB})}{I_{LSB}} = \frac{1}{2C\sqrt{2^N - 1}}, \tag{10.7}$$

where C depends only on the yield requirement and N is the number of bits. Various formulas for the C's dependence on the INL yield have been derived in the literature. Lakshmikumar has presented two rough bounds for INL yield estimation in binary-weighted D/A converters [Lak86], [Lak88]. The limitations of the two bounds were clearly shown in [Bos01a]. One is overly pessimistic, ignoring the strong correlation between different analog outputs, while the other is rather optimistic, considering only the contribution of the two mid-scale codes. Another formula proposed by van den Bosch for INL yield estimation is based on a nonstandard INL definition in which each current output is compared to the ideal value without correcting the gain error [Bos01a], although it is a well-known fact that a small gain error does not impact linearity. The formula therefore gives pessimistic results. The relationship between the INL yield requirement and the C for a fully segmented and binary weighted D/A converter, based on Monte Carlo simulations, has been presented [Con02]. For partially segmented architectures, corresponding values have not been presented. Monte Carlo simulations are therefore used to estimate the INL yield as a function of the unit current source standard deviation, as shown in Figure 10-3. In Figure 10-3, the INL performance, which is dependent on the segmentation level, is shown for four segmentation levels (0-12, 4-8, 8-4, 12-0), where the first number tells the number of the segmented bits. The fully segmented converter (12-0) does not give the worst yield shown in Figure 10-3; therefore, the thermometer and

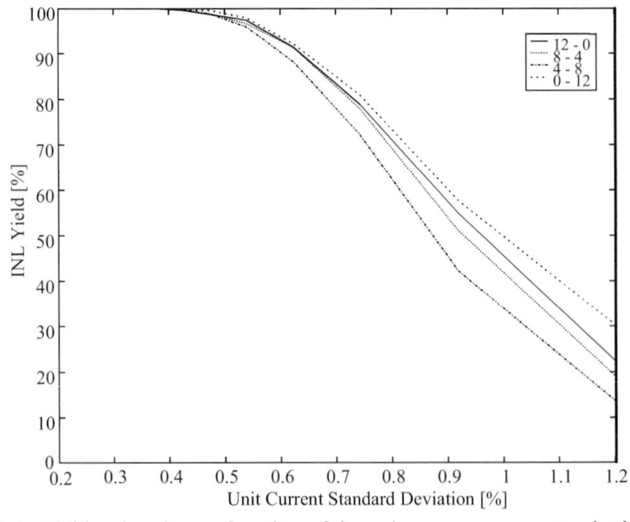

Figure 10-3. Yield estimation as function of the unit current source standard deviation (with zero mean).

binary weighted cases are not the extrema cases as stated in [Bas98]. It is shown in [Kos03] that the DNL yield also depends on the segmentation level, and that with more than two segmented bits, the INL yield is the limiting factor.

10.2.2 Systematic Errors

Gradient error distribution across a unary matrix can be approximated by a Taylor series expansion around the center of the unary array [Pla99]. The gradient error of the element located at (x,y) can be expressed as

$$g(x,y) = a_0 + a_{11}x + a_{12}y + a_{21}x^2 + a_{22}y^2 + a_{23}yx + \ldots \qquad (10.8)$$

It is generally assumed that the linear (first order) and quadratic (second order) terms are adequate to model gradient effects [Nak91], [Bas98]. That is, the error distribution is typically linear or quadratic or the superposition of both. For example, in a current source matrix, the doping and the oxide thickness over the wafer or the voltage drop along the power supply lines have been reported to cause approximately linear gradient errors [Pla99], [Bas98], [Nak91]. Temperature gradients and die stress may introduce approximately quadratic errors [Bas97]. The overall systematic error distribution is given by superposition of these error components [Nak91].

In a binary weighted D/A converter (Section 10.7.1), it is convenient

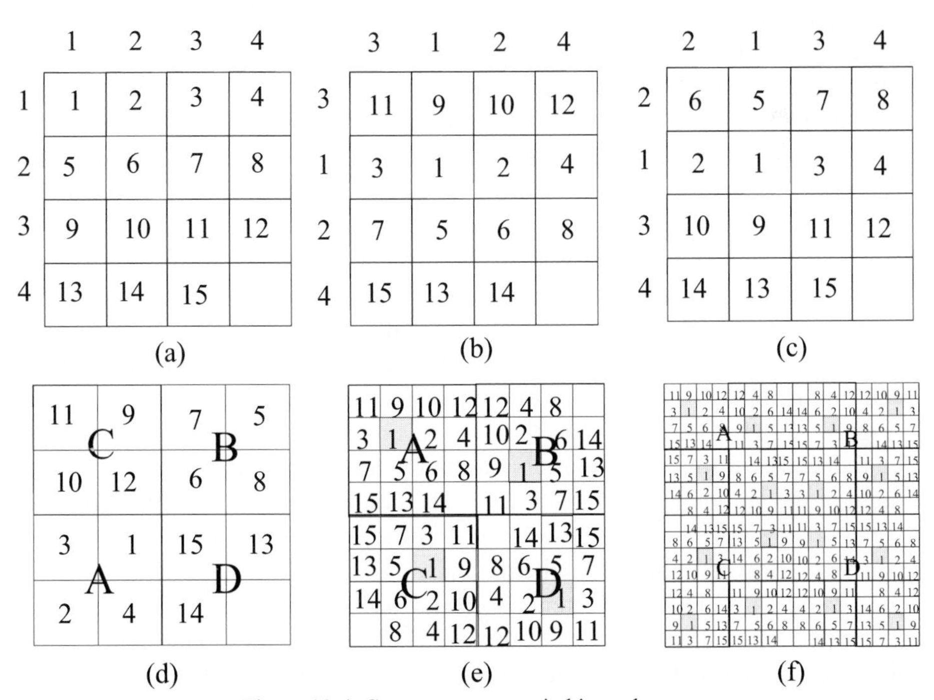

Figure 10-4. Current sources switching schemes.

from the matching point of view to generate the binary-weighted currents with equally sized current sources. Each binary-weighted current is constructed by connecting N unit current sources in parallel. The centroid of the composite sources is located at the center of the matrix [McC75], [Bas91]. This cancels the effects of linear process gradients across the matrix.

An unary decoded current sources are already equally sized, one can switch the current sources with a well-known symmetrical switching scheme [Mik86]. In this scheme, as shown in Figure 10-4(b), the current sources are arranged to average the spatial gradient errors in two directions, while the sequences for row and column selection are optimized independently. The switching optimization problem is thus reduced to a 1-D space. To compensate for both linear and quadratic errors, hierarchical symmetrical sequences were proposed (Figure 10-4(c)) [Nak91]. The switching optimization problem is also reduced to a 1-D space. The row-column and hierarchical symmetrical switching scheme are used in a 10 bit intrinsic accuracy D/A converter [Wu95], [Bav98]. The advantage of the row–column schemes are their simplicity for design and layout.

For the 2-D gradient error compensation, methods have been developed in which the current matrix is divided into regions (Figure 10-4(d)). The switching of regions is in the order of A, B, C, D; within the region the switching sequence is intended to compensate for quadratic and linear errors. The switching sequence could be determined by using algorithm techniques. Examples of methods are the "random walk" scheme [Pla99] and the INL-bounded scheme [Con00], which is developed for switching sequence optimization for unary D/A converter arrays.

The most efficient method to compensate process gradients is to divide current source transistors into four transistors; their centroid is located at the center of the matrix (Figure 10-4(e)) [McC75], [Bas98], [Pla99]. Linear errors are cancelled and quadratic errors are divided into the four regions (A, B, C, D). The residual error is ¼ of the original error. The residual errors in regions (A-D) are compensated using either the row–column switching scheme [Lin98], [Bas98] or the "random walk" scheme [Pla99]. The current source transistor inside the region (A-D) is divided to four transistors; their centroid is located at the center of the matrix in a quad quadrant (Q^2) scheme (Figure 10-4(f)) [Bos98], [Pla99]. Using this method, the residual error could be reduced four times. It is possible to divide the transistors further and place them in the common centroid scheme. The common centroid combined with some other switching scheme was used in the 10 and 12 bit D/A converters [Bas98], [Bos01]. The (Q^2) scheme and random walk scheme was used in the 12 and 14 bit D/A converters [Bos98], [Pla99], respectively.

Figure 10-4 shows current sources switching schemes. The numbers at the edge of the matrix refer to the row and columns switching order. The number inside the matrix refers to the current sources switching order. The letters in Figure 10-4 refer to the regions switching order. In the common centroid methods, the letters refer to the regions.

10.2.3 Calibration

The gradient and random errors can be compensated using calibration [Gro89], [Ger97], [Han99], [Bug00], [Rad00], [Tii01], [Con03], [Sch03], [Hyd03] and dynamic element matching [Rad00]. Two approaches can be taken to calibrate out the errors: mixed signal or analog. In the mixed signal calibration, the main D/A converter has an additional trim D/A converter in parallel, which is controlled by a calibration circuit [Ger97], [Con03], [Sch03]. For each of the MSB codes, there is a calculated correction term that is converted with the trim D/A converter. The analog output of the trim D/A converter compensates the linearity errors in the main D/A converter. However, the trim D/A converter must also be designed for dynamic linearity performance. In the analog calibration, the currents sources are trimmable [Bug00], [Tii01], [Hyd03]. The analog calibration needs a D/A converter for changing the digital calibration values to the analog domain in [Bug00]. A large number of the MSB cells require complex refreshing circuitry [Bug00].

The calibration cycle that determines the correction terms is typically performed at startup. This is not always sufficient, since the component values may drift over time and change with temperature and supply voltage. Thus, the calibration has to be repeated from time to time, which requires suspending the normal operation of the converter. In many systems, the input signal contains idle periods, which can be used for calibration. This is not always possible and the calibration has to be performed in the background [Bug00]. Proper calibration can dramatically reduce the area of the current source array and hence the parasitic capacitances. The calibration also reduces sensitivities to process, temperature and aging, thus providing high yields.

10.3 Finite Output Impedance

Any non-ideal current source has a finite output impedance and can be modeled as shown in Figure 10-5, where k is the digital code, R_{load} is the load resistance, C_{load} is the load capacitance, R_{lsb} is the LSB current source resistance, C_{out} is the parasitic capacitance and I_{lsb} is the LSB current. The output current is [Mik86]

$$I_{out} = \frac{I_{lsb}}{\frac{1}{k} + \frac{Z_{load}}{Z_{lsb}}}. \tag{10.9}$$

The output current does not depend linearly on the digital input code in (10.9), when the output impedance is finite. The relation between the impedance and INL could be derived using the (10.3) and (10.9) for a single-ended D/A converter

$$INL_{se}(k) = \frac{\frac{kZ_{load}}{1+\frac{kZ_{load}}{Z_{lsb}}} - \frac{kZ_{load}}{1+\frac{Z_{load}(2^N\text{-}1)}{Z_{lsb}}}}{\frac{Z_{load}}{1+\frac{Z_{load}(2^N\text{-}1)}{Z_{lsb}}}} I_{lsb} \approx \frac{Z_{load}}{Z_{lsb}} k(2^N\text{-}1\text{-}k)I_{lsb}, \tag{10.10}$$

and its maximum value is

$$INL_{\max} \approx \frac{Z_{load}}{Z_{lsb}} \frac{(2^N-1)^2}{4} I_{lsb}, \tag{10.11}$$

where N is the number of the bits. As the INL value should be below ½ I_{lsb}, from (10.11) we get for the required impedance ratio

$$\left(\frac{Z_{lsb}}{Z_{\text{load}}}\right)_{\text{se}} \approx \frac{(2^N\text{-}1)^2}{2}. \tag{10.12}$$

In the case of the differential output, the INL is

$$INL_{\text{dif}}(k) \approx \frac{Z_{load}^2}{2Z_{lsb}} k(2k^2 - 3(2^N-1)k + (2^N-1)^2) I_{lsb}, \tag{10.13}$$

and the required impedance ratio is

$$\left(\frac{Z_{lsb}}{Z_{load}}\right)_{\text{dif}} = \sqrt{\frac{2^N-1}{3\sqrt{3}}\left(\frac{Z_{lsb}}{Z_{load}}\right)_{\text{se}}}. \tag{10.14}$$

The output impedance variation causes mainly second order distortions

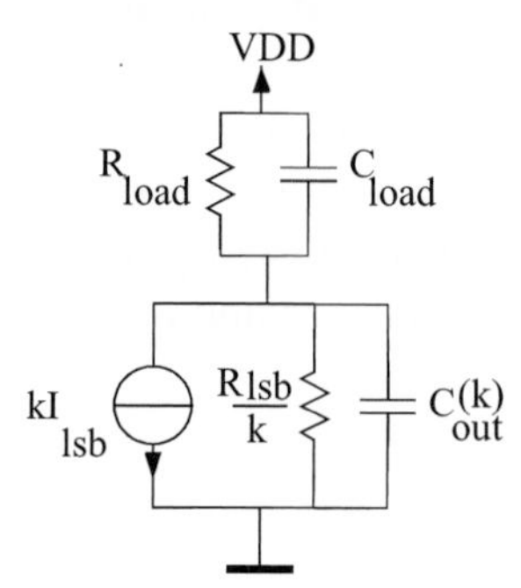

Figure 10-5. Small signal model of the D/A converter.

[Gus00], which are partly cancelled at the differential output. Therefore the differential output has output impedance specifications not as strict as the single-ended output. As the output frequency increases, it is more difficult to match the differential outputs.

The effect of the output current variations for the SFDR is analyzed in [Gus00]. The SFDR is [Gus00]

$$\mathrm{SFDR}_{\mathrm{maxse}} = (12.04 - 6.021\,N + 20\log(\frac{Z_{lsb}}{Z_{load}}))\ \mathrm{dB} \tag{10.15}$$

for differential signals

$$\mathrm{SFDR}_{\mathrm{maxdif}} = (24.08 - 12.042\,N + 40\log(\frac{Z_{lsb}}{Z_{load}}))\ \mathrm{dB}. \tag{10.16}$$

The finite output impedance also causes dynamic nonlinearities due to the code dependent capacitance. This error could be analyzed in the time or frequency domain. In the time domain, this capacitance increases the settling time and causes settling time variation according to the input code [Mik86]. In the frequency domain, it causes spurs [Bos98]. The LSB current source output impedance could be approximated by a small signal model (Figure 10-5)

$$Z_{lsb} = \frac{R_{lsb}}{1 + j\omega C_{lsb} R_{lsb}}, \tag{10.17}$$

where the pole frequency is

$$f_p = \frac{1}{2\pi C_{lsb} R_{lsb}}. \tag{10.18}$$

The output impedance reduces the 20dB decade after the pole frequency [Sed91]. Typical pole frequency is 10-10000 kHz. In Figure 10-6, the typical frequency response of the current source output impedance is shown. In output frequencies where the impedance ratio is reduced below the values of (10.12) and (10.14) the INL specification is not met. The D/A output impedance requirements should be higher than the DC requirements. The output impedance cannot be increased at high frequencies to increase the resistance in (10.17), because the pole frequency is decreased in (10.18). The output impedance could be increased by minimizing the parasitic capacitances. Different methods designed to increase the output impedance are presented in Section 10.10.

10.4 Other Systematic Errors

Edge effects
Voltage drop in the power supply lines
Temperature gradients

Temperature gradient and voltage drop can be modeled by (10.8), so their effect could be reduced in the same way as process gradients [Nak91], [Bas98], [Pla99], [Con00]. Sufficiently wide power supply lines minimize the voltage gradients. The temperature gradients cause symmetrical errors around the source, while voltage variations cause linear errors related to power supply line. The edge effects can be minimized by placing dummy transistors around the current source matrix and by avoiding the placement of all subtransistors to the center or edges of the matrix when using the common centroid placement.

10.5 Dynamic Errors

The dynamic nonlinearites describe the D/A converter performance at high output frequencies. As the update rate and/or code-to-code transition size increase, the error attributable to dynamic nonlinearities will constitute a greater fraction of the total error [Hen97].

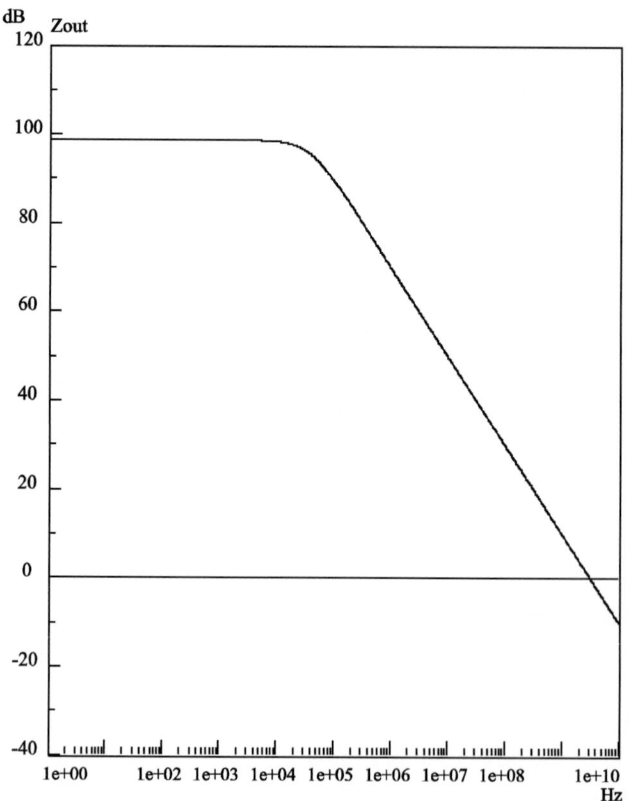

Figure 10-6. Typical frequency response of the current source output impedance.

10.5.1 Ideal D/A Converter

The ideal D/A converter cannot reconstruct the analog signal perfectly. The sampling theorem restricts the maximum output frequency to less than, or equal to, half the sampling frequency. The D/A converter output contains the original signal as well as images of it around every multiple of the sampling clock frequency, f_s, extending to infinity. The amplitudes of these components shown in Figure 10-7 are weighted by a function

$$\operatorname{sinc}(\pi \frac{f}{f_s}), \tag{10.19}$$

because of the sample and hold effect in the NRZ pulse.

This effect can be corrected by an inverse sinc($\pi f/f_s$) filter in the digital [Sam88] or analog domain [Dud97]. The filter that is after the D/A converter removes the high frequency images and provides a pure sine wave output. If the D/A generates frequencies close to $f_s/2$, the first image (f_s - f_{out}) becomes more difficult to filter. This results in a narrower transition band for the filter. The complexity of the filter is determined by the width of the transition band. Therefore, in order to keep the filter simple, the D/A converter frequency range is limited to less than 35 percent of the sampling frequency. As the D/A converter output is quantized with finite resolution, quantization

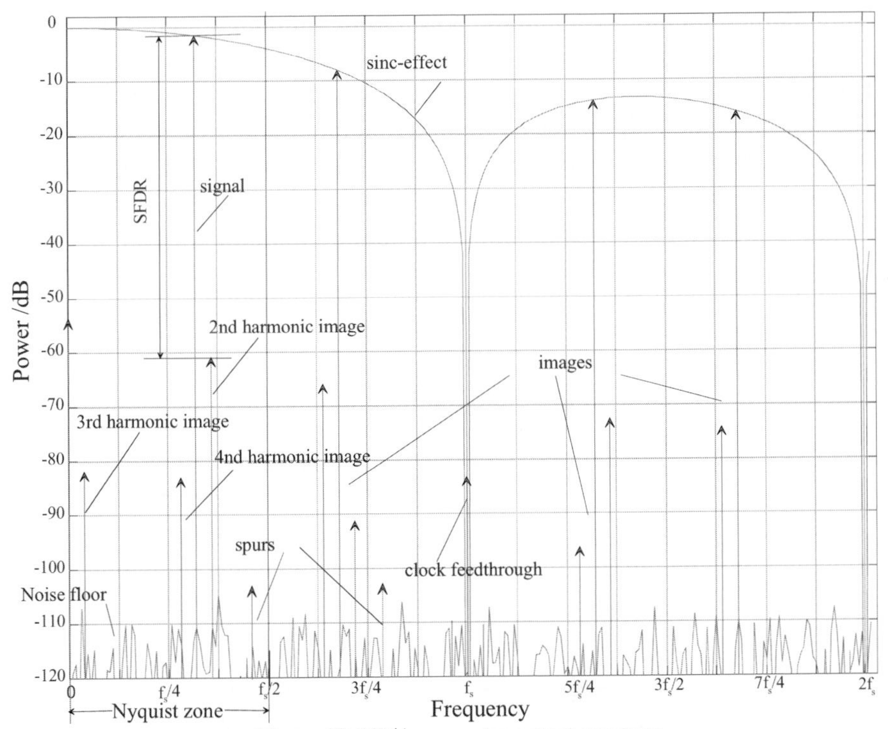

Figure 10-7. D/A converter output spectrum.

noise is introduced. The SNR for the full-scale sinewave is

$$\mathrm{SNR} = 6.02N + 1.76 + 10\log_{10}(f_s/(2B))\ \mathrm{dB}, \qquad (10.20)$$

where N is the number of the bits, f_s is the sampling frequency and B is the output bandwidth. The last term in (10.20) is the oversampling ratio.

10.5.2 Dynamic Performance Metrics

The dynamic performance specifications in the time domain are settling time, output slew rate, and glitch impulse. The settling time should be measured as the interval from the time the D/A-converter output leaves the error band around its initial value to settling within the error band around its final value. The slew rate is the maximum speed at which the D/A output is capable of changing. A difference between a rising and falling slew rate produces even harmonic distortion. The glitch impulse, often considered an important key in DDS and IF modulator applications, is simply a measure of the initial transient response (overshoot) of the D/A-converter between the two output levels, as shown in the grayed area in Figure 10-8. The glitches become more significant as the clock and/or signal frequency increases.

The most common dynamic performance metrics are signal-to-noise ratio (SNR), total harmonic distortion (THD), signal to noise and distortion ratio (SNDR), spurious free dynamic range (SFDR) [Hen97]. The SFDR specification defines the difference in the power between the signal of interest and the worst-case (highest) power of any other signal in the band of interest. The SFDR is perhaps the most often quoted D/A converter specification. The total harmonic distortion is a more complete specification of the D/A converter dynamic performance, since it includes most of the harmonically related spurs as well as aliased, i.e. folding back, harmonics within the D/A converter Nyquist bandwidth. The SNDR is the most encompassing specifi-

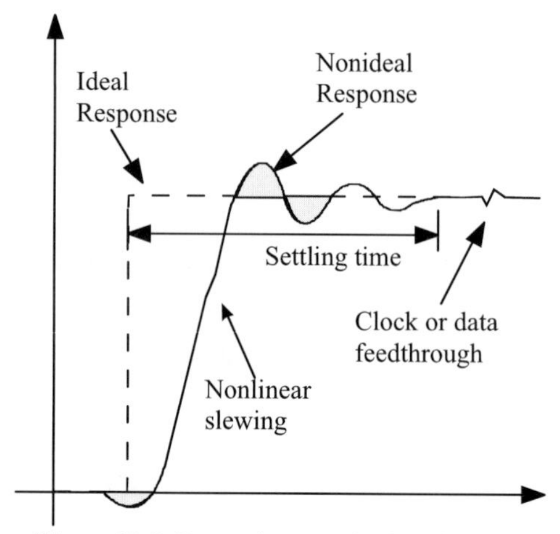

Figure 10-8. Dynamic errors in time domain.

cation since it includes all of the noise and distortion that falls within the Nyquist zone. Unfortunately, the SNDR is the most difficult parameter to measure and is therefore rarely quoted.

10.5.3 Dynamic Limitations

A key aspect of the performance of a D/A converter is its output glitch. The glitch is mainly due to the following effects [Cre89], [Wu95]:

1) imperfect synchronization of the control signals of the current switches
2) charge and discharge of parasitic capacitances associated with current sources
3) feedthrough of digital input data to the output
5) switching transistors being simultaneously in the off state

The imperfect synchronization can cause a high output glitch at the binary weighted D/A converter output. This problem is most severe at the midcode transition as all switches are switching simultaneously. At the midcode transition (0 111 111 111 $\rightarrow$ 1 000 000 000), if the current switches are not synchronized, the output have for a short period a current corresponding to the code 1 111 111 111. So one LSB change in the input code can cause one MSB output glitch. Because the timing differences are systematic, the same output glitch occurs periodically. The magnitude of the glitches depends on the change in digital input code and, as such, is a nonlinear function of the signal; therefore the output glitches cause harmonic distortions as well as noise.

The coupling of the control signals to the output lines through the parasitic gate to drain capacitance of the current switch transistors is also a source of glitches. The current through the parasitic capacitance is approximately

$$i = C\frac{\Delta v}{\Delta t}, \tag{10.21}$$

where Δv is the control signal swing and Δt is the switching time. At the high sampling frequencies, the portion of the glitch of the total signal energy increases.

The current source transistor source to drain voltage variation causes output current glitches due to the channel length modulation [All87]. The control signal feedthrough and supply voltage variations cause this voltage variation.

The fourth error source is that of switching transistors being turned off simultaneously. This causes the output node of the current source to rapidly discharge and the current source will turn off. To recover from this condition, the current source must progress through the linear region and back to the saturation. The recover time is proportional to the parasitic capacitance at the common source of the differential pair.

The output glitch influence to the D/A converter output can be analyzed by comparing the output glitch energy to the LSB bit energy. It is difficult to determine the output glitch effect in the frequency domain, because the glitch can be code related, partly clock related and partly voltage step related. Figure 10-9 shows the glitches that are related to the code, signal and

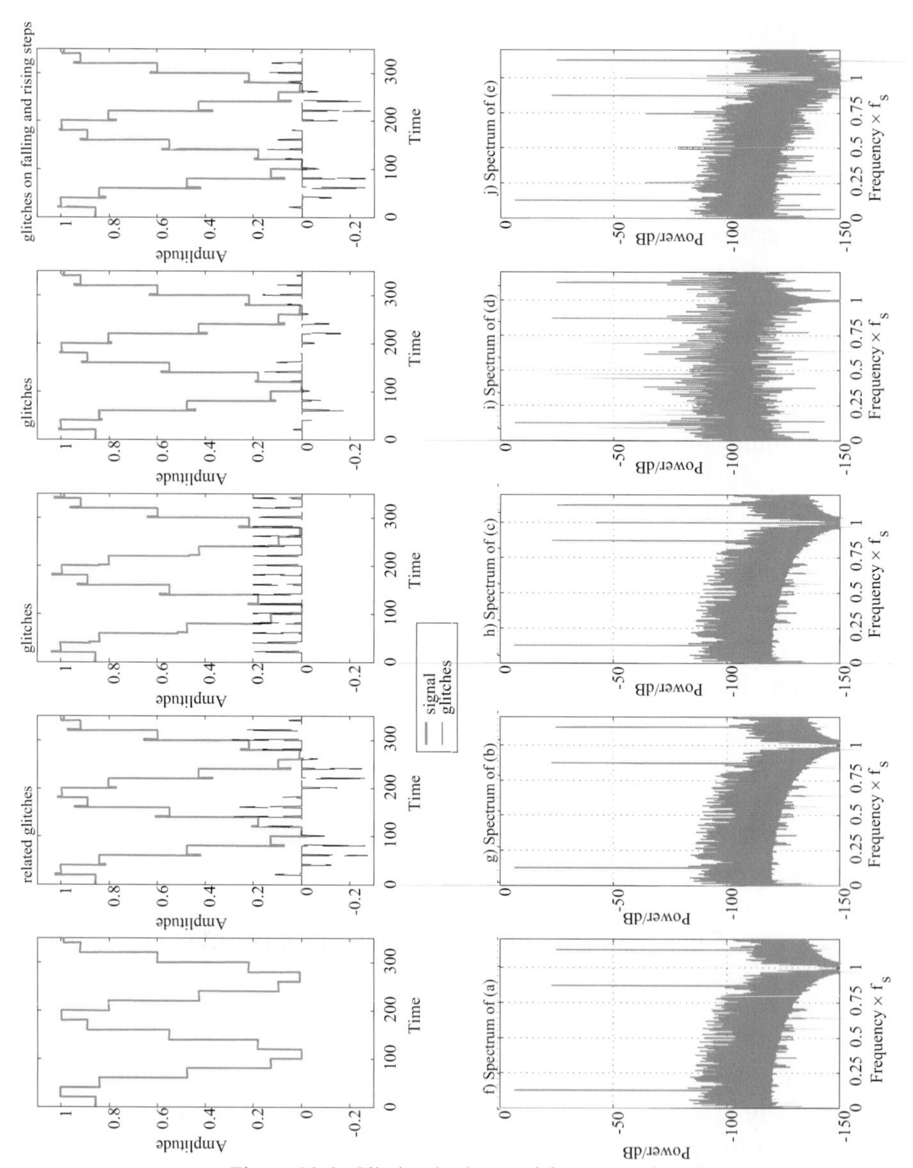

Figure 10-9. Glitches in time and frequency domain.

clock. The code related errors cause spurs in the Nyquist zone as shown in Figure 10-9(i). If the glitch energy is constant for each clock cycle, then this energy occurs at sampling frequency as shown in Figure 10-9(h). The glitches are out-of-band and do not cause an in-band spurious frequency problem. If the glitches could be transferred above the Nyquist band, then the reconstructing filter could remove them. If the glitch energy is proportional to the value of the output amplitude, or to the value of the step in the output amplitude, then the glitch energy is at the frequency of the desired output frequency as shown in Figure 10-9(g).

10.6 Inaccurate Timing of Control Signals

The D/A converter output pulse length variation with code and time is a timing related error. This timing difference causes pulse width modulation. The timing error can be global or local [Dor01].

In the global case, the system clock period exhibits fluctuations, but the D/A converter is ideal. The unit elements are synchronized with each other, but not with the ideal clock signal. This phenomenon is caused by external or internal mechanisms of the D/A converter. An example of the external physical mechanisms is the phase noise of an external clock driver, while internal mechanisms can be related with power supply jitter etc.

The local timing error can be spatially systematic or random [Dor01]. The systematic errors occur when the timing of a transistor or a group of transistors differs from the timing of other transistors (Figure 10-10). The random timing errors mean a situation in which the timing of each transistor timing differs randomly. Neither of these errors needs to be random in the time domain. The spatial error can be caused by, for example, current switch process variations, switch drivers mismatch and different wiring lengths of the control signals.

In Figure 10-10 the systematic and random timing error effects on the sine wave in the time domain are shown. The error is periodic to the signal in both cases. The error due to random spatial timing error has weaker harmonics than the error caused by systematic timing error. The level of harmonics is 20 dB higher in the systematic case than in the random case [Dor01]. Therefore the timing error should be randomized (spatially).

The power of the spurs caused by systematic timing difference can be calculated by [Dor01]

$$P_d(q, f_{out}, G) = 10\log_{10}\left(\frac{2J_{q-1}(4f_{out}\Gamma)}{2\pi q f_{out}}\sin(\pi q f_{out})\right) \text{ dBc}, \quad (10.22)$$

where J_q is the qth order Bessel function of the first kind, Γ the relative timing error, q the number of the harmonic, and f_{out} the normalized output frequency. Figure 10-11 shows the second and third harmonic as a function of the timing error, when the output frequency is 0.45 f_s. The second harmonic restricts the SFDR to 60 dB when the timing difference is about 0.1%. Furthermore, the second harmonic is much higher than the third. The second harmonic could be partly cancelled by the differential outputs.

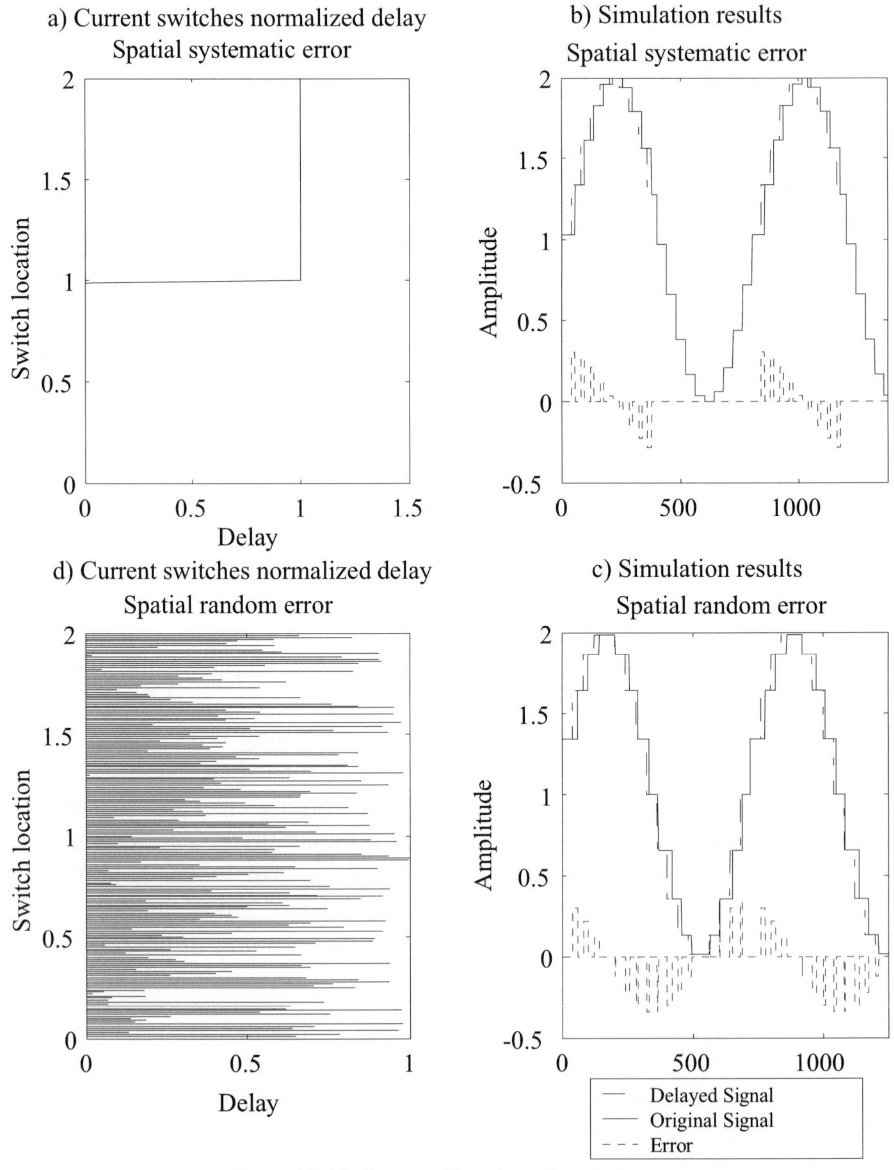

Figure 10-10. Systematic and random timing error.

10.6.1 D/A Converter Finite Slew Rate

In ideal D/A converters, the transition time from output level to another is zero. A transition time greater than zero is not problem in many applications if the transition time is constant for all step sizes and if output can settle during the clock period [Ess97], [Ess98]. The constant transition time is constant for all step sizes and the ramp is linear, the output frequency response roll-off follows

$$A(f) = \mathrm{sinc}(\pi\Delta\frac{f}{f_s})\mathrm{sinc}(\pi\frac{f}{f_s}), \tag{10.23}$$

where Δ is rise time normalized to the sampling period. The linear ramp causes 3dB more attenuation at the Nyquist frequency than the sinc effect in (10.19).

If the rise time depends on the output voltage step, then the output settling depends on the digital input code. If the slew rate (dV/dt) is constant and T_s is the clock period, then the difference between output voltage steps n and p (shadowed area in Figure 10-12) is

$$\Delta E = T_s\,(p-n) - \frac{1}{2}\frac{dt}{dV}(p^2-n^2), \tag{10.24}$$

where energy difference depends nonlinearly between voltage steps, which causes distortions. If the rise time is constant (t), the energy difference is

$$\Delta E = (p-n)\,(T_s - 0.5t), \tag{10.25}$$

where the energy difference depends linearly between voltage steps, which

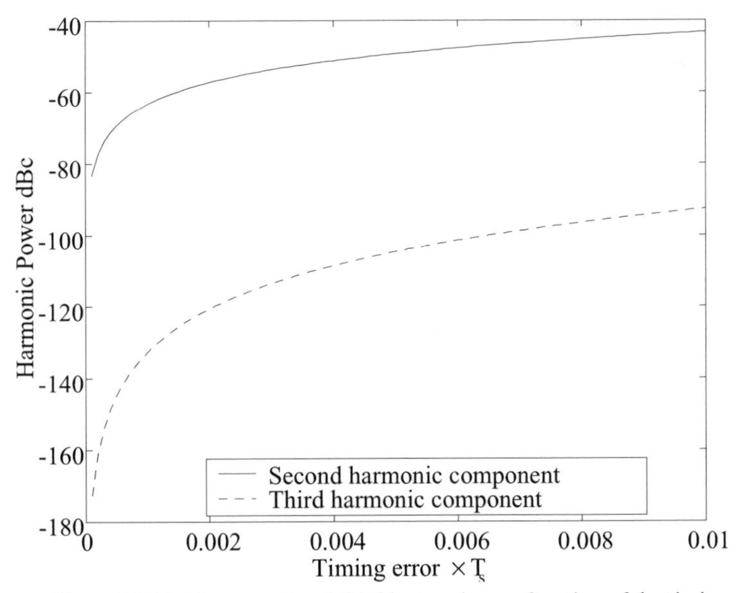

Figure 10-11. The second and third harmonic as a function of the timing error, when the output frequency is 0.45 f_s.

do not cause extra distortion. In reality, the rise time and slew rate depends on the voltage step, and the transition is not linear.

The asymmetrical rise and fall times generate harmonics. If errors are symmetrical in the rising and falling edges and are proportional to the given step size, then errors do not generate harmonics. The asymmetrical switching in the rising and falling characteristics tends to generate even-order harmonics [Cro98]. The opposite-polarity glitch components on the rising and falling steps will tend to generate even-order harmonics (see Figure 10-9(j)).

10.7 Different Current Steering D/A Converters Architectures

The core of a current-steering D/A converter consists of an array of current sources that are switched to the output according to the digital input code. This can be accomplished in several ways, resulting in three different architectures that will be briefly discussed.

10.7.1 Binary Architecture

The easiest solution can be found by a binary switched D/A converter. In this case, each digital input directly controls the number unit current sources (proportional to bit weight) that have to be switched to the output (Figure 10-13). This architecture can be implemented on a small silicon area and it minimizes the digital power consumption [Raz95], [Lin98], [Bos01]. The most significant bits control large currents that have to match the previous current with an accuracy of 0.5 LSB when the digital input value is increased by one. As the error is proportional to current, this becomes very hard to achieve. This can cause large DNL errors that lead to severe degradation of the static performance for this architecture.

The particular disadvantage of this architecture is glitches. The worst glitch in the output is most often found at the major carry of the D/A con-

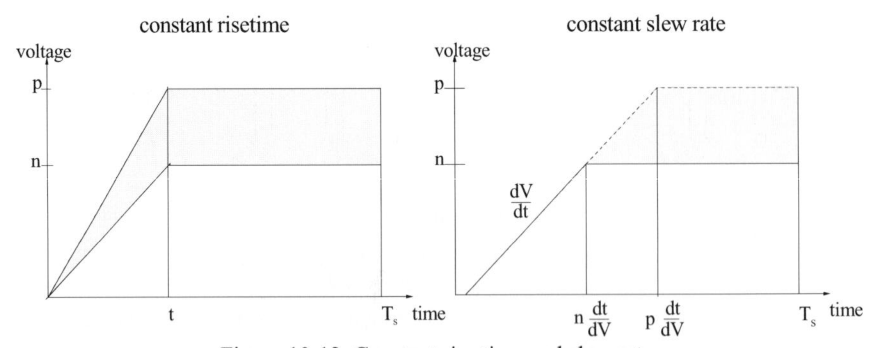

Figure 10-12. Constant rise time and slew rate.

verter (switching between 100...00 to 011...11). There will be a glitch in the output as each of the switches turn on and off at different times. The glitch is code dependent and thus produces harmonic distortion and even non-harmonic spurs. The current switches are usually size-scaled in binary-weighted design to account for the various current densities. So the parasitic capacitances are scaled by binary. This makes timing of the switches even harder.

10.7.2 Unary Architecture

In the unary architecture, the current source array consists of (2^N -1) unity current sources that can be accessed separately (Figure 10-14). The digital input code has to be converted to a thermometer code that determines the number of current sources that have to be switched to the output. The decoder consumes a lot of silicon area and has a large power consumption but this architecture has a good DNL specification since only one extra unity current source has to be switched to the output when the digital output is increased by one. Because in this D/A converter all the switches change in the same direction with unary code, the output is always monotonic. The timing error is smaller than in the binary weighted D/A converter, because all the current switches are equally sized, so the switch drivers have the same load. The decoder logic high current consumption causes supply voltage variation, which increases glitches at the D/A converter output.

Perhaps more importantly, a D/A converter based on a thermometer code, rather than on binary weighted architecture, greatly minimizes glitches. This is due to the fact that only one switch changes its state when the binary code changes by one. It should also be mentioned here that latches could be used in the binary-to thermometer code conversion such that no glitches occur in the digital thermometer-code words and that pipelining can be also used to maintain a high throughput speed. All the current switches in a thermometer-code approach are equal sizes, since they all pass equal currents.

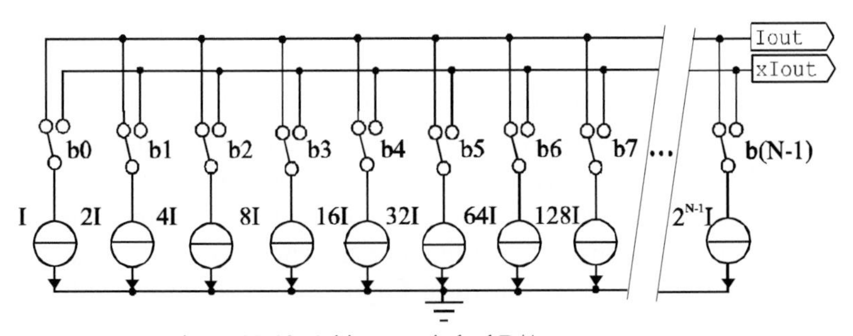

Figure 10-13. A binary switched D/A converter.

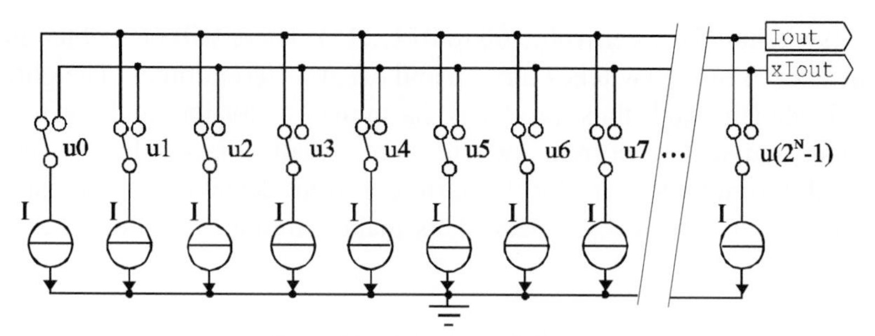

Figure 10-14. Unary architecture.

10.7.3 Segmented Architecture

To obtain architecture with a good DNL specification and an acceptable power and area consumption, a combination of the previous two architectures is usually chosen [Sch97]. This is called the segmented architecture (Figure 10-15). In the floorplan of this architecture, one can typically find two blocks. The first block consists of the binary conversion for the least significant bits and the second block contains the unary conversion for the remaining most significant bits. The number of unitary implemented bits is limited by the increased coding complexity and area constraints. The area constraints (routing of switch/latch and current source array) resulted in a 6-8 segmented architecture as a good trade-off [Lin98]. Timing error due to midcode transitions could be reduced by the MSB bits segmentation.

10.8 Methods for Reduction Dynamic Errors

10.8.1 Glitches Reduction

In principle, the differential current switch source voltage and current are constant, if the control voltages are

$$V_{ctrl} = \sqrt{\frac{2I_{ss}L_{sw}}{\mu C_{ox}W_{sw}}}\left(\sqrt{\frac{2}{T}t}-1\right), \quad \overline{V}_{ctrl} = \sqrt{\frac{2I_{ss}L_{sw}}{\mu C_{ox}W_{sw}}}\left(\sqrt{2-\frac{2}{T}t}-1\right), \quad (10.26)$$

where T is transition time. In practice, these waveforms are difficult to implement.

If both switching transistors are turned off, the output node of the current source will rapidly discharge and the current source will turn off [Tak91], [Wu95], [Koh95]. To recover from this condition, the current source must progress through the linear region and back into saturation. The situation can, however, be improved by setting the crosspoint of the V_{ctrl} and $\overline{V}_{ctrl}$ at the optimum high level, as shown in Figure 10-16. Then the voltage at the

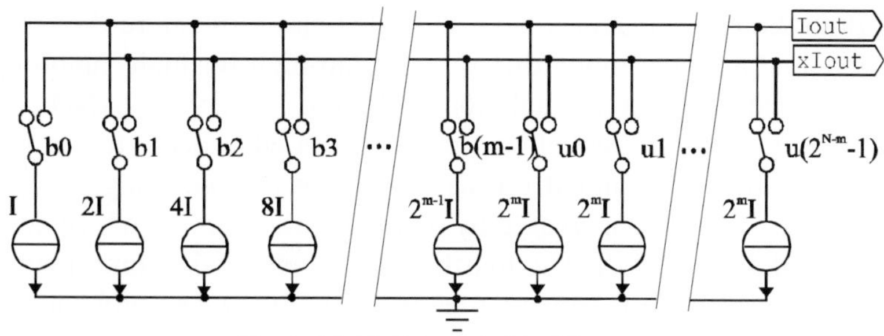

Figure 10-15. Segmented architecture.

differential current switch source does not need to decrease in order to deliver the current, and thus no glitches occur [Tak91], [Wu95]. If the cross-point is too high, the current is divided to both outputs, switch transistors V_{gs} voltage is reduced, source voltage increases, parasitic capacitances are charged and settling time is degraded. However, while the current sources are in saturation, their output current is changed. In practice, the source voltage is allowed to increase, because the settling time degradation is not as harmful as glitches (both switching transistors turning off simultaneously).

To shift the cross-point of the switch transistors control signals, two methods are usually used: the control signal falling edge is delayed (Figure 10-16 (a)) [Tak91], [Chi94], [Koh95], [Mar98], [Bav98], [Bos98] or fall time is increased (Figure 10-16 (b)) [Wu95], [Bos01] (in PMOS switches the control signal rising edge is delayed and rising time is increased).

The benefit of delaying control signal falling edge is that it is possible to set the crosspoint easily, even with large control voltage swings. The problem in this approach is that the current switches are not simultaneously off and on, therefore the output glitches do not occur simultaneously (output symmetry is destroyed). The simplest way to implement the edge delaying is to drive the inverter with the circuit, which has asymmetrical rise and fall times (Figure 10-17(a)). A short delay to the control signals can be implemented with this circuit [Ngu92]. If a longer delay time is needed, then the delay could be implemented with the NAND or NOR circuit in Figure 10-

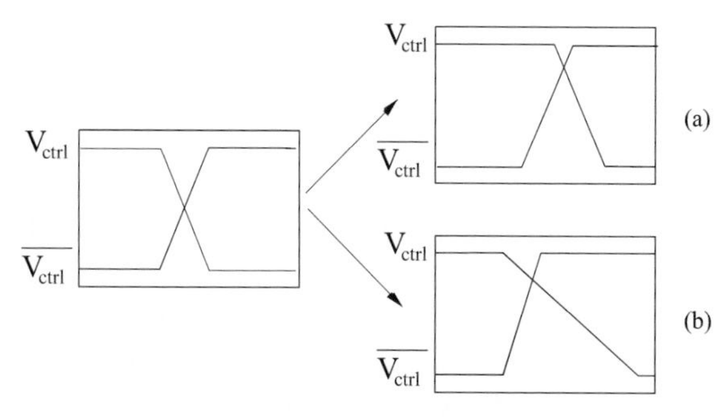

Figure 10-16. Setting the crosspoint of the V_{ctrl} and $\overline{V}_{ctrl}$.

17(b). Both the control signals have separate delays, therefore the circuit is not differential and the timing difference is too high for high sampling frequency applications. Another solution is the cross-coupled NAND (Figure 10-17(c)), which is differential but, nevertheless edge delay remains long. Furthermore, the crosscoupled transistors reduce the operating speed. The operation speed could be increased by removing transistor A, which does not affect to the delay [Ded99]. The fourth way is to delay the control signal of the NMOS transistor in the inverter (Figure 10-17(d)). The problem of this approach is that the delay can be too long, as the sampling frequency increases and the delay cannot be easily adjusted. The fifth approach is to use cross-coupled inverters (Figure 10-17(e)) [Bas98] (the area inside the dashed line is optional). The benefit of this approach is simple structure, built in clocking and differential outputs. The disadvantages are slow transitions due to restricted current steering capability of the cross-coupled inverters.

As the falling time is increased (Figure 10-16 (b)), control signal transitions occur simultaneously. The problem is that the rising and falling waveforms cause different glitches to the output according to (10.21). As the falling and rising waveforms are different [Bas98], the second harmonic is increased. By increasing the fall time, the crosspoint cannot be set as high as in the delay solutions, which cause problems in some applications. The simplest way to increase fall time is to change the buffer transistor sizes [Wu95] or to set the transistors, which operate in the linear region, between the output and the NMOS transistor (Figure 10-18) [Ngu92]. Another way is to remove the transistor B in Figure 10-17 (c) [Bos01]; the cross-coupled PMOS transistor will then decrease the rising time and increase the falling time.

10.8.2 Voltage Difference Between Control Signals

The problem of the switch transistor being simultaneously turned off could be overcome by the reduction of the voltage difference between the control signals. The minimum voltage necessary to completely steer the current from one output to the complementary output is less than the supply voltage (10.1) and not related to the absolute value. If the turn off voltage of the control signals is set so that the current sources are in the saturation region even when both signals are simultaneously off [Chi86]. However, the channel length modulation causes current variation, so this is a problem in high accuracy D/A converters. The problem of this approach is that the control voltage variation is so large that the switch transistors in the on-state should be in the linear region.

From (10.21) it may be seen that output current glitches are related to the control voltage. The coupling to the D/A converter output could be reduced by restricting the control voltage swing. Because switches source voltage follows during the transitions control voltage (not linearly), it is possible to reduce this variation by decreasing control voltage swing.

The problem with the control voltage swing reduction is that the switch driver current driving capability is decreased, thus the switches operation speed degrades. It is common to use cross-coupled or normal inverters, the power supply range of which is reduced [Bos98], [Bas98], [Bos01]. In NMOS current switches, if the positive supply voltage is set lower than the supply voltage, then the current switches are in the saturation region, and the negative voltage is the ground. The crosspoint of the control voltages must be set to a high level because the output voltage swing is minimized, and then voltage is near that at which the current sources transistors are in the saturation region. If the current switch need not operate in the saturation re-

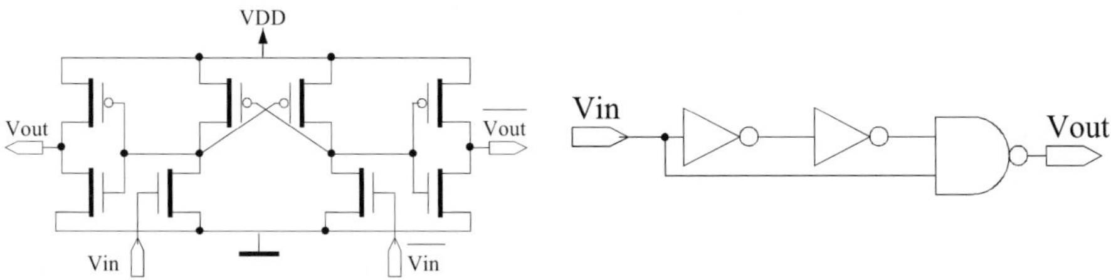

(a) Control signals with asymmetrical rise and fall times for inverters. (b) Edge delaying with NAND-gate.

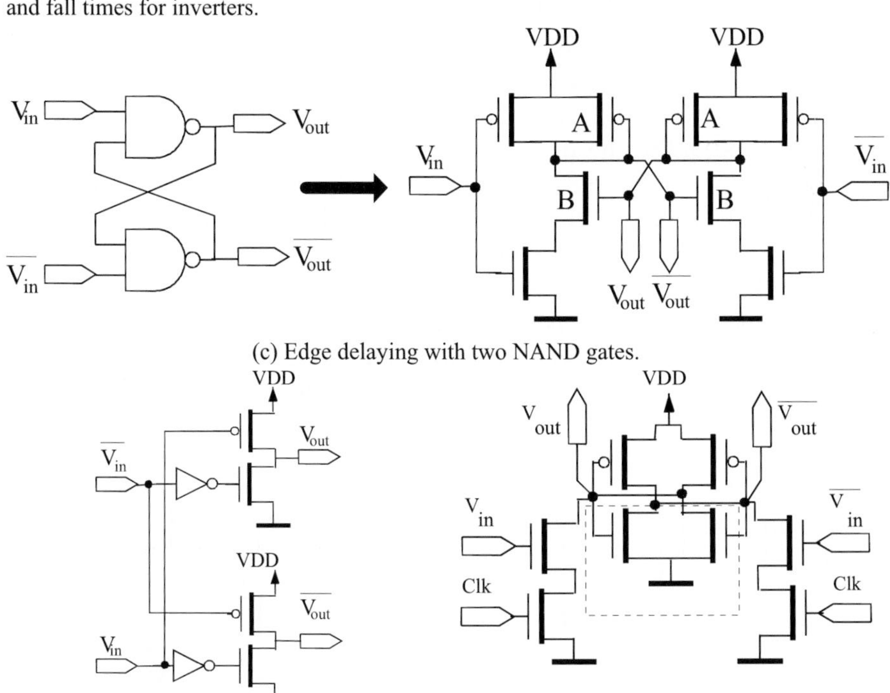

(c) Edge delaying with two NAND gates.

(d) Delaying the inverters control signals. (e) Cross-coupled inverters.

Figure 10-17. Circuits for delaying control signal falling edge.

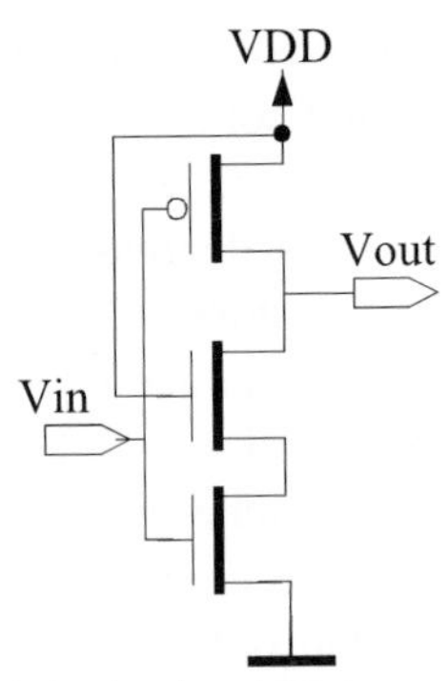

Figure 10-18. A circuit, which increases fall time.

gion or the output voltage range is low, the positive control voltage can be set to positive supply voltage and negative control voltage above the ground level. Then there can be more confidence that the current sources are in the saturation region.

The voltage range could be easily reduced by means of a normal differential pair [Bas91] by properly sizing the transistors. It is possible to restrict the supply voltage in both directions by setting constant current sources at the differential pair output [Bas91]. The differential pair performance degrades quickly as the sampling frequency is increased, as long as the bias current is not increased as well.

Another solution is to use the circuit shown in Figure 10-19(a) in which the output transistors are of the same type but the opposite of the current switches and driven by the complementary control signals. Then, in the case of the PMOS buffer, the minimum output voltage is the threshold voltage and is higher than the negative power supply when the lower transistor turned off. The benefit of this circuit is that the turning on by the control signal occurs as quickly as in the inverter topology, which sets the crosspoint of the control signals high. Adjusting the buffer transistor or the sizing of their drivers could set the crosspoint higher. Reducing the power supply range could restrict the output swing or setting the diode connected transistor between the buffer transistor and output [Chi86]. The problem in this topology

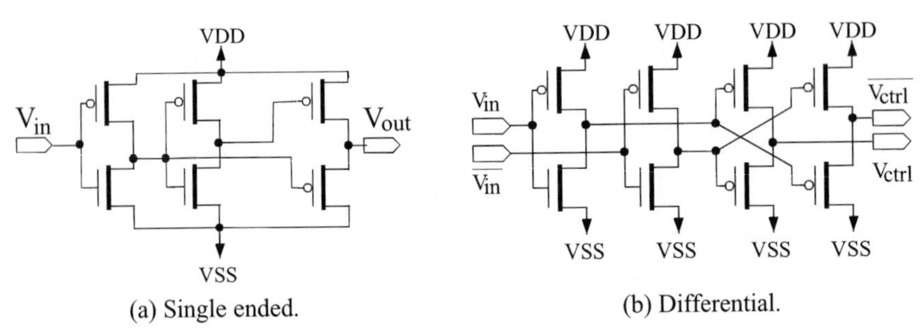

(a) Single ended. (b) Differential.

Figure 10-19. PMOS buffers for reducing control signals swing.

is that both switch driver buffers needs a two inverters chain, which easily destroy the differential output if their sizing is changed due to the process variations. By connecting cascade transistors biased in the saturation region between the buffer transistors and the output [Fal99], it is possible to reduce the extent to which the control signals of the inverter couple to the output with the penalty of reduced switching speed.

A differentially controlled version of the previous buffer pair is shown in Figure 10-19(b) [Mer00]. Because both buffers are steered with the same control signals, the transitions occur simultaneously. The problem in this topology is the limited possibility of adjusting the crosspoint; this is why the minimum value of the control signals should be set as high as possible, thus reducing the supply range, which degrades the operation speed.

A simple differential voltage swing restricting circuit is shown in Figure 10-20(a). The signals V_{in} and $\overline{V}_{in}$ set if the output is connected to positive supply or the drain voltage of the diode connected transistor. The ability of the current to steer to negative directions is not great because there are two transistors in the current path, which raises the crosspoint of the control signals.

A switch driver in Figure 10-20(b) is a combination of the inverter and differential pair properties [Ded02]. It generates the switches difference voltage instead of two control voltages and follows the process variations at the switches difference voltage. In Figure 10-20(b), the current path is shown in black, when V_{in} is off and $\overline{V}_{in}$ is on. The transistors A and B are biased in the linear region, so that the node voltage between them keeps the transistor G at the saturation region. Furthermore, this voltage adjusts the transistor G source node voltage. The transistors A and B can be replaced with resistors,

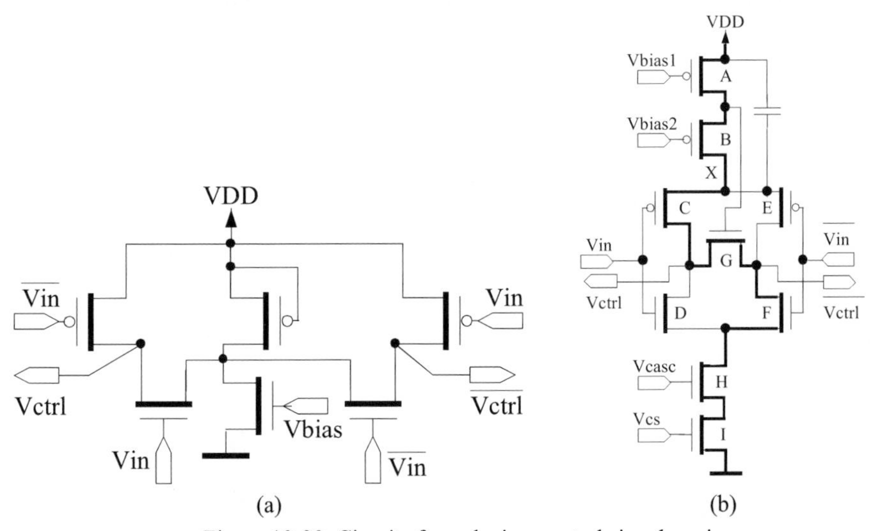

Figure 10-20. Circuits for reducing control signals swing.

but, in the transistor implementations, the bias circuit can be used to assure that the voltage at node X follows the process variations in the threshold voltages of the switch transistors. The positive control voltage depends on the drain to source current of the transistors A, B and C. Because the current is generated by the transistors I and H (located near the current source matrix) in the same way as in the current switches, the negative voltage follows the process variation changes by altering the negative control voltage according to the current switch source voltage. The switch driver is fast because the drain to source voltage of the transistor G achieves its maximum voltage quickly due to the steep current to voltage curve of the transistor. Furthermore, the difference voltage could be set near the minimum voltage because the safety margin can be reduced due to the process variation compensation.

10.8.3 Current Switch Sizing

The control signals of the current switch coupling to the D/A converter output is related to the current switch gate capacitance [Tak91], $C_g \sim WL$. Therefore the size of the current switches should be minimized. If the current in the source is constant, then $V_{GS} \sim 1/\text{sqrt}(W)$. The source voltage of the current switches is the difference between the control voltage and V_{GS}. When the size of the transistor is decreased, the required V_{GS} grows so high that the current sources remain no longer in the saturation region. The control voltage should be increased or the output current decreased to keep the current sources in the saturation region. Both approaches increase distortions; increasing the control voltage might set the current switches in the linear region. The size of the current switches should be set so that the current sources drain to source voltage is above saturation voltage when the output voltage is at the minimum.

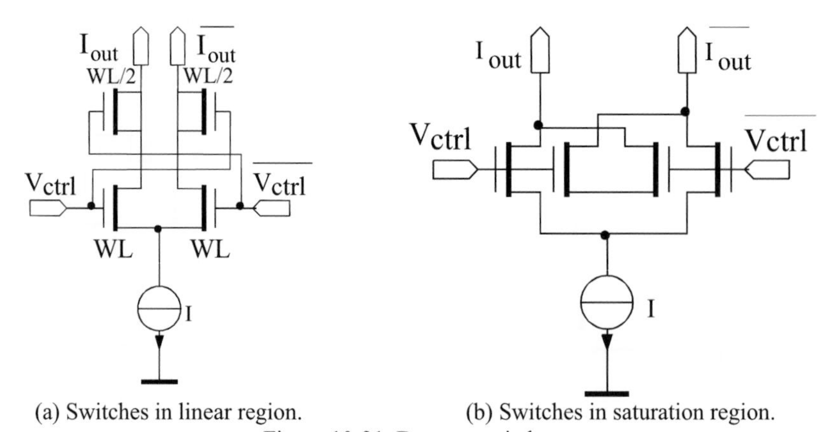

(a) Switches in linear region. (b) Switches in saturation region.
Figure 10-21. Dummy switches.

10.8.4 Dummy Switches

Adding to the output equal but opposite sign glitch could compensate the glitch caused by the switch control signal feedthrough. If the current switches in the on-state are in the linear region, then the gate capacitance is divided approximately equally between the drain and source [All87]. The glitch due to the control signal feedthrough could be compensated by a dummy transistor, which has a size half of the switch transistor size and its source and drain connected together (Figure 10-21(a)) [Sch88]. The gate capacitance to the output is the same as that in the switch transistor. When the dummy switch is driven by the switch complementary control signals, the equal but opposite sign glitch cancels the original spike. In practice, the switches capacitance is not divided equally, and thus the cancellation is not perfect. If the glitches have a timing difference, there will be two glitches. Furthermore, the transistor cannot be minimum sized while the dummy transistors add capacitive load to the output, which makes the converter slower.

If the current switches in on state are in the saturation region, the gate capacitance to the output is the same as in the cut-off region [All87]. This being so, glitches could be compensated by adding equal sized dummy transistors parallel to current switches, so that dummy switches sources are cross-coupled and the drains are connected together (Figure 10-21(b)) [Luh00]. The dummy transistors are in the cut-off region, the voltage coupling from the gate to the output is complementary to the switches feedthrough. As above, the dummy switches increase output load capacitance and imperfect synchronization causes extra glitches.

10.8.5 Removing Spurs from Nyquist Band

The D/A converter spectrum is periodical with sampling frequency. The reconstructing filter must remove the images. Therefore the D/A converter spectral purity is important in the Nyquist band. If the glitches could be

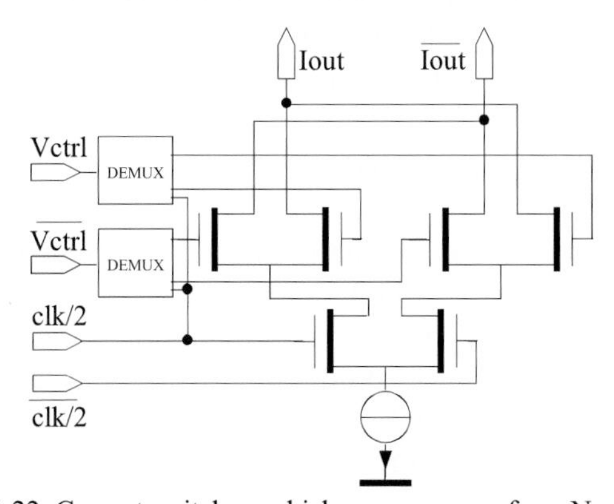

Figure 10-22. Current switches, which remove spurs from Nyquist band.

transferred above the Nyquist band, then the reconstructing filter could remove them.

One method for this is shown in Figure 10-22 [Her91]. Two differential pairs are driven by alternating the control signals. Both differential pairs have a common current source, which is connected alternatively to the differential pairs at half the clock frequency. The outputs of the differential pair are connected together. Then one differential pair conducts at every even clock period, while the other at every odd period. The control signals are divided by the DEMUX to the two differential pairs so that the control signals cannot change its state when its differential pair conducts. Then the glitches due to the transitions cannot affect the output and consequently cause no harmonic distortions. Instead, the switches that control the current driven to the differential pairs change their states, which will cause a glitch at the output. Although this glitch is still generated when the switches are opened, this glitch occurs regularly, at the half sampling frequency, with amplitude that is dependent only on the value of the output. Therefore, the glitches contribute frequency components at the desired output frequency and at the half sampling frequency.

The circuit in Figure 10-22 does not reduce code related errors due to control signals feedthrough to the output. These code related errors could be cancelled by parallel dummy transistors (see Section 10.8.4).

The problem in this circuit is that the part of glitches caused by clock driven switches is not related to the output current. Therefore, part of the glitches is code related and causes distortion to the Nyquist band. The half clock related spurs are at the Nyquist frequency, therefore there might be difficulties in removing them. The half clock related spurs could be transferred to the sampling frequency by driving the lower switches at the clock frequency. In this case, the upper switches synchronization is more difficult. The circuit in Figure 10-22 increases complexity, because two extra demultiplexers and four switch transistors are needed. The extra switches between the output and ground reduce the output voltage swing range. Furthermore, the other differential pair source node is floating every other clock period, which increases the settling time.

10.8.6 Sample and Hold

One conceptual solution to the dynamic linearity problem is to eliminate the dynamic non-linearities of the D/A converter, all of which are associated with the switching behavior, by placing a track/hold circuit at the D/A converter output. The track/hold would hold the output constant while the switching is occurring, and track once the outputs have settled to their dc value. Thus, only the static characteristics of the D/A converter would show up at the output, and the dynamic ones would be attenuated or eliminated. The problem with this approach is that the track/hold circuit in practice introduced dynamic non-linearities of its own. A different approach, employing a return-to-zero (RZ) circuit at the output, is proposed in [Bug99], [Hyd03]. The output stage implements a RZ action, which tracks the D/A converter once it has settled, and then returns to zero. The problems of this approach are: large voltage steps cause extreme clock jitter and waveform fall/rise asymmetry sensitivity, large steps cause problems for the analog lowpass filter, and the output range (after filtering) is reduced by a factor of 2. To remedy these problems, current transients could be sampled to an external dummy resistor load, and settled current to external output resistor loads, by multiplexing two D/A converters [Bal87], [Van02].

The rise time is typically longer for a large step size than for a small step size, which causes distortions. This problem could be alleviated by the fractional-order-hold, where the transition is a fraction of the clock period and linear. It is easier to model this by real circuits than step functions. When the transition time is a fractional of the clock period, then output current transition time is constant [Ess97], [Ess98] and transition is linearly independent of the parasitic capacitances. Using this method, it is possible to get a 20 dB improvement at a 13 bit D/A converter distortion level at $0.4f_s$ [Ess98]. The problem of the fractional-order-hold is that it is complex, and therefore requires extra chip area and consumes more power.

10.9 Timing Errors

There are three methods of reducing timing errors: the control signals synchronization, matching switch drivers loads and layout techniques.

10.9.1 Control Signals Synchronization

The control signals synchronization is needed even at low sampling frequencies, because the signal paths have different lengths even in the binary weighted D/A converter, where there is no longer additional logic. In the segmented D/A converters, the decoding delays make synchronization necessary. However, 8-bit D/A converters with a 65 MHz sampling frequency have been implemented without synchronization [Kim98]. The synchronization is implemented by the D flip-flop or latch, which are placed symmetrically near the switches. In [Mik86], [Bos98], [Pla99], [Bos01] the last latches are with current switches in the same matrix. This can cause the digital noise to be coupled to the output. The switch drivers can also do single ended to differential conversion, crosspoint setting and control voltage reduction [Tak91], [Bas98], [Bos98], [Pla99], [Bos01]. As the standard cell libraries there are usually not fast enough D-flip-flop or latches that have differential outputs and predistort the control signal, the latch or D-flip-flop is full-custom design.

In its simplest form, the latch can be implemented with a switch and capacitor (Figure 10-23(a)) [Mer97]. The problem of this approach is that the switch transistor is not ideal so the capacitor may not be fully discharged even when the switch is closed. Furthermore, as the output node is floating when the switch is open, the external and internal disturbances may change its state. This solution, therefore, could be used in the applications where the capacitance charge is frequently refreshed. In these applications, the problem is that the capacitance is not charged enough and sharp transitions cause voltage fluctuations. If the switch is only off when the control signal changes, the time that capacitor is floating is shorter and so less susceptive to disturbances [Mer93], [Mer94]. The crosspoint of the control signals can be only set at a higher level by changing rise and fall times.

Two control signals are fed to the both the switches of the differential pair in Figure 10-23(b). The control signal, which is settled, is selected. The circuit removes the timing differences in the transitions, but most of the timing difference is actually generated in the buffers before the differential pair, so the circuit does not affect them. Furthermore, the circuit needs four different phase clock signals, which are hard to implement at high frequencies and which double the number of the D-flip-flops.

A cross-coupled inverter is a more reliable storage element (Figure 10-23(c)) than the previous one [Mer97]. The cross-coupled inverters prevent the outputs from floating. Furthermore, the output signals are automatically differential. The problem is that cross-coupled inverters try to keep output constant and inhibit transitions. Furthermore, the inverters increase the ca-

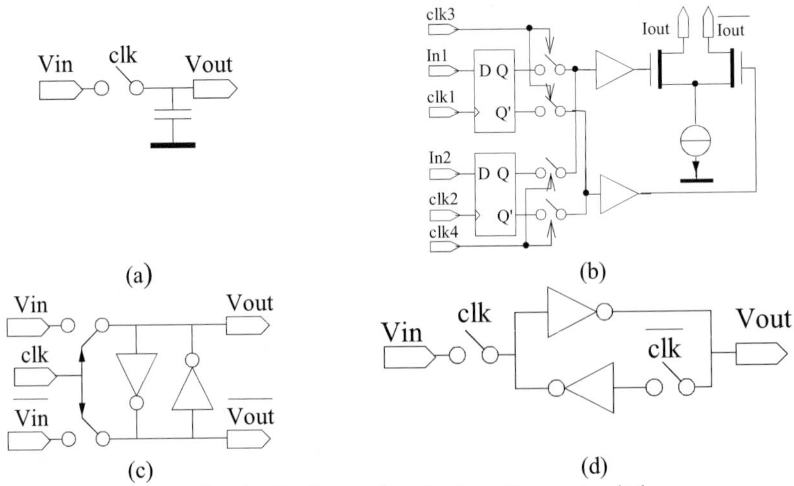

Figure 10-23. Circuits for the synchronization of current switches.

pacitive load for the control signals. This increases the output settling time. The inverters could be used at the output of the cross-coupled inverters to provide sufficient drive current. This, however, can degrade the timing of the differential outputs.

If the feedback of cross-coupled inverter is connected only when the input switch is on, the settling time decreases (Figure 10-23(d)). The problem is that the circuit is not differential because both the control signals require their own storage element.

In single-ended D/A converters, it is possible to use a current switch, which is driven by only one control signal (Figure 10-24) [Kas95]. When V_{ctrl} is on, the transistor C is on and transistor B is off. When V_{ctrl} is off, the same amount of current goes through the transistors. The benefit of this current switch is that the source voltage of transistor B is stabilized by the feedback so the control signals cannot directly feedthrough to the output.

10.9.2 Switch Driver Load Matching

If the D/A converter is not totally segmented, the synchronization of the control signals cannot guarantee a small enough timing error at the high sampling frequencies, because the size of the current switches is scaled to according to the various current densities. This causes the capacitive loads of the switch drivers to be different. One way to match the loads is to scale the switch driver transistors according to the switches (Figure 10-25(a)). This solution is suitable for low sampling rates and moderate resolution, where timing errors due to different loads are not significant. For an 8-bit binary weighted D/A converter, the largest buffer dimensions should be 128 times the smallest. The required current and dimension grows large, because the transistor minimum dimension restricts the smallest transistor size. Furthermore, accurate buffer matching is difficult due to different sizes, so timing errors occur.

Another way to match the loads is to use a dummy load parallel to the switch transistor. This can be implemented with the transistor, which has source and drain connected to the ground (Figure 10-25(b)). The dimensions

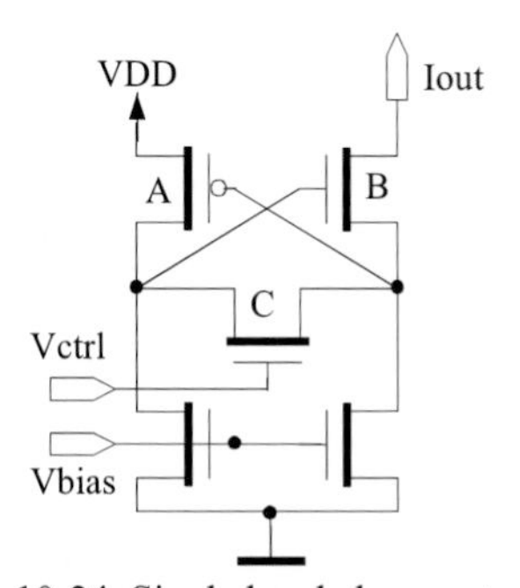

Figure 10-24. Singled ended current switch.

of the dummy transistors could be selected by rule that the combined size of the switch and dummy switch is equal to the MSB current switch size. The transistor gate capacitance depends on the transistor operation region; furthermore, the voltage over C_{GS} and C_{GD} varies in the switch transistor differently from in the dummy transistors. Therefore, the timing error can not made small enough using this method with high sampling rates. Therefore the transistor dimensions should be determined by simulations. The matching accuracy could be improved by dividing the dummy transistor into two half sized transistors connected to VDD and GND as shown in Figure 10-25(c) [Vol02]. This corresponds better to the switch transistors average load. For fast sampling rates, this is not enough, because the dummy and current switch transistors still have different operation points.

Third method is to use a dummy current switch with the current source so that the sum of the actual current source and dummy current is equal to the MSB current so the sum of their sizes is equal to the size of the MSB switch (Figure 10-25(d)) [Tei02]. This arrangement makes the capacitive loads equal, due to the gate capacitance seen by the drivers. The problem is the power consumption, because the dummy transistor current is wasted. The second problem is extra area, but the dummy transistors in the current matrix could be used for this purpose.

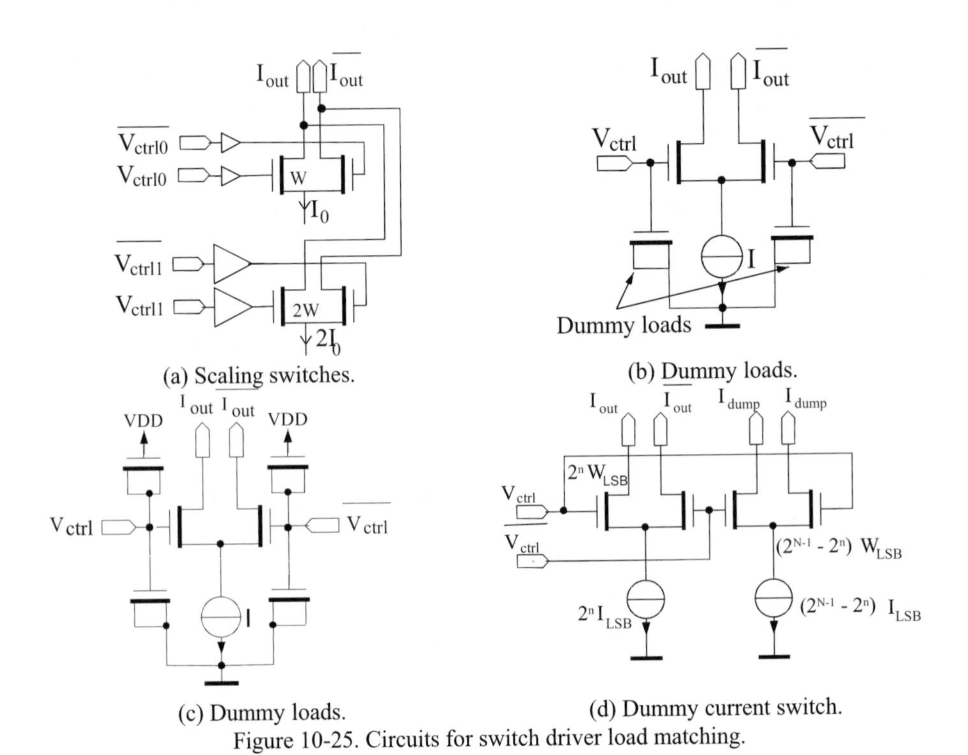

Figure 10-25. Circuits for switch driver load matching.

10.9.3 Layout

As the sampling frequency increases, the symmetry of the interconnections and cross coupling become important issues. For the methods presented in the previous sections to be useful, the signal paths from the D-flip-flop inputs to the current switches should be designed to be similar, while, to minimize the parasitic effects, the D-flip-flops should be located near the switches. The switches can be placed in a matrix [Mik86], [Tak91, [Bas98], [Pla99], [Bos01] such as the current sources, as this improves the matching between the switches. If the timing error minimization is not possible, the timing error should be made as random as possible between the switches, so that the error spectrum becomes noise [Dor01].

The supply fluctuations and clock jitter inside the chip cause current switches transition errors. Wider metal lines could minimize the supply voltage drops. Avoiding extensive buffering could reduce the clock jitter. The clock line sees a variable capacitive load, because the gate to source capacitance of the input transistors varies according to the states of switch drivers. If one global clock buffer is used, this will result in large code-dependent timing errors. The local clock buffers are used to reduce these code-dependent timing errors [Fal99]. Dummy switch drivers were used in [Sch03], [Sch04] to ensure that same number of data transitions occur every clock cycle. Then the power supply of the switch drivers delivers a constant data-independent charge at the clock rate, and the clock line sees a fixed capacitive load. This was a key feature in achieving the exceptional performance levels in [Sch03], [Sch04].

10.10 Cascode Transistor

The output resistance of the output current source transistor is approximately [All87]

$$R_{cs} = \frac{1}{I_{cs}\lambda_{cs}}, \tag{10.27}$$

where $\lambda \sim 1/L$. The current source output resistance is related to the transistor channel length. The output current of the current source is inversely proportional to the channel length.

If the current switch transistors are in saturation region at the on-state, the output impedance of the current source is multiplied by the current switch voltage gain

$$g_{msw} = \sqrt{\frac{2\mu C_{ox} W_{sw}}{I_{cs}\lambda_{sw}^{2} L_{sw}}} \sim \sqrt{W_{sw} L_{sw}}. \tag{10.28}$$

The current switches gain typically improves the impedance ratio about 20-30 dB [Tei02]. However, the current switches should be minimum sized in order to reduce data feedthrough. Therefore, without cascode transistors, it is possible to achieve an impedance ratio of about 130dB [Tei02]. This is not enough to fulfill the 12 bit single-ended D/A converter impedance requirements [Tei02]. However, for 12 bit differential output this is enough, therefore 12 bit D/A converters have been implemented without the cascode transistors [Bos98].

The current sources output impedance could be increased by using cascode transistors (Figure 10-26(a)) [Chi86], [Bos99]. The cascode transistor multiplies the current source output impedance by their gain (10.28). The dimension of the cascode transistors are restricted by the W/L ratio and bias voltage range, where the transistors are in saturation range. The output impedance can be increased 50dB [Tei02]. The other advantage is that the cascode transistors isolate the current switches and current sources, and the effect of the switching noise is minimized [Chi86], [Bos99]. It is possible to add two cascode transistors to enhance the output impedance [Bav98], but this requires additional bias voltage and reduces the output voltage range.

The cascode transistor effect could be interpreted in the time domain. The D/A converter settling time depends on the source capacitance of the current switches. When a cascode transistor is inserted between the current switches and current sources so that the most of the wiring stray capacitances are below the cascode transistor, the cascode transistor isolate this stray capacitance, causing the output settling to be faster.

Cascode transistors could also be added between the current switches and the output (Figure 10-26(b)) [Tak91]. The cascode transistors multiply the output impedance by their gain. Furthermore the cascode transistors turn off, when the switches turn off, therefore the switching noise feedthrough to the output is reduced. The cascode transistors turn on later than the current switches, so they reduce the switching transients. The high output resistances

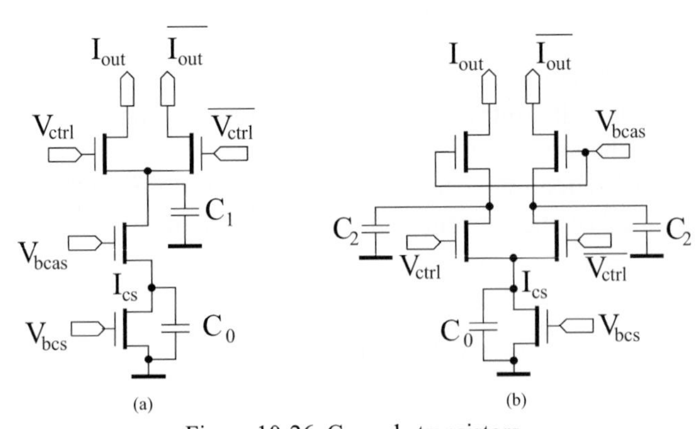

Figure 10-26. Cascode transistors.

of the cascode transistors attenuate the high frequency noise when the cascode is on [Mar98]. The high resistance also increases the output settling time; therefore the channel length should be minimum. This restricts the cascode transistors voltage gain to about 30 dB [Tei02], therefore it improves the linearity less than the cascoded transistor between the current switches and current sources. When the current switch is turned off, the cascode is turned off. The parasitic capacitance at the cascode transistor source is discharged through the cascode transistor, which reduces the output settling time at the positive voltage steps. This output rise and fall waveform asymmetry causes even harmonics, but this is partly cancelled at the differential output. However, the cascode transistors of the switches are used in [Tak91], [Bas98], [Mar98], [Zho01], [Xu99].

The cascode switch and current source transistors can be used together. The two cascode transistors reduce the output voltage range. Biasing the cascode transistors in the linear region could alleviate this problem, if the output impedance is not a limiting factor [Tak91]. It is quite easy to achieve DC impedance requirements, as the limiting factor is the output impedance degradation at high frequencies. If the current switches effect is taken into account, then the output impedance is

$$Z_{imp} = R_{sw}(1 + g_{msw}(R_{cs} \,||\, C_0)) \approx R_{sw} g_{msw} R_{cs} \frac{1 + \frac{j\omega C_0}{g_{msw}}}{1 + j\omega C_0 R_{cs}}, \qquad (10.29)$$

where R_{sw}/R_{cs} is current switches/source resistance. According to the equation, the switch transistors add one zero at $f_z = g_{msw}/(2\pi C_0)$, but not affect the poles. The output impedance corner frequency does not depend on parasitic capacitance (C_0) only, but also on the switches transconductance or current source resistance. As stated before, switch transistor size cannot be set arbitrarily. The zero is normally at high frequency, so it does not have a strong effect.

If the cascode transistors are between the current sources and switches as shown in Figure 10-26(a), then the output impedance becomes

$$Z_{swcas} = R_{sw} R_{cscas} R_{cs} g_{msw} g_{mcscas} \frac{(1 + \frac{j\omega C_1}{g_{msw}})(1 + \frac{j\omega C_0}{g_{mcscas}})}{(1 + j\omega R_{cs} C_0)(1 + j\omega R_{cscas} C_1)}, \qquad (10.30)$$

where R_{cscas} and g_{mcscas} are cascode transistor resistance and transconductance, respectively. The cascode transistors add a pole at the frequency

$$f_p = \frac{1}{2\pi R_{cscas} C_1} \qquad (10.31)$$

and zero at

$$f_z = \frac{g_{mcscas}}{2\pi C_0}. \tag{10.32}$$

In principle, it is possible to compensate the zeros and poles by sizing the cascode transistors. In practice, the transistor minimum size is dependent on the process. As the transistor dimensions increase, the transistor parasitic capacitances become more significant than the wiring stray capacitances. In spite of that the corner frequency of the output impedance could be increased by cascode sizing [Bos99].

If the cascode transistors are added above the current switches, as shown in Figure 10-26(b), the output impedance will then be

$$Z_{imp} = R_{sw} R_{swcas} R_{cs} g_{msw} g_{mswcas} \frac{(1+\frac{j\omega C_0}{g_{msw}})(1+\frac{j\omega C_2}{g_{mswcas}})}{(1+j\omega R_{cs} C_0)(1+j\omega R_{sw} C_2)}. \tag{10.33}$$

where R_{swcas} and g_{mswcas} are cascode transistor resistance and transconductance, respectively. According to the above equation, the poles do not depend on the cascode transistor size. Furthermore the cascode transistor above the current switches should be minimum size, which reduces the possibility of the compensating the poles and zeros. It is possible to use both the switch and current source cascodes, if the chip area and power supply are not criti-

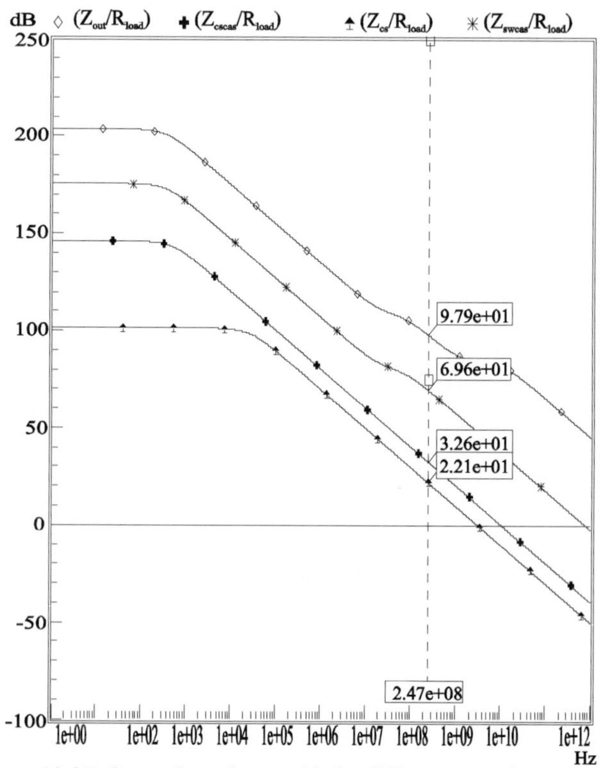

Figure 10-27. Output impedance with the different cascode structures.

cal factors. The output impedance would then be

$$Z_{out} = R_{swcas} R_{sw} R_{cscas} R_{cs} g_{mswcas} g_{msw} g_{mcscas} \frac{(1+\frac{j\omega C_0}{g_{msw}})(1+\frac{j\omega C_1}{g_{mcscas}})(1+\frac{j\omega C_2}{g_{mswcas}})}{(1+j\omega R_{cs} C_0)(1+j\omega R_{sw} C_2)(1+j\omega R_{cas} C_1)}, \tag{10.34}$$

where half of the zeros and poles could be tuned by cascode transistors dimensions. In Figure 10-27, the 12bit D/A converter output impedance is shown with the different cascode structures (the wiring stray capacitances are taken into account). The curves from down to top are: current source output impedance (Z_{cs}), cascode current source output impedance (Z_{cscas}), cascode current source with current switch output impedance (Z_{swcas}) and cascode current source with cascode switch output impedance (Z_{out}). The current source transistor and cascode transistor has little difference at high frequencies due to lack of the wiring capacitance in the current source simulations. In Figure 10-27 the effect of the poles and zeros on the impedance can clearly be seen.

REFERENCES

[All87] P. Allen, and D. Holberg, "CMOS Analog Circuit Design," Oxford University Press, 1987.

[Bal87] G. Baldwin, et al., "Electronic Sampler Switch," U. S. Patent 4,639,619, Jan. 27, 1987.

[Bas91] C. Bastiaansen, D. Wouter, J. Groeneveld, H. Schouwenaars, and H. Termeer, "A 10-b 40 MHz 0.8-µm CMOS Current-Output D/A Converter," IEEE Journal of Solid State Circuits, Vol. 26, No. 7, pp. 917-921, July 1991.

[Bas97] J. Bastos, M. Steyaert, A. Pergoot, and W. Sansen, "Influence of Die Attachment on MOS Transistor Matching," IEEE Trans. Semiconduct.Manufact., Vol. 10, pp. 209–217, May 1997.

[Bas98] J. Bastos, A. Marques, M. Steyaert, and W. Sansen, "A 12-bit Intrinsic Accuracy High-Speed CMOS DAC," IEEE J. Solid-State Circuits, Vol. 33, No. 12, pp. 1959–1969, Dec. 1998.

[Bav98] N. van Bavel, "A 325 MHz 3.3V 10-Bit CMOS D/A Converter Core With Novel Latching Driver Circuit," Proceedings of the IEEE Custom Integrated Circuits Conference 1998, pp. 245-248.

[Bos01] A. van den Bosch, M. Borremans, M. Steyaert, and W. Sansen, "A 10-bit 1G Samples Nyquist Current-Steering CMOS D/A Converter," IEEE Journal of Solid State Circuits, Vol. 36, No. 3, pp. 315-324, Mar. 2001.

[Bos01a] A. Van den Bosch, M. Steyaert, and W. Sansen, "An Accurate Statistical Yield Model for CMOS Current-Steering D/A Converters,"

Analog Integrated Circuits and Signal Processing, Vol. 29, No. 3, pp. 173-180, Dec. 2001.

[Bos98] A. van den Bosch, M. Borremans, J. Vandenbussche, G. Van der Plas, A. Marques, J. Bastos, M. Steyaert, G. Gielen, and W. Sansen, "A 12 bit 200MHz Low Glitch CMOS D/A Converter," Proceedings of the IEEE Custom Integrated Circuits Conference 1998, pp. 249-252.

[Bos99] A. van den Bosch, M. Steyaert, and W. Sansen, "SFDR-Bandwidth Limitations for High Speed High Resolution Current Steering CMOS D/A Converters," Proceedings of the 6th IEEE International Conference on Electronics, Circuits and Systems 1999, Vol. 3, pp. 1193-1196.

[Bug00] A. Bugeja, and B.-S. Song, "A Self-Trimming 14-b 100-MS/s CMOS DAC," IEEE Journal of Solid State Circuits, Vol. 35, No. 12, pp. 1841-1852, Dec. 2000.

[Bug99] A. R. Bugeja, B. S. Song, P. L. Rakers, and S. F. Gillig, "A 14-b, 100-MS/s CMOS DAC Designed for Spectral Performance," IEEE J. Solid-State Circuits, Vol. 34, No. 12, pp. 1719-1732, Dec. 1999.

[Bur01] M. Burns, and G. Roberts, "An Introduction to Mixed-Signal IC Test and Measurement," Oxford University Press, 2001.

[Chi86] K. Chi, C. Geisenhainer, M. Riley, R. Rose, P. Sturges, B. Sullivan, R.Watson, R. Woodside, and M. Wu, "A CMOS Triple 100-Mbit/s Video D/A Converter with Shift Register and Color Map," IEEE Journal of Solid-State Circuits, Vol. SC-21, pp. 989-995, Dec. 1986.

[Chi94] S.-Y. Chi, and C.-Y. Wu, "A 10-b 125-MHz CMOS Digital-to-Analog Converter (DAC) with Threshold-Voltage Compensated Current Sources," IEEE Journal of Solid State Circuits, Vol. 29, No. 11, pp. 1374-1380, Oct. 1994.

[Con00] Y. Cong, and R. Geiger, "Switching Sequence Optimization for Gradient Error Compensation in Thermometer-Decoded DAC Arrays," IEEE Transactions on Circuits and Systems-II: Analog and Digital Signal Processing, Vol. 47, No. 7, pp. 585-595, July 2000.

[Con02] Y. Cong, and R. L. Geiger, "Formulation of INL and DNL Yield Estimation in Current-Steering D/A Converters," in Proc. IEEE International Symposium on Circuits and Systems (ISCAS), May 2002, pp. III.149-III.152.

[Con03] Y. Cong, and R. Geiger, "A 1.5V 14b 100MS/s Self-Calibrated DAC," Digest of Technical Papers of the IEEE International Solid-State Circuits Conference 2003, pp. 128-129.

[Cre89] A. Cremonesi, F. Maloberti, and G. Polito, "A 100-MHz CMOS DAC for Video-Graphic Systems," IEEE Journal of Solid State Circuits, Vol. 24, No. 3, pp. 635-639, June 1989.

[Cro97] D. Crook, and E. Stroud, "Drive Circuit and Method for Controlling the Crosspoint Levels of a Differential CMOS Switch Drive Signal," U. S. Patent 5,703,519, Dec. 30, 1997.
[Cro98] D.Crook, and R. Cushing, "Sources of Spurious in a DDS/DAC System," RF Design, pp. 28-42, April 1998.
[Ded02] I. Dedic, "Switch Driver Circuitry Having First and Second Output Nodes with a Current-Voltage Converter Connected There Between Providing Current Paths of First and Second Directions There Between and Switching Circuitry Connected Therewith," U. S. Patent 6,340,939, Jan. 22, 2002.
[Ded99] I. Dedic, "Differential Switching Circuitry," U. S. Patent 6,100,830, Aug. 8, 2000.
[Dor01] K. Doris, D. Leenaerts, and A. van Roermund, "Time Non Linearities in D/A Converters," Proceedings of the European Conference on Circuit Theory and Design, Vol. III, pp. 353-356, 2001.
[Dud97] F. Dudek, B. M. Al-Hashimi, and M. Moniri, "CMOS Equaliser for Compensating sinc(x) Distortion of Video D/A Converters," Electron. Lett., Vol. 33, No. 19, pp. 1618-1619, Sep. 1997.
[Ess97] K. Essenwanger, "Digital-to-Analog Converted and Method that Set Waveform Rise and Fall Times to Produce an Analog Waveform that Approximates a Piecewise Linear Waveform to Reduce Spectral Distortion," U. S. Patent 5,663,728, Sept. 2, 1997.
[Ess98] K. Essenwanger, "Slewer Fractional-Order-Hold: The Ideal DAC Response For Direct Digital Synthesis," Proceedings of the IEEE International Frequency Control Symposium 1998, pp. 379-389.
[Fal99] K. Falakshahi, C.-K. Yang, and B. Wooley, "A 14-bit 10-Msamples/s D/A Converter Using Multibit ΣΔ Modulation," IEEE Journal of Solid State Circuits, Vol. 34, No. 5, pp. 607-615, 1999.
[Ger97] J. Gersbach, "Self Calibrating Segmented Digital-to-Analog Converter," U. S. Patent 5,642,116," June 24, 1997.
[Gro89] D. W. J. Groeneveld, H. J. Schouwenaars, H. A. H. Termeer, and C. A. A. Bastiaansen, "A Self-Calibration Technique for Monolithic High-Resolution D/A Converters," IEEE J. Solid-State Circuits, Vol. 24, No. 6, pp. 1517-1522, Dec. 1989.
[Gus00] M. Gustavsson, J. Wikner, and N. Tan, "CMOS Data Converters for Communications," Kluwer Academic Publishers, 2000.
[Han99] J. Hanna, "Circuit and Method for Calibrating a Digital-to-Analog Converter," U. S. Patent 5,955,980, Sept. 21, 1999.
[Hen97] P. Hendriks, "Specifying Communications DACs," IEEE Spectrum, Vol. 34, No. 7, pp. 58-69, July 1997.
[Her91] R. Herman, A. McKay, and A. Chao, "Synchronizing Switch Arrangement for a Digital-to-Analog Converter to Reduce in-Band Switching Transients," U. S. Patent 5,059,977, Oct. 22, 1991.

[Hyd03] J. Hyde, T. Humes, C. Diorio, M. Thomas, and M. Figueroa, "A 300-MS/s 14-bit Digital-to-Analog Converter in Logic CMOS," IEEE Journal of Solid State Circuits, Vol. 38, No. 5, pp. 734-740, May 2003.

[Kas95] K. Kasai, and K. Matsuo, "Differential Current Source Circuit in DAC of Current Driving Type," U. S. Patent 5,406,135, Apr. 11, 1995.

[Kim98] J. Kim, and K. Yoon, "An 8-bit CMOS 3.3-V-65MHz Digital-to-Analog Converter with a Symmetric Two-Stage Current Cell Matrix Architecture," IEEE Transactions on Circuits and Systems-II: Analog and Digital Signal Processing, Vol. 45, No. 12, pp. 1605-1609, Dec. 1998.

[Koh95] H. Kohno Y. Nakamura, A. Kondo, H. Amishiro, T. Miki, and K. Okada, "A 350-MS/s 8-bit CMOS D/A-Converter Using Delayed Driving Scheme," Proceedings of the IEEE Custom Integrated Circuits Conference 1995, pp. 211-214.

[Kos03] M. Kosunen, J. Vankka, I. Teikari, and K. Halonen, "DNL and INL Yield Models for a Current-Steering D/A Converter," in Proc. ISCAS'03, Vol. I, May 2003, pp. 969-972.

[Lak86] K. Lakshmikumar, and et al., "Characterization and Modeling of Mismatch in MOS Transistors for Precision Analog Design," IEEE Journal of Solid State Circuits, Vol. 21, pp. 1057–1066, Dec. 1986.

[Lak88] K. Lakshmikumar, and et al., "Reply to 'A Comment on: Characterization and Modeling of Mismatch in MOS Transistors for Precision Analog Design," IEEE Journal of Solid State Circuits, Vol. 23, p. 296, Feb. 1988.

[Lin98] C-H. Lin, and K. Bult, "A 10b 500MSample/s CMOS DAC in 0.6 mm^2," IEEE Journal of Solid State Circuits, Vol. 33, No. 12, pp. 1948-1958, Dec. 1998.

[Luh00] L. Luh, J. Choma Jr., and J. Draper, "A High-Speed Fully Differential Current Switch," IEEE Transactions on Circuits and Systems-II: Analog and Digital Signal Processing, Vol. 47, No. 4, pp. 358-363, Apr. 2000.

[Mar98] A. Marques, J. Bastos, A. van den Bosch, J. Vandenbussche, M. Steyaert, and W. Sansen, "A 12b Accuracy 300MSample/s Update Rate CMOS DAC," Digest of Technical Papers of the IEEE International Solid-State Circuits Conference 1998, pp. 216 -217.

[McC75] J. McCreary, and P. Gray, "All-MOS Charge Redistribution Analog-to-Digital Conversion Techniques-Part I," IEEE Journal of Solid State Circuits, Vol. 10, No. 6, pp. 371-379, Dec. 1975.

[Mer00] D. Mercer, "Differential Current Switch," U. S. Patent 6,031,477, Feb. 29, 2000.

[Mer93] D. Mercer, "Two Approaches to Increasing Spurious Free Dynamic Range In High Speed DACs," Proceedings of the IEEE Bipolar Circuits and Technology Meeting 1993, pp. 80-83.

[Mer94] D. Mercer, "A 16-b D/A Converter with Increaced Spurious Free Dynamic Range," IEEE Journal of Solid State Circuits, Vol. 29, No. 10, pp. 1180-1185, Oct. 1994.

[Mer97] D. Mercer, D. Reynolds, D. Robertson, and E. Stroud, "Skewless Differential Current Switch and DAC Employing the Same," U. S. Patent 5,689,257, Nov. 18, 1997.

[Mik86] T. Miki, Y. Nakamura, M. Nakaya, S. Asai, Y. Asaka, and Y. Horiba, "An 80-MHz 8-bit CMOS D/A-Converter," IEEE Journal of Solid State Circuits, Vol. 21, No. 6, pp. 983-988, Dec. 1986.

[Nak91]Y. Nakamura, T. Miki, A. Maeda, H. Kondoh, and N. Yazawa, "A 10-bit 70 MS/s CMOS D/A converter," IEEE J. Solid-State Circuits, Vol. 26, pp. 637–642, Apr. 1991.

[Ngu92] T. Nguyen, "Spike Current Reduction in CMOS Switch Drivers," U. S. Patent 5,089,728, Feb. 18, 1992.

[Pel89] M. Pelgrom, A. Duinmaijer, and A. Welbers, "Matching Properties of MOS Transistors," IEEE Journal of Solid State Circuits, Vol. 24, No. 5, pp. 1433-1440, Oct. 1989.

[Pla99] G. Van der Plas, J. Vandenbussche, W. Sansen, M. Steyaert, and G. Gielen, "A 14-bit Intrinsic Accuracy Q2 Random Walk CMOS DAC," IEEE J. Solid-State Circuits, Vol. 34, No. 12, pp. 1708–1718, Dec. 1999.

[Rad00] R. Radke, and A. Eshraghi, "A 14-Bit Current Mode SD-DAC Based Upon Rotated Data Weighted Averaging," IEEE Journal of Solid State Circuits, Vol.35, No. 8, pp. 1074-1084, Aug. 2000.

[Raz95] B. Razavi, "Principles of Data Conversion System Design," IEEE Press, 1995.

[Sam88] H. Samueli, "The Design of Multiplierless FIR Filters for Compensating D/A Converter Frequency Response Distortion," IEEE Trans. Circuits and Syst., Vol. 35, No. 8, pp. 1064-1066, Aug. 1988.

[Sch03] W. Schofield, D. Mercer, and L. St. Onge, "A 16b 400MS/s DAC with -80dBc IMD to 300MHz and -160dBm/Hz Noise Power Spectral Density," Digest of Technical Papers of the IEEE International Solid-State Circuits Conference 2003, pp. 126-127.

[Sch04] B. Schafferer, and R. Adams, "A 3V CMOS 400mW 14b 1.4GS/s DAC for Multi-Carrier Applications," ISSCC Digest of Technical Papers, February 2004, San Francisco, USA, pp. 360-361.

[Sch88] H. Schouvenaars, D. Wouter, J. Groeneweld, and H. Termeer, "A Low-Power Stereo 16-bit CMOS D/A Converter for Digital Audio," IEEE Journal of Solid State Circuits, Vol. 23, No. 6, pp. 1290-1297, Dec. 1988.

[Sch97] W. Schnaitter, "Switchable Current Source for Digital-to-Analog Converter (DAC)," U. S. Patent 5,598,095, Jan. 1997.

[Sed91] A. Sedra, and K. Smith, "Microelectronic Circuits," 3rd edition, Saunders College Publishing, 1991.

[Tak91] H.Takakura, M. Yokoyama, and A. Yamaguchi, "A 10 bit 80MHz Glitchless CMOS D/A Converter," Proceedings of the IEEE Custom Integrated Circuits Conference 1991, pp. 26.5/1 -26.5/4.
[Tei02] I. Teikari, "High-speed Nyquist-rate D/A Converter for Telecommunications Applications," Master thesis, Helsinki University of Technology, Electronic Circuit Design Laboratory, Oct. 2002.
[Tii01] M. Tiilikainen, "A 14-bit 1.8-V 20mW 1-mm^2 CMOS DAC," IEEE Journal of Solid State Circuits, Vol. 36, No. 7, pp. 1144-1147, July 2001.
[Van02] J. Vankka, J. Pyykönen, J. Sommarek, M. Honkanen, and Kari Halonen, "A Multicarrier GMSK Modulator with on-chip D/A converter for Base Station," IEEE Journal of Solid-State Circuits, Vol. 37, No. 10, pp. 1226-1234, Oct. 2002.
[Vol02] A. Volk, "Method To Reduce Glitch Energy in Digital-to-Analog Converter," U. S. Patent 6,507,295, Jan. 14, 2003.
[Wu95] T.-Y. Wu, C.-T. Jih, J.-C. Chen, and C.-Y. Wu, "A Low Glitch 10-bit 75-Hz CMOS Video D/A Converter," IEEE Journal of Solid State Circuits, Vol. 30, No. 1, pp. 68-72, Jan. 1995.
[Xu99] Y. Xu, and H. Min, "A Low-Power Video 10-Bit CMOS D/A Converter Using Modifed Look-Ahead Circuit," IEEE Transactions on Consumer Electronics, Vol. 45, No. 2, pp. 295-298, May 1999.
[Zho01] Y. Zhou, and J. Yuan, "An 8-Bit 100-MHz Low Glitch Interpolation DAC," Proceedings of the 2001 IEEE International Symposium on Circuits and Systems, Vol. 4, pp.116 -119.

Chapter 11

11. PULSE SHAPING AND INTERPOLATION FILTERS

Different methods of designing the pulse shaping filters are reviewed in this chapter. In the digital modulators, phase distortion cannot be tolerated, thus the filters are required to have a linear phase response. A FIR filter can be guaranteed to have an exact linear phase response if the coefficients are either symmetric or antisymmetric about the center point. Three FIR filter structures (direct form, transposed direct form and hybrid form) are presented. The quantization effects and scaling methods within the fixed point FIR architectures are reviewed. Using canonic signed digit (CSD) coefficients, the FIR filtering operation can be simplified to add and shift operations. The well known carry save (CS) numbers are very attractive for VLSI implementation since the basic building block for arithmetic operations is a simple full adder. The multirate signal processing is particularly important in the digital modulator, where sample rates are low initially and must be increased for efficient subsequent processing. The efficient filter structures for the multirate signal processing (polyphase filters, halfband filters and comb filters) are presented. Taking advantage of the fact that in the modulator data streams in the I and Q paths are processed with the same functional blocks (see Figure 16-6), a further hardware reduction can be achieved by pipeline interleaving techniques.

11.1 Pulse Shaping Filter Design Algorithms

The pulse shaping filter design has two main objectives: minimization of the inter symbol interference (ISI) and maximization of the adjacent channel leakage power ratio (ACLR) [Che82], [Sam88], [Sam91], [Mor95]. [Sam88] and [Sam91] present two iterative algorithms that allow to the design of an overall filter of a given order N (an even number) with zero ISI and linear phase. The attenuation in the stopband is minimized. In [Sam88], only equiripple filters are considered and linear programming is used for determining the filter taps. [Sam91] introduces a ripple weighting vector that allows arbitrary magnitude transfer functions to be designed, and uses equations instead of inequalities. Both [Sam88] and [Sam91] illustrate how to design a Nyquist filter that can be subsequently split into a transmitter and a receiver filter. [Sam88] and [Sam91] provide zero-ISI solutions but nonlinear phase characteristics. In [Mor95], the matched filter condition, i.e. identical transmit and receive filters with time reversal, is relaxed in order to obtain two linear phase transmit and receive filters and zero ISI in the composite filter. The transmit and the receive filters have different lengths. In [Che82], Chevillat and Ungerboeck present nonlinear optimization techniques for designing transmit and receive filters that result linear phase solutions with a non-zero-ISI [Che82].

The N-tap transmit filter is characterized by the coefficient vector $h = (h_0, h_1, ..., h_{N-1})^T$, which is clocked at the rate M/T corresponding to an over-sampling ratio M. The receive filter (hr) is a K tap filter, which is M times over-sampled from the root raised cosine function. The transmit filter is convolved with the receive filter. Ideally, the result of the convolution will be an ideal raised cosine filter. There will be an EVM due to the truncation of the receive filter impulse response, if the length of the receive filter is short. It is therefore better to use a long receive filter so that the transmit filter will dominate the EVM.

The transmit and receive filter lengths are assumed to be either even or odd, so as to have one middle sample for decision in the composite pulse $RC(n)$. The convolution of the transmitter and receiver filters should satisfy the zero inter-symbol interference constraint:

$$RC(n)=0, \quad n=n_c \pm lM, \quad l=1,2,...,L, \tag{11.1}$$

where n_c is the center tap and M is the over-sampling ratio. The center tap is $(N+K-2)/2$. The total number of the terms in (11.1) is $2L$, where $L = \lfloor n_c/M \rfloor$, and $\lfloor x \rfloor$ denotes the integer part of x. The equation (11.1) can be written as

$$RC(n_c + Ml) = \sum_{i=0}^{N-1} h_i \, hr_{i+Ml} = h^T \, S_l \, hr, \quad l = \pm 1, \pm 2, ..., \pm L, \tag{11.2}$$

where the elements of the "shift" matrices S_l are zero, except $s_{i,k}(l) = 1$ for $i - k = (N-K)/2 + Ml$ [Che82]. The "shift" matrices S_l are $N \times K$ matrices.

The passband ripples of the linear phase half-band filters (interpolation filters in Figure 16-6) cause EVM as well, which could be partly compensated for by predistortion of the pulse shaping filter. The receive filter (hr) could be convolved with the interpolation filters. This convolution could be calculated with the noble identities [Vai93]. The result is decimated back to the M over-sampled ratio and convolved with the transmit filter in (11.2).

One code channel is transmitted, when the EVM is measured. The EVM consists of two components, which are mutually uncorrelated:

$$\sigma_{EVM}^2 = \sum_{\substack{l=-L \\ l\neq 0}}^{L} (h^T S_l\, hr\,)^2 + \delta_e^2, \tag{11.3}$$

where δ_e^2 is the quantization noise due to finite word length effects. The D/A converter dominates this quantization noise, because it is the most critical component. The ISI term is

$$\begin{aligned} \delta_{ISI}^2 &= \sum_{\substack{l=-L \\ l\neq 0}}^{L} (h^T\, S_l\, hr)^2 = h^T\, W\, h, \\ \text{where } W &= \sum_{\substack{l=-L \\ l\neq 0}}^{L} S_l\, hr\,(S_l\, hr)^T, \end{aligned} \tag{11.4}$$

and W is a $N \times N$ matrix. A linear constraint is added to guarantee proper scaling of the pulse peak

$$RC(n_c) = h^T\, S_0\, hr = 1. \tag{11.5}$$

The lowpass channel energy (E_c) from dc to f_b (the cut-off frequency of the lowpass channel) is

$$\begin{aligned} E_c &= \int_{f=-f_b}^{f=f_b} |H(f)|^2\, df = \sum_{i=0}^{N-1} \sum_{k=0}^{N-1} h_i\, h_k \int_{f=-f_b}^{f=f_b} e^{-j2\pi f(i-k)T/M}\, df \\ &= \sum_{i=0}^{N-1} \sum_{k=0}^{N-1} h_i\, h_k\, r_{ik} = h^T\, R\, h, \end{aligned} \tag{11.6}$$

where R is a $N \times N$ matrix with elements

$$\mathrm{r}_{\mathrm{ik}} = \begin{cases} 2\, f_b & i = k \\ \dfrac{\sin(2\pi\, f_b\, (i-k)T/M)}{\pi\,(i-k)T/M} & i \neq k. \end{cases} \tag{11.7}$$

The stopband energy (E_s) from f_s (stopband corner frequency) to M/(2*T)* is

$$\begin{aligned} \mathrm{E_s} &= 2 \int_{f=f_s}^{f=M/(2T)} |H(f)|^2\, df = 2\sum_{i=0}^{N-1} \sum_{k=0}^{N-1} h_i\, h_k \int_{f=f_s}^{f=M/(2T)} e^{-j2\pi f(i-k)T/M}\, df \\ &= \sum_{i=0}^{N-1} \sum_{k=0}^{N-1} h_i\, h_k\, v_{ik} = h^T\, V\, h, \end{aligned} \tag{11.8}$$

where V is a $N \times N$ matrix with elements

$$v_{ik} = \begin{cases} M/T - 2f_s & i = k \\ \dfrac{\sin(\pi(i-k))}{\pi(i-k)T/M} - \dfrac{\sin(2\pi\, f_s\,(i-k)T/M)}{\pi(i-k)T/M} & i \neq k. \end{cases} \tag{11.9}$$

The ISI can be traded off against the power ratio of the main channel power to the power of the adjacent channels. The ISI performance decreases while the power ratio of the main channel power to the power of the adjacent channels increases. The cost function, which should be maximized, is written as

$$E = a \times E_c - b \times E_s - c \times \delta_{ISI}^2 . \qquad (11.10)$$

The objective is to maximize the ratio of the main channel power to the power of the adjacent channels power under the constraint that the ISI is below 2%. Therefore weighting terms, *a, b* and *c* are added. No well-developed method exists for choosing the weighting terms, *a, b* and *c*. Suitable values have to be found by trial and error. Employing the Lagrangian method for the maximization of (11.10), subject to (11.5), the objective function is

$$\begin{aligned} \Phi(h, \lambda) &= a \times h^T R h - b \times h^T V h - c \times h^T W h - \lambda \ (h^T S_0 hr - 1) \\ &= h^T D h - \lambda \ (h^T S_0 hr - 1), \end{aligned} \qquad (11.11)$$

where $D = a \times R - b \times V - c \times W$. The solution is found with the standard Lagrange multiplier techniques (by setting the derivatives with respect to $h(0),...,h(N\text{-}1)$ and λ to zero) to be

$$h = \frac{D^{-1} S_0 hr}{(S_0 hr)^T D^{-T} S_0 hr} . \qquad (11.12)$$

The main shortcoming of this algorithm is that the effect of the weighting

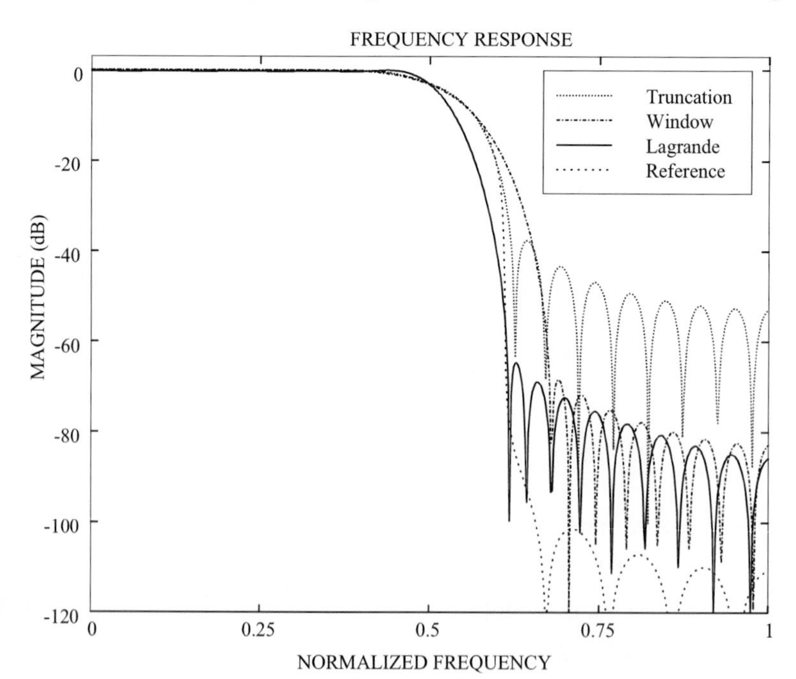

Figure 11-1. Comparison of filter design methods.

Table 11-1. Comparison of filter design methods.

Method	ACLR	ISI
Truncation	45.30 dB	-59.21 dB
Window with Kaiser, β = 4	36.15 dB	-40.07 dB
Lagrange	73.38 dB	-45.08 dB
Root raised cosine with 1001 coefficients	71.2 dB	-106.10 dB

factors *a*, *b* and *c* has to be found by trial and error. Results of different filter design methods are compared in Table 11-1. The number of the filter coefficients is 37 for each filter. The oversampling ratio is 2 and the sample frequency is normalized to that. The passband is defined to be from 0 to 0.61 Hz and the stopband (adjacent channel) from 0.61 Hz to 1 Hz. In Table 11-1, it can be seen that the ACLR value of the filter designed with the window method suffers from the increased width of the passband. The frequency responses of the filters are presented in Figure 11-1.

11.2 Direct Form Structure of FIR Filter

The structure of the folded direct form FIR filter is presented in Figure 11-2. If the FIR filter coefficients are symmetrical or anti-symmetrical, i.e. the filter is phase linear, the advantage of folding the taps can be applied. Folding the taps is hardware efficient, because only about a half of the taps need be realized. More accurately, the number of taps to be realized is $\lfloor N/2 \rfloor + 1$ if

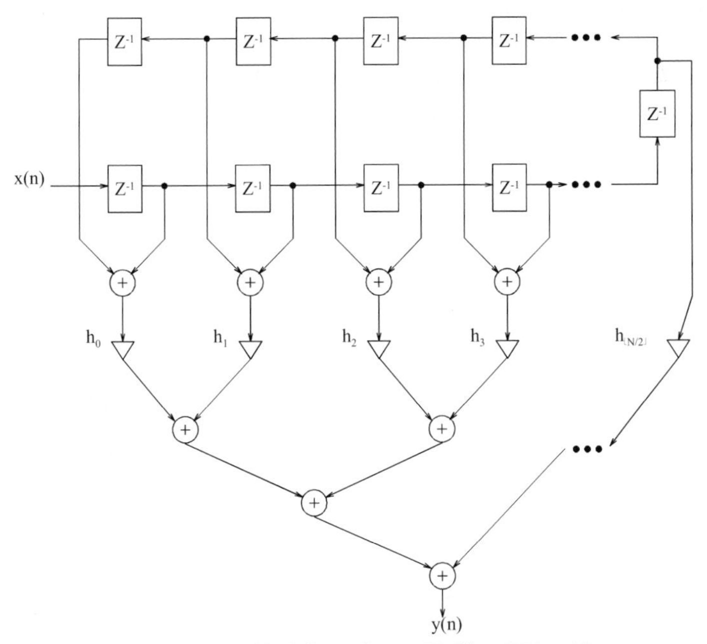

Figure 11-2. Folded direct form FIR filter (*N* is odd).

the number of taps N is odd and $\lfloor N/2 \rfloor$ if N is even. $\lfloor x \rfloor$ stands for the integer part of x. In the folded regular direct form, the number of bits needed in the delay elements is defined by the number of input bits rather than the required word length of the filter, which may lead to a reduced amount of hardware. This structure is suitable for the programmable filter designs, because the effect of changing the filter coefficients is seen immediately or simultaneously at the filter output.

The long critical path causes problems in the high-speed systems when a short cycle-time is desired. However, this can be overcome by pipelining at the expense of an increased amount of hardware. The subexpression sharing method [Har96] can be applied easily, which reduces the amount of hardware.

11.3 Transposed Direct Form Structure of FIR Filter

The structure of the folded transposed direct form FIR filter is presented in Figure 11-3. In this case, as well as in the folded direct form filter described in Section 11.2, folding the taps reduces the amount of hardware. As can be clearly seen from Figure 11-2 and Figure 11-3, the maximum delay path is considerably shorter in the transposed form realization, resulting in a faster performance. The subexpression sharing method [Har96] can be applied easily, which reduces the amount of hardware.

A shortcoming of this structure is that if the filter coefficients are changed, the effect is spread over the multiple number of clock cycles, resulting in this structure being unsuitable for the programmable filter applications, when an immediate response to the change of the coefficients is required. Another problem of the transposed direct structure relates to the location of the delay elements in the output path of the filter. Because of the ad-

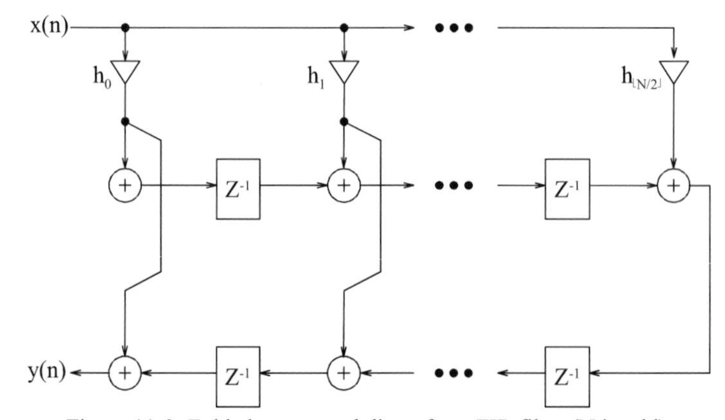

Figure 11-3. Folded transposed direct form FIR filter (N is odd).

ditions, the delay elements might have to be implemented with a larger word length than in the direct structure counterpart [Haw96]. Hence, they require a larger area and consume more power. Anyhow, the necessity of enlarging the word length depends on the filter coefficients and is not always necessary.

In the folded transposed direct structure, a large loading occurs on the input data bus since all taps are fed in parallel, as shown in Figure 11-3. If the taps are realized as the multiplierless signed digit representation described in Section 11.6, this problem becomes even worse, due to the sign bit extensions in the shift operations. Fortunately, this sign bit load can be avoided by using the constant vector addition method proposed in [Haw96].

11.4 Hybrid Form

It is also possible to combine the filter structures described above into a so-called hybrid form filter [Lee96], [Kho01], [Hat01]. The purpose of this structure is to trade off the cycle-time requirement of an FIR filter with its area requirement. The hybrid form, depicted in Figure 11-4, can be thought of as being obtained from the direct structure by moving a minimum number of registers from the input path to the output path to satisfy the cycle-time requirement. Then the word length, and thus the required area of only a few registers, has to be increased. The cycle-time of a subsection, illustrated by a dash line in Figure 11-4, can be determined independently from the other subsections. An optimal design method of the hybrid form FIR filters is presented in [Kho01].

There are some restrictions on the advantages of the hybrid form. The benefit of using the hybrid form instead of the transposed direct form is not very large if the filter coefficients can be chosen such that the increase of the word length in the output path is not necessary and if the sign bit reduction technique [Haw96] is used to reduce the loading in the input path of the transposed direct form filter. The hybrid form can not be used in the programmable filters where an immediate or simultaneous response to the change of the filter coefficient is required. It is also very complicated to write a reusable hardware description of the hybrid form filter where the length of the filter is parameterizable. In case of a linear phase FIR filter,

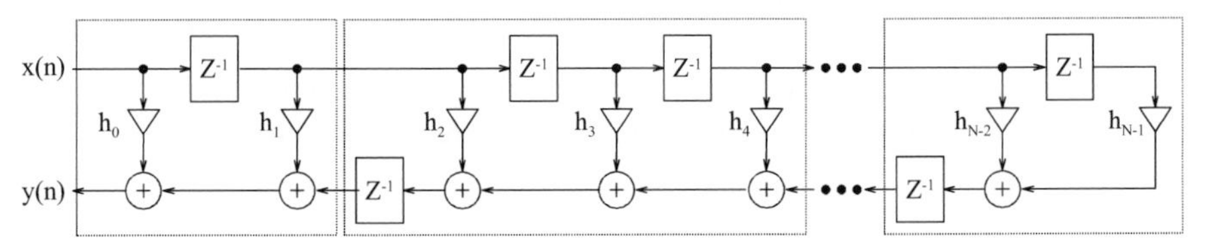

Figure 11-4. Hybrid form FIR filter.

folding the taps is possible only by introducing an extra delay element for every delay element moved to the output path. Anyhow, this increase of hardware is, in most cases, acceptable; also, the benefits of the hybrid form are more important.

11.5 Word Length Effects and Scaling

In order to get the best possible trade off between the signal to noise ratio at the output of the digital filter and the amount of hardware, the minimum accuracy needed has to be found. The noise sources are quantization errors and overflows in the FIR filters.

The mean and variance of the two's complement truncation error are [Con99]

$$m_e = -\frac{2^{-Bt} - 2^{-Bo}}{2}, \tag{11.13}$$

$$\delta_e^2 = \frac{2^{-2Bt}}{12} - \frac{2^{-2Bo}}{12}, \tag{11.14}$$

where Bt and Bo are the LSBs of the truncated word and the original, respectively. The DC offset at the FIR filter output due to the internal word length truncation is

$$m_{ne} = m_e \sum_{n=0}^{N-1} h(n), \tag{11.15}$$

where $h(n)$ is the filter coefficient. The DC offset can be removed by adding an appropriate offset at the output. The noise variance at the FIR filter output due to the internal word length quantization is

$$\delta_{ne}^2 = \delta_e^2 \sum_{n=0}^{N-1} h(n)^2. \tag{11.16}$$

The overflows in the filter output may be avoided, if the filter coefficients are scaled according to rule

$$s < \frac{1}{x_{\max} \sum_{n=0}^{N-1} |h(n)|}, \tag{11.17}$$

where x_{max} is the maximum of the input signal and s is the scaling factor. This scaling method is suitable for short FIR filters because the probability for overflow is high for a filter with a short impulse response. However, the above equation usually gives pessimistic values for the bandlimited signals. For narrowband signals, the filter coefficients are scaled according to rule

$$s < \frac{1}{x_{\max} \max[|H(e^{j\omega T})|]}. \tag{11.18}$$

The two scaling methods just discussed are the ones used in practice. However, they lead only to a reasonable scaling of the filter. In practice, simulations with actual input signals should be performed to optimize the dynamic range.

A common scheme to reduce the size of overflow errors and their harmful influence is to detect numbers outside the normal range and limit them to either the largest or smallest representable number. This scheme is referred to as saturation arithmetic [Hwa75].

11.6 Canonic Signed Digit Format

A direct implementation of an N-tap FIR filter requires N multipliers, which consume a substantial amount of the chip area and power. Furthermore, the maximum speed of operation would be severely limited without pipelining. Implementing a fixed coefficient with a multiplier is very inefficient. In order to achieve the required throughput rate and to save hardware, a multiplierless filter coefficient architecture based on the canonic signed-digit (CSD) representation of coefficients is used. A coefficient can be recoded from a binary code to a signed-digit code containing the digits {-1, 0, 1}. In the signed-digit representation, the number is represented as the sums and differences of powers-of-two having the form

$$h(n) = \sum_{i=1}^{L} c_i 2^{-i}, c_i \in \{-1, 0, 1\}, \tag{11.19}$$

where L is the number of ternary bits in the signed-digit representation. The

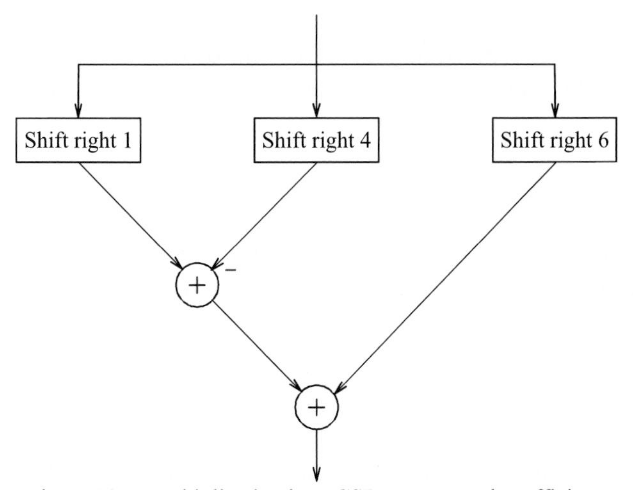

Figure 11-5. Multiplication by a CSD-represented coefficient.

benefit of the CSD representations is that the multiplication operation can be realized by using only adders/subtractors and shift operations that can be realized with hardwired shifts. The minimum signed-digit representation refers to a code requiring the minimum number of non-zero digits. There may be more than one choice of minimum signed-digit representation. A CSD representation is defined as the minimum signed-digit representation for which no two adjacent non-zero digits c_i exists. The CSD representation is unique. As an example, a multiplication by a CSD-represented coefficient 0.100$\underline{1}$01, where $\underline{1}$ denotes -1, equals to $2^{-1} - 2^{-4} + 2^{-6} = 0.4531$. The implementation of this is illustrated in Figure 11-5. The programmable CSD coefficients have been presented in [Hau93], [Tat90], [Oh95], [Kho96], [Zha01], [Dua00], [Gra01].

The advantage of the CSD representation is that most of the numbers may be represented with many fewer non-zero digits than in a conventional binary code. The number of adders or subtractors required to realize a CSD coefficient is one fewer than the number of non-zero digits in the code. The accuracy of a CSD coefficient is controlled by the number of ternary bits L and by the maximum number of non-zero digits in the representation. Several algorithms have been presented for mapping the floating point or regular two's complement presentation to the CSD presentation [Zha88], [Sam89], [Li93], [Yag96], [Che99].

11.7 Carry Save Arithmetic

The well known carry save (CS) numbers are very attractive for VLSI implementation since the basic building block for arithmetic operations is a simple full adder. The CS-principle was developed in the late 40's and early 50's for fast digital computers. The basic idea is to postpone the time consuming carry propagation (CP) from a number of multiple additions. From the view of number representation the saving of the carry word results in a redundant number representation because of the multiple alternative repre-

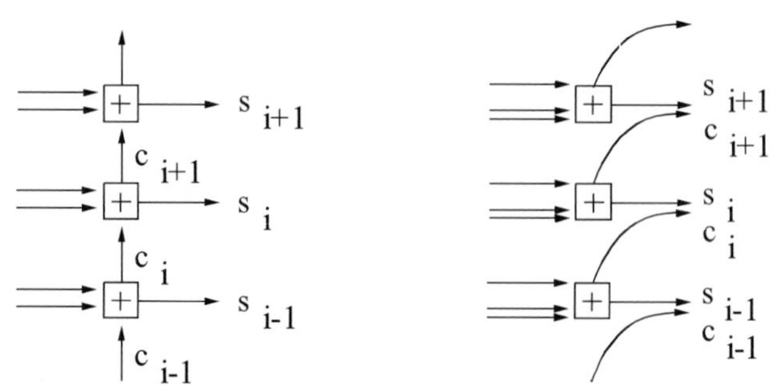

Figure 11-6. Carry ripple addition (left hand side) and 3-2 Carry Save addition (right hand side).

sentations of any given algebraic value.

In order to represent the CS digits two bits are necessary which are called c_i and s_i. The two vectors C and S given by c_i and s_i can be considered to be two two's complement numbers (or binary numbers, if only unsigned values occur). All rules for two's complement arithmetic (e.g. sign extension) apply to the C and the S number. An important advantage of CS numbers is the very simple and fast implementation of the addition operation. In Figure 11-6, a two's complement carry ripple addition and a CS addition is shown. Both architectures consist of W full adders for a wordlength of W digits. The carry ripple adder exhibits a delay corresponding to W full adder carry propagations while the delay of the CS adder is equal to a single full adder propagation delay and independent of the wordlength. The CS adder is called 3-2 adder since 3 input bits are compressed to 2 output bits for every digit position. This adder can be used to add a CS number represented by two input bits for every digit position and a usual two's complement number. Addition of two CS numbers is implemented using a 4-2 adder as shown in Figure 11-7. The CS subtraction is implemented by negation of the two two's complement numbers C and S in the minuend and addition as for two's complement numbers. It is well known that due to the redundant number representation pseudo overflows [Man89], [Nol91] can occur. A correction of these pseudo overflows can be implemented using a modified full adder cell in the most significant digit (MSD) position. For a detailed explanation the reader is referred to [Man89], [Nol91].

The conversion from the CS to two's complement numbers is achieved using a so called Vector-Merging adder (VMA) [Nol91]. This is a conventional adder adding the C and the S part of the CS number and generating a two's complement number. Since this conversion is very time consuming compared to the fast CS additions it is highly desirable to concatenate as many CS additions as possible before converting to two's complement representation.

The combination of the CS-principle with bit-level pipelining offers high computing speed, limited only by the delay time of one single (three bit) full adder. This limit can be seen as an upper throughput bound for a given tech-

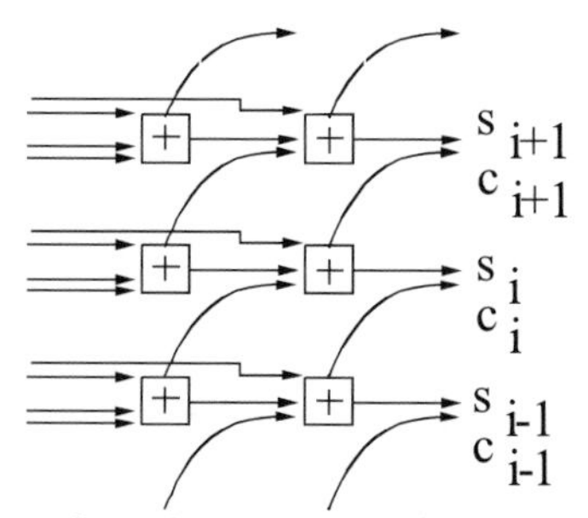

Figure 11-7. Addition of two Carry Save numbers (4-2 Carry Save addition).

nology. In today's CMOS technologies the CS-arithmetic offers the potential of clock frequencies up to several hundred MHz. But in practice the maximum frequency is limited by clocking and I/O-bandwidth problems.

In recursive structures simple pipelining is not applicable. Here the CS-arithmetic is a very attractive number representation, in order to deal with the fixed delay of the time critical recursive loops. The basic idea is to postpone the CP and therefore move the CP-path out of the recursive loops into parts of the structure where the timing restrictions are relaxed or where pipelining can be used.

A typical disadvantage of all redundant number representations is the problem of the sign detection, because a complete word level carry-propagation is still required in order to determine the sign of a redundant number. The same problem as with the sign detection occurs in saturation control. Difficulties with the sign detection and saturation control can nullify some, or all of the speed advantages of the CS-number representations (see section 6.3).

11.8 Polyphase FIR filters in Sampling Rate Converters

A FIR filter with a transfer function H(z) can be decomposed into its polyphase components as follows [Fli94]:

$$H(z) = \sum_{l=0}^{L-1} z^{-l} H_l(z^L), \tag{11.20}$$

where L is the number of polyphase components and, at the same time, the interpolation rate. One realization of a polyphase composed interpolation filter using two subfilters is shown in Figure 11-8. In this structure, the commutator takes care of the upsampling and right timing, so the separate up-samplers and the delay elements are not necessary. Using the polyphase decomposition, a lower clock rate can be used for the computation. The benefit of a lower clock rate appears as a possibility of using the pipelining/interleaving technique described in Section 11.11 or as a possibility of reducing the supply voltage in order to minimize the power dissipation. The

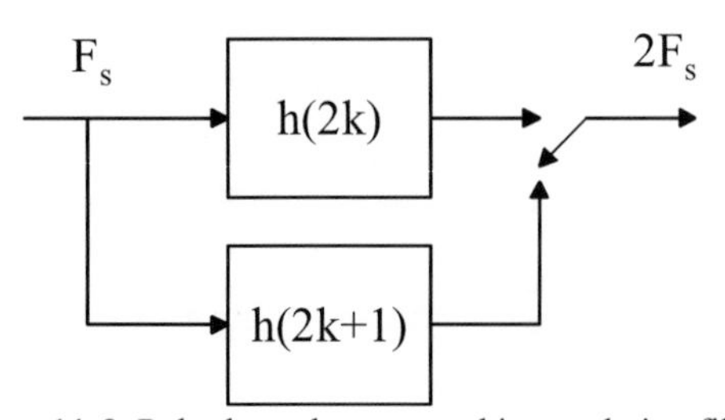

Figure 11-8. Polyphase decomposed interpolation filter.

use of a lower clock rate also relaxes the timing constraints when synthesizing the design. As described in Section 11.2, in case of the linear phase FIR filters, the symmetry of the coefficients can be exploited using the folding method to reduce the area. If the subfilters are required to be symmetrical in order to make use of folding, the maximum number of the subfilters is two, when interpolating by the factor $L = 2$ and the number of filter taps is odd [Haw96]. In case of a larger number of subfilters, the subfilter coefficients are, in fact, symmetric or mirror image pairs, i.e. $h_0(n) = h_1(N - 1 - n)$, and these symmetries can be exploited to implement efficient calculation structures, similar to the folding method [Whi00].

11.9 Half-Band Filters for Interpolation

The interpolation filters are usually implemented with multirate FIR structures. There exists an well-known multirate architecture for implementing very narrow-band FIR filters, which consists of a programmable coefficient FIR filter, and half band filters followed by the cascaded-integrator-comb (CIC) structure [Hog81]. Unfortunately, the CIC structure is not particularly suitable for implementing wideband filters because the frequency response of the CIC filter does not have satisfactory stopband attenuation. Furthermore, the CIC filter introduces droop in the passband. The problems could be alleviated by increasing the ratio of the sampling frequency to the signal bandwidth. The half-band filters have frequently been applied to increase the pre-oversampling ratio. The half band filters have the center of their transition band at the quarter of the sampling frequency. They can be designed with the filter design algorithms with the following constraints

$$\begin{aligned} \delta_S &= \delta_P, \\ \omega_S &= \pi - \omega_{P,} \end{aligned} \tag{11.21}$$

where δ_S is the ripple in the stopband from ideal response, δ_P the ripple in the passband from ideal response, ω_S the stopband edge frequency and ω_P the passband edge frequency. A "trick" for their design has been introduced in [Vai87]. The half band filters have the property that every second of their coefficients, except the center coefficient (odd N), has zero value. Therefore, these filters can be implemented with approximately half the number of multiplications than arbitrary choices of filter designs.

11.10 Cascaded Integrator Comb (CIC) Filter

In [Hog81], an efficient way of performing decimation and interpolation was introduced. Hogenauer devised a flexible, multiplier-free filter suitable for hardware implementation that can also handle arbitrary and large rate changes. These are known as cascaded integrator comb filters, or CIC filters for short. Figure 11-9 shows the basic structure of the CIC interpolation filter. A CIC interpolator would be N cascaded comb stages with a differential delay of M samples per stage running at f_s/R, followed by a zero-stuffer, followed by N cascaded integrator stages running at f_s. The differential delay is a filter design parameter used to control the filter's frequency response. In practice, the differential delay is usually held to $M = 1$ or 2. The transfer function for a CIC filter at f_s is

$$H(z) = \frac{(1-z^{-RM})^N}{(1-z^{-1})^N} = \left[\sum_{k=0}^{RM-1} z^{-k}\right]^N. \tag{11.22}$$

This equation shows that even though a CIC has integrators in it, which by themselves have an infinite impulse response, a CIC filter is equivalent to N FIR filters, each having a rectangular impulse response. Since all of the coefficients of these FIR filters are unity, and therefore symmetric, a CIC filter also has a linear phase response and constant group delay.

The power response at the output of the filter can be shown to be

$$H(f) = \left[\frac{\sin(\pi M f)}{\sin(\pi f / R)}\right]^{2N}. \tag{11.23}$$

An example power response is given in Figure 11-10 for $N = 4$ stage CIC filter with a differential delay of $M = 1$ and a rate change factor of $R = 16$. The passband cutoff is at $f_c = 0.19$ with the aliasing/imaging bands centered around the nulls at frequency of 1, 2, and 3 relative to the low sampling rate.

For a CIC interpolator, the gain, G, at the ith stage is

$$G_i = \begin{cases} 2^i & i = 1,2,\ldots,N \\ \dfrac{2^{2N-i}(RM)^{i-N}}{R} & i = N+1,\ldots,2N. \end{cases} \tag{11.24}$$

As a result, the register width, W_i, at ith stage is

$$W_i = \lceil B_{in} + \log_2 G_i \rceil, \tag{11.25}$$

where B_{in} is the input register width. The analysis of this is beyond the scope of this book, but is fully described in [Hog81].

In many CIC designs, the rate change R is programmable. Since the bit growth is a function of the rate change, the filter must be designed to handle both the largest and smallest rate changes. The largest rate change will dic-

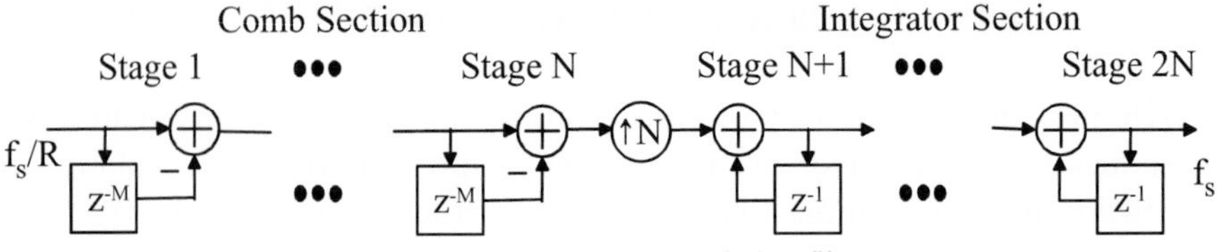

Figure 11-9. CIC interpolation filter.

tate the total bit width of the stages, and the smallest rate change will determine how many bits need to be kept in the final stage. In many designs, the output stage is followed by a shift register that selects the proper bits for transfer to the final output register.

Rounding or truncation cannot be used in CIC interpolators, except for the result, because the small errors introduced by rounding or truncation can grow without bound in the integrator sections.

Because of the passband droop, and therefore narrow usable passband, many CIC designs utilize an additional FIR filter at the low sampling rate. This filter equalizes the passband droop and performs a low rate change, usually by a factor of two to eight.

The economics of CIC filters derives from the following sources: 1) no multipliers are required; 2) no storage is required for filter coefficients; 3) intermediate storage is reduced by integrating at the high sampling rate and differentiating at the low sampling rate, compared to the equivalent implementation using cascaded uniform FIR filters; 4) the structure of comb-filters is very "regular" consisting of two basic building blocks; 5) little external control or complicated local timing is required; 6) the same filter design can easily be used for a wide range of rate change factors, N, with the addition of a scaling circuit and minimal changes to the filter timing. Some problems encountered with CIC filters include the following. 1) Register widths can become large for large rate change factors, R. 2) The frequency response is fully determined by only three integer parameters (R, M, and N),

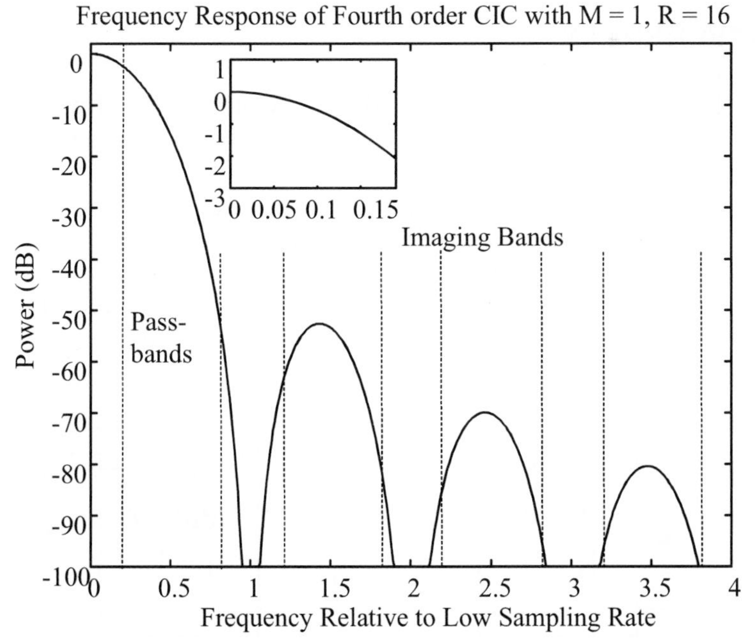

Figure 11-10. Example frequency response for $N = 4$, $M = 1$, $R = 16$.

resulting in a limited range of filter characteristics.

Filter sharpening can be used to improve the response of a CIC filter. This technique applies the same filter several times to an input to improve both passband and stopband characteristics. The interested reader is referred to [Kwe97] for more details.

11.11 Pipelining/Interleaving

The pipelining/interleaving (P/I) technique can be used to reduce the amount of hardware when the same filtering operation is performed for several independent data streams [Par89], [Jia97]. The idea is to use the same hardware for all K data streams, as shown in Figure 11-11. This may be achieved with some additional logic. The data streams are interleaved in a serial form in time. The filtering operations are made with a K times higher clock frequency; after filtering, the data stream is de-interleaved back into K parallel data streams. This serial-in-time computation requires K register elements instead of one register in the filter structure. In addition to that, extra logic is needed in the interleave/de-interleave interfaces. The expense of this technique is the increased power consumption. This increase accrues because the number of delay elements remains same, but the clock frequency is doubled. The increased power consumption due to the doubled clock frequency in the other elements such as in the taps and adders is compensated by the halved number of elements.

REFERENCES

[Che82] P. R. Chevillat, and G. Ungerboeck, "Optimum FIR Transmitter and Receiver Filters for Data Transmission over Band-Limited Channels," IEEE Trans. Commun., Vol. 30, No. 8, pp. 1909-1915, Aug. 1982.

[Che99] C. Chen, and A. Willson, "A Trellis Search Algorithm for the Design of FIR Filters with Signed-Powers-of-Two Coefficients," IEEE Transactions on Circuits and Systems-II, Vol. 46, No. 1, pp. 29-39, Jan. 1999.

[Con99] G. A. Constantinides, P. Y. K. Cheung, and W. Luk, "Truncation noise in Fixed-Point SFGs [digital filters]," Electron. Lett., Vol. 35, No. 23,

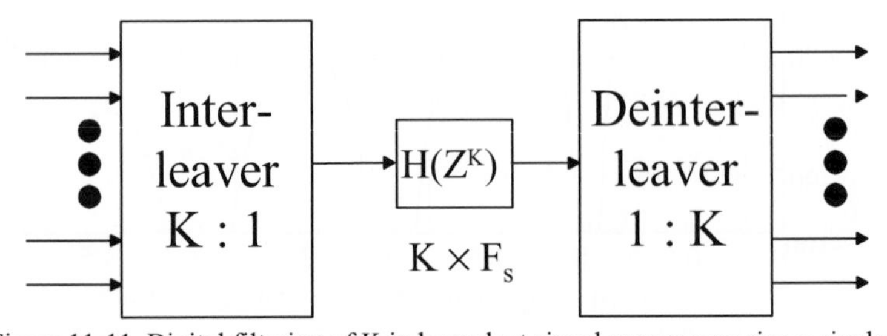

Figure 11-11. Digital filtering of K independent signal sequences using a single filter.

pp. 2012 -2014, Nov. 1999.
[Dua00] J. N. Duan, L. Ko, and B. Daneshrad, "A Highly Versatile Beam-forming ASIC for Application in Broad-Band Fixed Wireless Access Systems," IEEE J. of Solid State Circuits, Vol. 35, No. 3, pp. 391-400, Mar. 2000.
[Fli94] N. J. Fliege, "Multirate Digital Signal Processing," John Wiley & Sons, 1994.
[Gra01] E. Grayver, and B. Daneshrad, "VLSI implementation of a 100-μW Multirate FSK Receiver," IEEE J. of Solid State Circuits, Vol. 36, No. 11, pp. 1821-1828, Nov. 2001.
[Har96] R. Hartley, "Subexpression Sharing in Filters Using Canonic Signed Digit Multipliers," IEEE Transactions on Circuits and Systems-II, Vol. 43, No. 10, pp. 677-688, Oct. 1996.
[Hat01] M. Hatamian, "Efficient FIR Filter for High-Speed Communication," U. S. Patent 6,272,173, Aug. 7, 2001.
[Hau93] J. C. Hausman, R. R. Harnden, E. G. Cohen, and H. G. Mills, "Programmable Canonic Signed Digit Filter Chip," U. S. Patent 5,262,974, Nov. 16, 1993.
[Haw96] R. Hawley, B. Wong, T. Lin, J. Laskowski, and H. Samueli, "Design Techniques for Silicon Compiler Implementations of High-Speed FIR Digital Filters," IEEE Journal of Solid-State Circuits, Vol. 31, No. 5, pp. 656-667, May 1996.
[Hog81] E. B. Hogenauer, "An Economial Class of Digital Filters for Decimation and Interpolation," IEEE Trans. Acoust., Speech, Signal Process., Vol. ASSP-29, No. 2, pp. 155-162, Apr. 1981.
[Hwa75] S. Hwang, "On Monotonicity of L_p and l_p Norms," IEEE Transactions on Acoustics, Speech, and Signal Processing, Dec. 1975, pp. 593-594.
[Jia97] Z. Jiang, and A. Willson, "Efficient Digital Filtering Architectures Using Pipelining/Interleaving," IEEE Transactions on Circuits and Systems-II, Vol. 44, No. 2, pp. 110-119, Feb. 1997.
[Kho01] K. Y. Khoo, Z. Yu, and A. Willson, "Design of Optimal Hybrid Form FIR Filter," IEEE International Symposium on Circuits and Systems 2001, Vol. 2, pp. 621-624.
[Kho96] K. Y. Khoo, A. Kwentus, and A. N. Willson Jr., "A Programmable FIR Digital Filter using CSD Coefficients," IEEE J. of Solid State Circuits, Vol. 31, No. 6, pp. 869-874, June 1996.
[Kwe97] A. Y. Kwentus, Z. Jiang, and Alan N. Wilson, Jr. , "Application of Filter Sharpening to Cascaded Integrator-Comb Decimation Filters," IEEE Transactions on Signal Processing, Vol. 45, No. 2, pp. 457-467, 1997.
[Lee96] H. R. Lee, C. W. Jen, and C. M. Liu, "A New Hardware-Efficient Architecture for Programmable FIR Filters," IEEE Transactions on Circuits and Systems-II, Vol. 43, No. 9, pp. 637-644, Sept. 1996.

[Li93] D. Li, J. Song, and Y. Lim, "Polynomial-Time Algorithm for Designing Digital Filters with Power-of-Two Coefficients," In Proc. IEEE International symposium on Circuits and Systems 1993, Vol. 1, May 1993, pp. 84-87.

[Man89] E. D. Man, and T. Noll, "Arrangement for Bit-Parallel Addition of Binary Numbers with Carry-Save Overflow Correction," U. S. Patent 4,888,723, Dec. 19, 1989.

[Mor95] F. Moreau de Saint-Martin, and P. Siohan, "Design of Optimal Linear-Phase Transmitter and Receiver Filters for Digital Systems," IEEE International Symposium on Circuits and Systems, 1995, Vol. 2, pp. 885 -888.

[Nol91] T. Noll, "Carry-Save Architectures for High-Speed Digital Signal Processing," Journal of VLSI Signal Processing, Vol. 3, pp. 121-140, June 1991.

[Oh95] W. J. Oh, and Y. H. Lee, "Implementation of Programmable Multiplierless FIR Filters with Powers-of-Two Coefficients," IEEE Trans. Circuits and Systems II, Vol. 42, No. 8, pp. 553 - 556, Aug. 1995.

[Par89] K. Parhi, and D. Messerschmitt, "Pipeline Interleaving and Parallelism in Recursive Digital Filters-Part I: Pipelining Using Scattered Look-Ahead and Decomposition," IEEE Transactions on Acoustics, Speech, and Signal Processing, Vol. 37, No. 7, pp. 1099-1117, July 1989.

[Sam88] H. Samueli, "On the Design of Optimal Equiripple FIR Digital Filters for Data Transmission Applications," IEEE Transactions on Circuits and Systems, Vol. 35, No. 12, pp. 1542-1546, Dec. 1988.

[Sam89] H. Samueli, "An Improved Search Algorithm for the Design of Multiplierless FIR Filters with Powers-of-Two Coefficients," IEEE Transactions on Circuits and Systems, Vol. 36, No. 7, pp. 1044-1047, Jul. 1989.

[Sam91] H. Samueli, "On the Design of FIR Digital Data Transmission Filters with Arbitrary Magnitude Specifications," IEEE Transactions on Circuits and Systems, Vol. 38, No. 12, pp. 1563-1567, Dec. 1991.

[Tat90] L. R. Tate, "Truncated Product Partial Canonical Signed Digit Multiplier," U. S. Patent 4,967,388, Oct. 30, 1990.

[Vai87] P. Vaidyanathan, and T. Nguyen, "A "Trick" for the Design of FIR Half-Band Filters," IEEE Transactions on Circuits and Systems, Vol. 34, No. 3, pp. 297-300, Mar. 1987.

[Vai93] P. P. Vaidyanathan, "Multirate Systems and Filter Banks," Prentice-Hall, 1993.

[Whi00] B. A. White, and M. I Elmasry, "Low-Power Design of Decimation Filters for a Digital IF Receiver," IEEE Transactions on Very Large Scale Integration (VLSI), Vol. 8, No. 3, pp. 339-345, June 2000.

[Yag96] M. Yagu, A. Nishihara, and N. Fujii, "Design of FIR Digital Filters Using Estimates of Error Function over CSD Coefficient Space," IEICE Transactions on Fundamentals, Vol. E79-A, No. 3, pp. 283-290, Mar. 1996.

[Zha01] T. Zhangwen, Z. Z. Zhanpeng, Z. Jie, and M. Hao "A High-Speed, Programmable, CSD Coefficient FIR Filter," in Proc. 13th Annual IEEE International ASIC/SOC Conference, Sept. 2001, pp. 397-400.
[Zha88] Q. Zhao, and Y. Tadokoro, "A Simple Design of FIR Filters with Powers-of-Two Coefficients," IEEE Transactions on Circuits and Systems, Vol. 35, No. 5, pp. 566-570, May 1988.

Chapter 12

12. RE-SAMPLING

The multi-standard modulator has to be able to accept data with different symbol rates. This fact leads to the need for a variable interpolator that performs a conversion between variable sampling frequencies. Furthermore, because the symbol rates in the different systems are not multiples of each other, the interpolation ratio has to be a rational number. An interpolator that is used for the sampling rate conversion is often called a re-sampler.

For example, the symbol rate of the GSM is 270.833 ksym/s and the symbol rate of the WCDMA is 3.84Msym/s. These rates are not multiples of each other. If these rates need to be converted to the sample rate 76.8MHz, the interpolation ratios for the GSM and WCDMA modes have to be 283.56 and 20, respectively. The ratio of the sampling frequency to the signal bandwidth determines, in part, the complexity of an interpolator with a rational number [Ves99]. Therefore, in these particular cases, an integer rate interpolator should be used to increase the pre-oversampling ratio [Ves99], [Sam00], [Cho01], [Ram84]. The sampling rate can first be raised using power-of-two ratios, for example. An efficient integer ratio interpolator is fairly easy to implement using a polyphase decomposition, as discussed in Section 11.8.

There are several methods of realizing a re-sampler with an arbitrary sampling rate conversion. In this chapter, the design of the polynomial-based interpolation filter using the Lagrange method is presented. Some other polynomial-based methods are also discussed.

Even though the concentration in this chapter is focused on the polynomial-based approaches, these are not the only ways to implement an interpolator with a rational number ratio. One well-known approach presented in most of the digital signal processing (DSP) books, in [Mit98], for example, is to cascade an interpolator of an integer ratio L and a decimator of an integer ratio M to form an interpolator with a rational number ratio L/M. Usually the multistage techniques are most efficient for these kind of solutions [Mit98]. In [Gra01], a similar approach in which the decimation ratio M can be a mixed integer/fractional number is used. In this system, the nearest interpolated value is selected as an output value and the exact value at the time instant is not calculated, which introduces some timing jitter. The methods described in [Che93] and [Che98] allow the decimation and interpolation factors to be prime numbers while preserving the multistage efficiency. Programmable cascaded integrator-comb filters are used for interpolation with a ratio of variable rational numbers in [Hen99] and [Ana01].

12.1 Interpolation for Timing Adjustment

When a sampling rate is to be converted, new samples using the known input samples need to be calculated. An interpolator calculates one interpolant $y(kT_{out})$ at a time using N input samples. Let's first consider a continuous-time signal that is sampled with a sampling frequency $F_s = 1/T_s$. The output interpolants of an ideal interpolator are the same as if the original continuous-time signal had been sampled at the sampling frequency $F_{out} = 1/T_{out}$ [Ves96a].

The time of the interpolant is governed by the fractional interval μ_k in Figure 12-1 and defined by

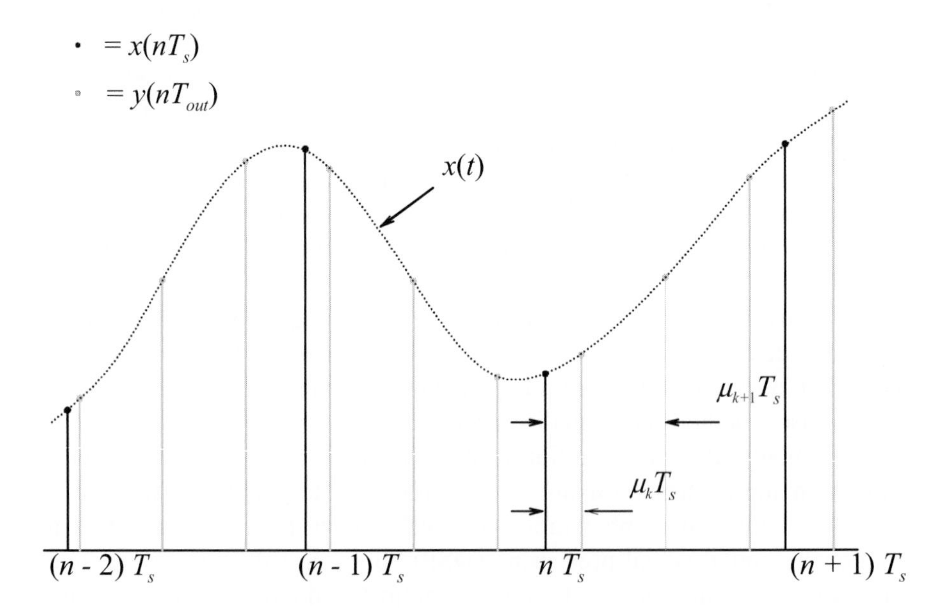

Figure 12-1. Interpolation in time-domain.

$$\mu_k = \frac{(kT_{out} - nT_s)}{T_s}, \tag{12.1}$$

where the fractional interval is $\mu_k \in [0,1)$, T_s the input sample interval, T_{out} the output sample interval and n the largest integer for which $nT_s \leq kT_{out}$. The output of the interpolator is

$$y(kT_{out}) = \sum_{i=-N/2+1}^{N/2} x[(n-i)T_s]\, h_a[(i+\mu_k)T_s], \tag{12.2}$$

where $x[(n - i)T_s]$ and $h_a[(i+ \mu_k)T_s]$ for $i = -N/2+1,...,N/2$ are the N input samples, i.e. basepoints, and the N samples of the continuous impulse response of the interpolation filter, respectively.

12.2 Interpolation Filter with Polynomial-Based Impulse Response

As can be seen from Figure 12-1, in case of the arbitrary re-sampling rate, the fractional interval μ_k has to be controlled on-line during the computation. Equation (12.2) states that the continuous-time impulse response $h_a(t)$ is a function of μ_k. This means that the filter coefficients of the interpolation filter must change for every new value of μ_k. This can be achieved by calculating the coefficients for all possible values of μ_k and storing these into a lookup table. A new set of coefficients is then read into the filter structure every time μ_k changes. Of course, this leads to a need for a huge coefficient memory, particularly when using long filters and a high resolution of the fractional interval. To avoid the very inefficient implementation using the coefficient memory, the filter coefficients can be calculated on-line using a polynomial-based interpolation.

The idea of the polynomial-based interpolator is that the continuous-time impulse response $h_a(t)$ of the interpolation filter is expressed in each interval of length T_s by means of polynomial [Ves99]

$$h_a[(i+\mu_k)T_s] = \sum_{m=0}^{M} c_m(i)(\mu_k)^m \tag{12.3}$$

for $i = -N/2+1, -N/2+2,..., N/2$ and for $\mu_k \in [0,1)$. $c_m(i)$ denotes the polynomial coefficients for $h_a[(i + \mu_k)T_s]$ and M is the degree of polynomial. The coefficients $c_m(i)$ are constants and independent of μ_k. This leads to the most attractive feature of the polynomial-based interpolation filters; they can be implemented very efficiently using the so-called Farrow structure [Far98] described in Section 12.3.

12.2.1 Lagrange Interpolation

One of the best known polynomial-based interpolation methods is the Lagrange interpolation. An advantage of the Lagrange interpolation is that the polynomial coefficients can be determined in a closed form, and therefore no optimization is needed [Ves99]. On the other hand, this leads to a fixed frequency response that cannot be controlled by any other design parameter than the degree of the polynomial M. All the same, the Lagrange interpolators provide a fairly good approximation of the continuous-time signal at low frequencies, i.e. far away from half the sampling rate [Ves99]. It is also a maximally flat design of the interpolation filters or the fractional delay finite impulse response (FIR) filters [Väl95].

The approximating polynomial of degree M, denoted by $y_a(t)$, can be obtained by using the following $M+1$ time-domain conditions

$$y_a(iT_s) = x(i), \quad \text{for } n - \frac{N}{2} + 1 \leq i \leq n + \frac{N}{2}, \tag{12.4}$$

which means that the reconstructed signal value is exactly the value of the input signal at the sampling instants. To minimize the approximation error, only the middle interval $nT_s \leq \mathrm{t} < (n+1)T_s$ of the approximating polynomial is normally used. In order to uniquely determine the middle interval, the length of the filter $N = M + 1$ has to be even. Even though not recommended in [Eru93], for instance, it is also possible to use the Lagrange interpolator with an odd N. In this case, the locations of the filter taps have to be changed when μ_k passes the value 0.5 to maintain μ_k within half a sample from the center point of the filter [Väl95]. This is practically difficult to implement and it also causes a discontinuity to the output signal [Väl95].

The impulse response $h_a(t)$ of the interpolation filter based on the Lagrange interpolation can be derived by using the conditions of Equation (12.4). It is a piecewise polynomial and can be expressed as (12.3),

$$h_a[(i+\mu_k)T_s] = \sum_{m=0}^{M} c_m(i)(\mu_k)^m \tag{12.5}$$

for $i = -N/2+1, -N/2+2, \ldots, N/2$. The so-called Lagrangian coefficient functions can be determined from the following equation

$$\sum_{m=0}^{M} c_m(i)x^m = \prod_{\substack{j=-N/2+1 \\ j \neq i}}^{N/2} \frac{j-x}{-i+j}. \tag{12.6}$$

One of the possible derivations of (12.6) is shown in Appendix [Väl95]. The coefficients $c_m(i)$ of a cubic, i.e. third order $M = 3$, Lagrange interpolator are calculated in Table 12-1. The impulse and magnitude responses of the linear (M = 1) and cubic (M = 3) Lagrange interpolators are depicted in

Table 12-1. Filter coefficients for a cubic Lagrange interpolator.

	$m = 0$	$m = 1$	$m = 2$	$m = 3$
$c_m(-1)$	0	-1/3	1/2	-1/6
$c_m(0)$	1	-1/2	-1	1/2
$c_m(1)$	0	1	1/2	-1/2
$c_m(2)$	0	-1/6	0	1/6

Figure 12-2 and Figure 12-3, respectively. One important property of the interpolator, when using it as a sampling rate converter, is the phase linearity. Even though the underlying continuous time impulse response of the Lagrange interpolator is symmetric and thus linear phase, the phase response of the interpolator is not linear for all μ_k. Fortunately, it is phase linear on average. The average phase linearity can also be seen from the phase responses of the cubic Lagrange interpolation filter for μ_k =0, 0.1, 0.2, 0.3, 0.4, 0.5, 0.6, 0.7, 0.8, 0.9, 1 depicted in Figure 12-4.

12.3 Farrow Structure

The Farrow structure is a very efficient implementation structure for a polynomial-based interpolator. It suits well in real-time systems because it is fast and it has fixed coefficient values. A Farrow structure can be found for any polynomial-based filter [Ves99].

Substituting (12.5) into (12.2), the following is obtained

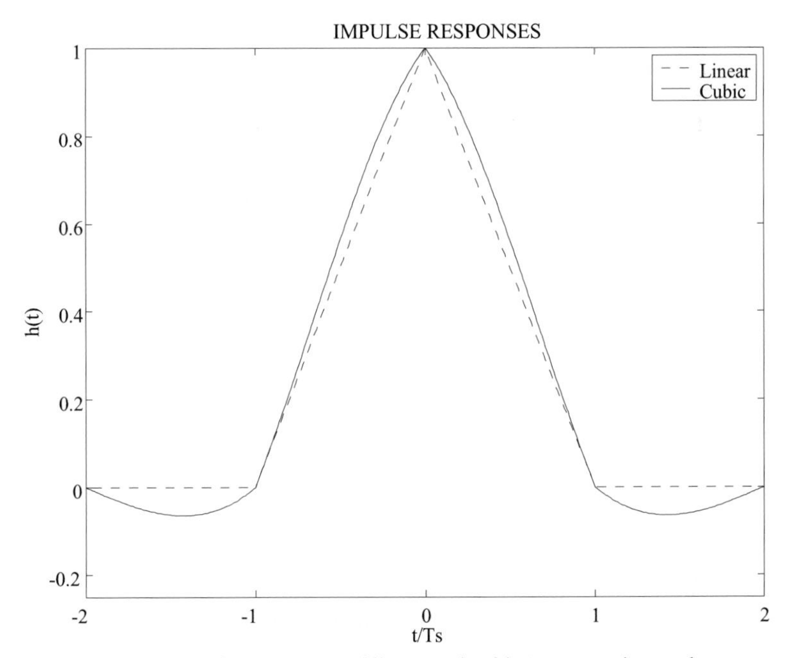

Figure 12-2. Impulse responses of linear and cubic Lagrange interpolators.

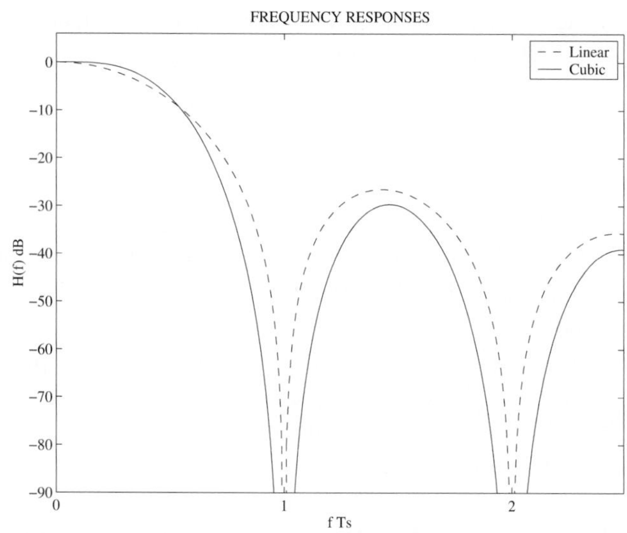

Figure 12-3. Magnitude responses of linear and cubic Lagrange interpolators.

$$y(kT_{out}) = \sum_{i=-N/2+1}^{N/2} x[(n-i)T_s] \sum_{m=0}^{M} c_m(i)(\mu_k)^m = \sum_{m=0}^{M} v_m(n)(\mu_k)^m, \quad (12.7)$$

where

$$v_m(n) = \sum_{i=-N/2+1}^{N/2} x[(n-i)T_s] c_m(i). \quad (12.8)$$

(12.7) and (12.8) indicate that the polynomial-based interpolation filter

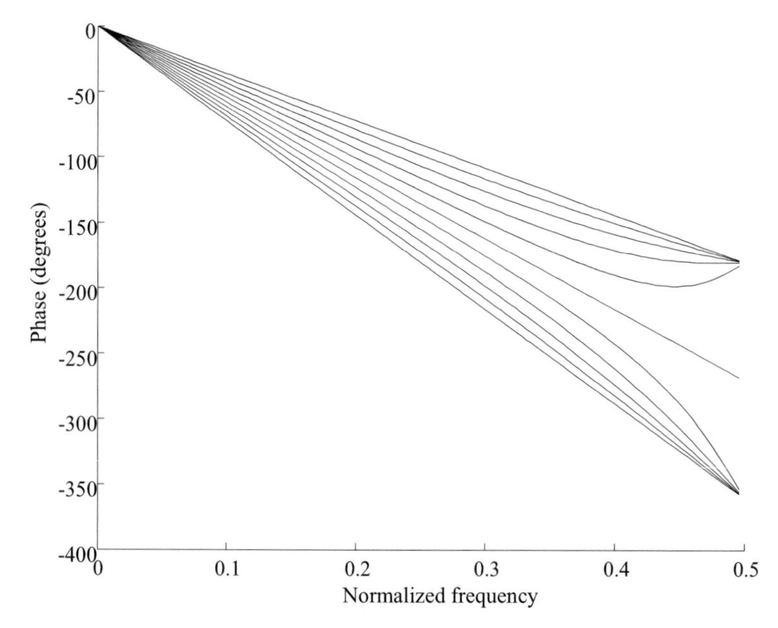

Figure 12-4. Phase response of cubic Lagrange interpolator for different μ_k values.

can be realized as $M+1$ parallel FIR filters of length N having the coefficient values $c_m(i)$. The transfer functions of these filters are given by

$$C_m(z)=\sum_{i=0}^{N-1} c_m(-i+N/2)\,z^{-i}, \text{ for } m=0,1,\ldots,M. \tag{12.9}$$

The interpolant $y(kT_{out})$ is obtained by multiplying the output samples of the parallel filters $v_m(n)$ by $(\mu_k)^m$ for $m = 0, 1,\ldots, M$. The Farrow structure realization of the interpolator described above is depicted in Figure 12-5. This efficient structure is implemented using the Horner's method [Väl95]

$$\sum_{m=0}^{M} v_m(n)(\mu_k)^m = v_0(n)+[v_1(n)+[v_2(n)+\ldots+[v_{M-1}(n)+v_M(n)\mu_k]\mu_k]\mu_k\ldots]\mu_k \tag{12.10}$$

The number of the FIR filter coefficients in the Farrow structure is $N(M+1)$. These coefficients are fixed and the multiplications can be efficiently implemented as, for example, Canonic Signed Digit taps described in Section 11.6. In addition to these, M multiplications by the fractional interval μ_k are needed. This results in the total number of multiplications $M+N(M+1)$. This is a maximal number; usually some of the coefficients are 0 or ±1 and they do not have to be implemented, or two coefficients may have the same value and thus the hardware can be shared [Väl95]. The unit delay elements of the filters $C_m(z)$ can also be shared in order to reduce the amount of hardware.

Some modifications to the Farrow structure are presented in [Ves99] and [Väl95]. The modification in [Ves99] results in the symmetric or antisymmetric impulse responses of the FIR filters $C_m(z)$ in the Farrow structure if $h_a(t)$ is symmetrical, i.e. if they are linear-phase Type II or Type IV filters.

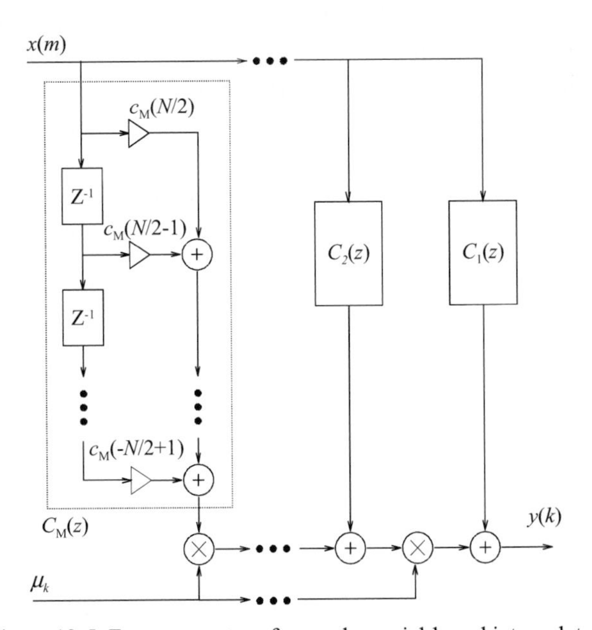

Figure 12-5. Farrow structure for a polynomial-based interpolator.

This is generally not the case in the Farrow structure. The modification in [Väl95] aims for a more efficient implementation by means of changing the range of the fractional interval.

12.4 Alternative Polynomial Interpolators

There are many polynomial-based interpolation methods other than the classical Lagrangian method described in Section 12.2.1. Actually, the Lagrange interpolator is optimal only in the maximally flat sense [Ves99]. In this section, a number of other polynomial-based interpolation methods are presented briefly, and some of their pros and cons are discussed. The advantages of the Farrow structure in a very large scale integration (VLSI) implementation gives grounds for concentration only in the polynomial-based methods.

In addition to the Lagrange interpolation, another well-known time-domain approach is the B-spline interpolation (see Figure 12-6). This approach, together with some modifications of it, is described in, for example, [Hou78], [Ves99], [Ves95], [Uns91], [Egi96] and [Got01]. In this approach, the B-spline functions are used as the interpolation functions. As a time-domain approach, the B-spline interpolation suffers from the same lack of flexibility as the Lagrange interpolation. After giving the degree of interpolation M, the polynomial coefficients are uniquely determined. On the other hand, this can be considered as an advantage, because no optimization is needed. The maximal total number of multiplications in the Farrow structure is $M + N(M + 1) = M + (M + 1)^2$ [Ves99], as is the case also with the Lagrange interpolation, but a prefilter is needed before the Farrow structure in order to calculate the B-spline coefficients. The non-stable and non-causal IIR prefilter can be in most cases approximated using a causal FIR filter of

Table 12-2. Comparison of computational complexity.

	Cubic Lagrange (Figure 16-7)	Cubic B-spline (Figure 12-6)	Vesma-Saramäki (Figure 12-7)
Delay Elements	3	11	3
Scale by constant	6	11	7
Scale by constant (not powers of two)	3	8	6
Add/Subtract	11	19	16
Multiple/Divide	3	3	3
Passband Ripple (dB)	0.003	0.048	0.001
Image (dB) Stopband	-77.94	-94.28	-92.86

length 9 [Ves95]. The B-spline interpolators have a better stopband attenuation than the Lagrange interpolators of the same degree [Ves99]. Actually, the stopband attenuation of the Lagrange interpolator cannot be greatly improved by increasing the degree of interpolation. The main problems of both the Lagrange and B-spline interpolations are a wide transition band and fairly poor frequency selectivity for the wideband signals.

A better frequency-domain behavior for the wideband signals can be achieved by using a frequency-domain approach. The polynomial coefficients of the impulse response $h_a(t)$ can be optimized directly in the frequency domain. These methods have usually at least three design parameters, which makes them more flexible compared to the time-domain methods with only one design parameter. These parameters include the length of the filter N, the degree of the interpolation M, which is not tied to the length of the filter as is the case with the time-domain methods, and the passband edge ω_p. Some methods also allow the definition of the passband and stopband regions. In [Ves96a], [Far98], [Laa96] and [Ves98], for example, several frequency domain design methods are discussed. All these interpolators are mentioned as offering a better frequency response for the wideband signals than the filters based on the Lagrange or B-spline interpolations [Ves99]. An interpolator designed using the method presented in [Ves96a] is taken here as an example. It is referred to here as the Vesma-Saramäki interpolator (see Figure 12-7). The output instant for the interpolant $y(kT_{out})$ in Figure 12-7 is

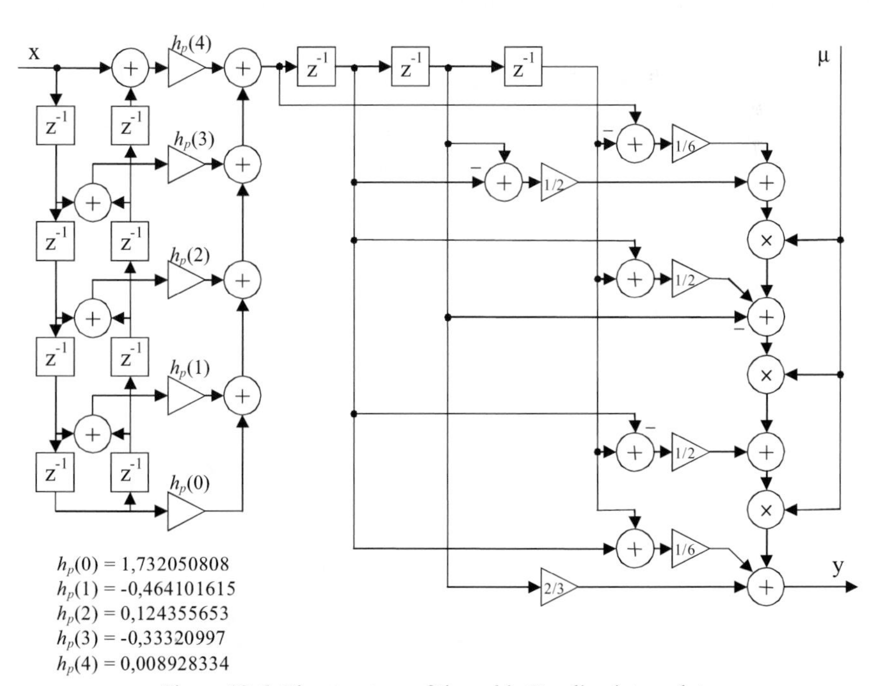

Figure 12-6. The structure of the cubic B-spline interpolator.

controlled by $(2\mu_k - 1)$, and not by μ_k like for the Lagrange and B-spline interpolators. The filter coefficients in Figure 12-7 were designed so that the frequency-domain error function is minimized in the minimax sense [Ves96a], [Ves00].

The Computational Complexity and the maximum level of the attenuated images of the cubic Lagrange (see Figure 16-7), cubic B-spline (see Figure 12-6) and Vesma-Saramäki (see Figure 12-7) interpolators are compared in Table 12-2. The interpolators are used as a resampler in GSM/EDGE/WCDMA modulator in Figure 16-6. The image rejection specification is more than 75 dB in WCDMA mode. The interpolation filter specifications in Table 12-2 are as follows: the passband is $[0\ 0.0625F_s]$, the stopbands are $[0.9375\ F_s,\ 1{,}0625\ F_s]$ and $[1.9375\ F_s,\ 2{,}0625\ F_s]$, the maximum passband deviation is 0.05 dB, and the minimum stopband attenuation is 75 dB. The magnitude responses of the cubic Lagrange (see Figure 16-7), cubic B-spline (see Figure 12-6) and Vesma-Saramäki (see Figure 12-7) interpolators are depicted in Figure 12-8. One important property of the interpolator, when using it as a sampling rate converter, is the phase linearity. The continuous time impulse responses of the cubic Lagrange, cubic B-spline and Vesma-Saramäki interpolators are symmetric around $t = 0$ (see Figure 12-8) and thus they have linear phase. The elements of the prefilter are added to the number of elements of the B-spline interpolator in Table 12-2. The number of operations only gives an overview, because they depend on the implementation of the interpolator and thus there is no unique way to calculate

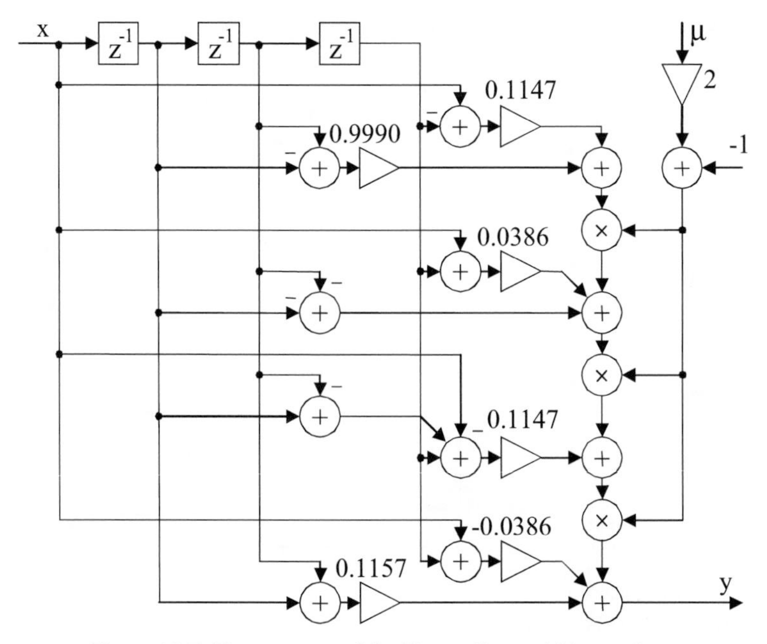

Figure 12-7. The structure of the Vesma-Saramäki interpolator.

them. Scaling by a constant is considered a fairly simple operation that can be performed using the multiplierless coefficients described in Section 11.6. The passband ripple is here the maximum value of the frequency response minus unity. The values in Table 12-2 indicate that the cubic Lagrange interpolator is the most appropriate method of these with a sufficient performance and an efficient realization. It is therefore used in Figure 16-6.

Even though the results in some articles, in [Ves96b], for example, indicate that the frequency domain approaches are better in performance and in computational complexity, this was not the result in this case. The values in Table 12-2 indicate that the time-domain approach, the B-spline interpolation and the Lagrange interpolation give performance similar to the frequency domain approach Vesma-Saramäki interpolation. A reason for this discrepancy in the results is that, in this case, the sampling rate is increased before the re-sampler by an integer ratio in Figure 16-6 and so the ratio of the signal bandwidth to the sampling rate is smaller ($0.0625F_s$).

Some other polynomial interpolation methods worthy of a mention are the piece-wise parabolic explained and used in [Eru93], [Cho01], for example, and the use of the trigonometric functions as interpolation functions in, for example, [Fu99]. The disadvantage of the interpolation filters based on

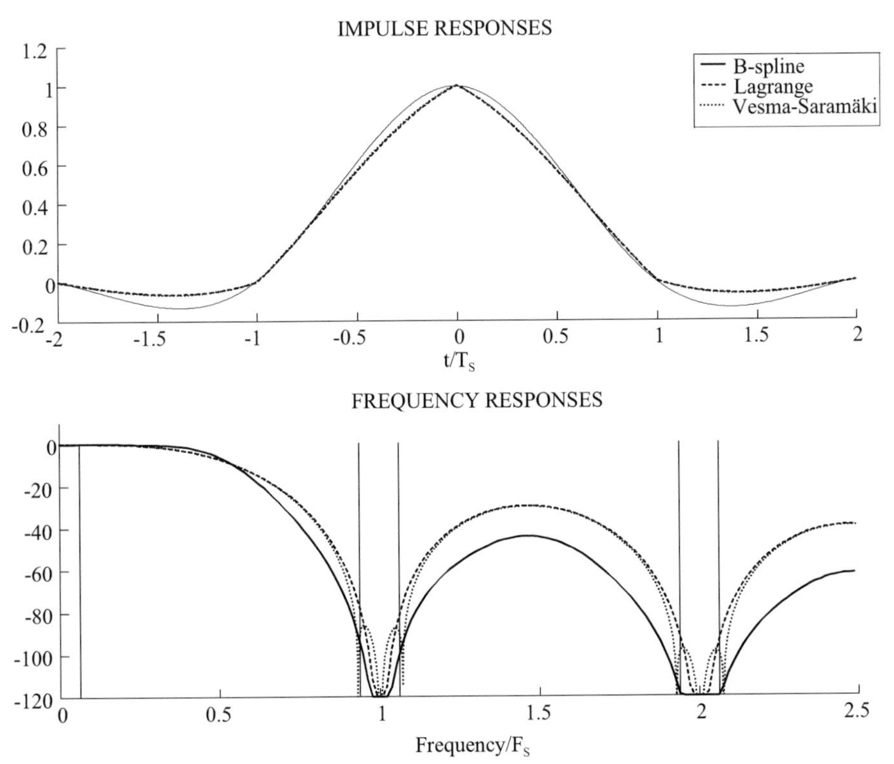

Figure 12-8. Impulse and magnitude responses of cubic Lagrange (see Figure 16-7), cubic B-spline (see Figure 12-6) and Vesma-Saramäki (see Figure 12-7) interpolators.

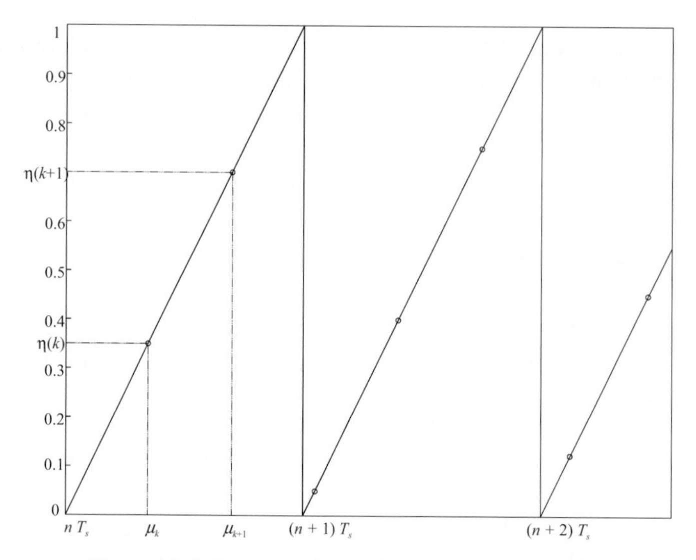

Figure 12-9. Output of the NCO for control word 0.35.

the trigonometric interpolation is that they do not offer good frequency selectivity for wideband signals, i.e. for signals with the highest baseband frequency component close to half the sampling rate [Ves99]. In [Sne99], an average of the overlapping interval of multiple polynomials fitted to different input samples is used for the interpolation. This way lower-order polynomials can be used to calculate the interpolant using the same number of input samples that is used by one higher-order polynomial. This approach is mentioned to provide a better aliasing rejection with only slightly more complex coefficient calculation than using only one lower-order polynomial.

12.5 Calculation of Fractional Interval μ_k Using NCO

As shown by (12.3), in case of a polynomial-based interpolation, there is no need to calculate the polynomial coefficients $c_m(i)$ on-line. The only variables in the interpolation (12.2) are the fractional interval $\mu_k \in [0,1)$ and the index n of the input samples. The approximations of these variables can be calculated using a numerically controlled oscillator (NCO) [Gar93]. The output of the NCO, denoted as $\eta(k)$, is defined

$$\eta(k+1) = \text{mod}[\eta(k) + \Delta\phi, 1], \tag{12.11}$$

where the positive $\Delta\phi$ is the NCO control word and mod[] denotes the modulus operation. The NCO is clocked with the output sampling rate F_{out} of the interpolator. The output of the NCO, which is a positive fraction, will be incremented by an amount $\Delta\phi$ at every T_{out} seconds and the register will overflow every $1/\Delta\phi$ clock ticks, on average. The control word $\Delta\phi$ can be calculated using the equation

$$\Delta\phi = \frac{F_s}{F_{out}}. \tag{12.12}$$

Figure 12-9 shows the output η of the NCO for a control word $\Delta\phi = 0.35$. When an overflow occurs, the MSB of the binary NCO output word $\eta(k)$ switches from the state '1' to the state '0'. As mentioned, this happens on average at intervals

$$\frac{1}{\Delta\varphi} T_{out} = \frac{F_{out}}{F_s} T_{out} = T_s. \tag{12.13}$$

(12.13) implies that the MSB of the NCO output $\eta(k)$ can be used as the sampling clock of the interpolator (it must be inverted when a rising edge logic is used). This clock controls the shift registers in the Farrow structure shown in Figure 12-5 and therefore takes care of the index n. Because the output of the NCO can change only at the clock tick, (12.13) holds only on average, introducing some jitter to the clock derived from the MSB [Gar93].

From the triangles in Figure 12-9 the following equation can be derived

$$\frac{\mu_k T_s}{\eta(k)} = \frac{\mu_{k+1} T_s}{\eta(k+1)} = \frac{\mu_{k+1} T_s}{\eta(k) + \Delta\phi}, \tag{12.14}$$

where

$$\eta(k) = \frac{\mu_k \Delta\phi}{\mu_{k+1} - \mu_k}. \tag{12.15}$$

By substituting Equation (12.12) and (12.1) into the denominator, it follows

$$\eta(k) = \frac{\mu_k F_s / F_{out}}{T_{out} / T_s} = \mu_k. \tag{12.16}$$

(12.16) shows that the NCO output $\eta(k)$ is a direct approximation of the

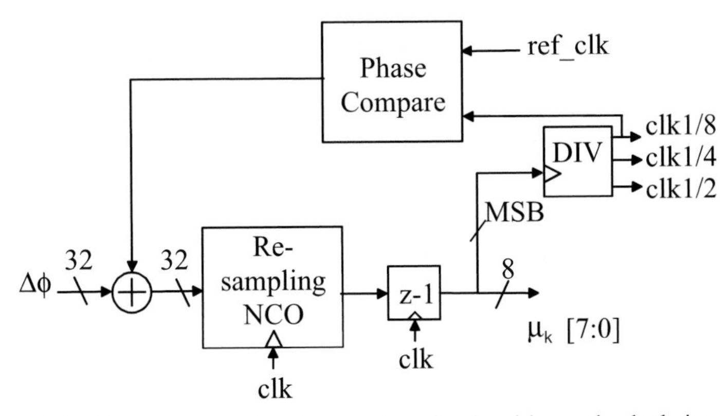

Figure 12-10. NCO with the synchronizer used for fractional interval calculation and clock generation.

fractional interval μ_k. Since it is not possible to represent all frequencies exactly, the accuracy of the approximation depends on the word length of the NCO [Int01]. The maximum error of the control word $\Delta\phi$ is half the LSB. This error can be corrected by using a timing recovery loop and adding the derived error signal to the control word $\Delta\phi$.

12.5.1 Synchronization of Resampling NCO

The infinite word length of the digital control word $\Delta\phi_{bin}$ makes it not always possible to have a binary control word representing the ideal $\Delta\phi$ exactly. This means that the input registers of the interpolator are sampling either a little too fast or a little too slow. The error of the closest binary control word is

$$\varepsilon_{\Delta\phi} = \left|\Delta\phi - \Delta\phi_{bin}\right|, \tag{12.17}$$

where

$$\Delta\phi_{bin} = \text{round}(\Delta\phi \cdot 2^j)/2^j, \tag{12.18}$$

and j is the number of bits in the binary control word. The error will occur in the MSB used as the sampling clock for the interpolator after $n_{cyc} = 2^{-1}/\varepsilon_{\Delta\phi}$ clock cycles and after

$$t_{err} = \frac{2^{-1}}{F_{out}\,\varepsilon_{\Delta\phi}} \tag{12.19}$$

seconds. For 32-bit NCO with $\varepsilon_{\Delta\phi} = 0.00000000009313$ and $F_{out} = 80\text{MHz}$, the error will occur after 67.1 s. This is a rather long time and if the operation of the interpolator can be interrupted and the device reset before the error occurs, there is no need for compensation. Of course, according to (12.17) and (12.18) the error will occur even less frequently if the word length of the control word is increased. Naturally there are also cases where $\Delta\phi_{bin} = \Delta\phi$ and then (12.17) equals zero and no compensation is needed. In the devices that require continuous-time operation, the error due to the infinite length control word causes sampling errors. To prevent this, the NCO

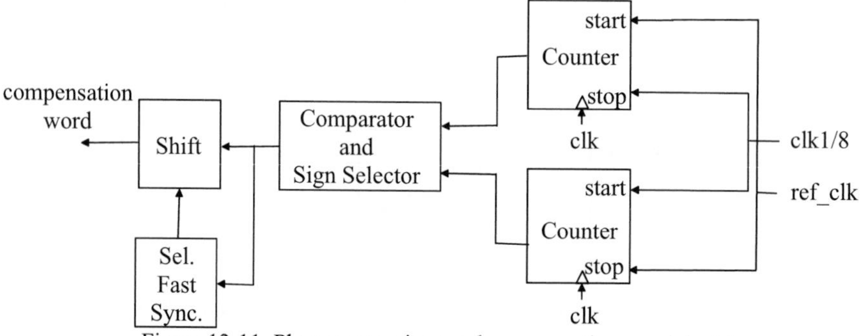

Figure 12-11. Phase comparison and compensation word decision.

has to be synchronized to the input data. This can be done by varying the control word between values that are a little too large and a little too small, such that on average the control word is equivalent to the accurate value. A phase comparator can be used to compare the phase of the generated clock and the reference clock that implies the input data sample periods. If the generated clock has been drifted due to a too small control word, a compensation value is added to the control word and the phase of the generated clock drifts back to the optimum phase. The compensation word can be chosen so small that it has a negligible effect on the much shorter fractional interval μ_k and on interpolation result. At smallest, the compensation word might be no larger the size of the LSB of the control word. For example, if $\Delta\phi = 0.4$ and $j = 32$, we get $\Delta\phi_{bin} = 0.39999999990687$, and by adding one LSB we get $\Delta\phi_{bin} + \text{LSB} = 0.40000000013970$. Such a synchronizer is actually a simple phase locked loop, where the loop filter is of the first order. The block diagram of the NCO with the synchronizer is depicted in Figure 12-10.

After the device is reset, the phase of the generated clock can be far from the optimum. In this case, with a very small compensation word the drifting to optimum phase can take a very long time, depending on how accurate the frequency control word is. To prevent this long initialization time, we can use a larger compensation value for fast drifting and change the value to the

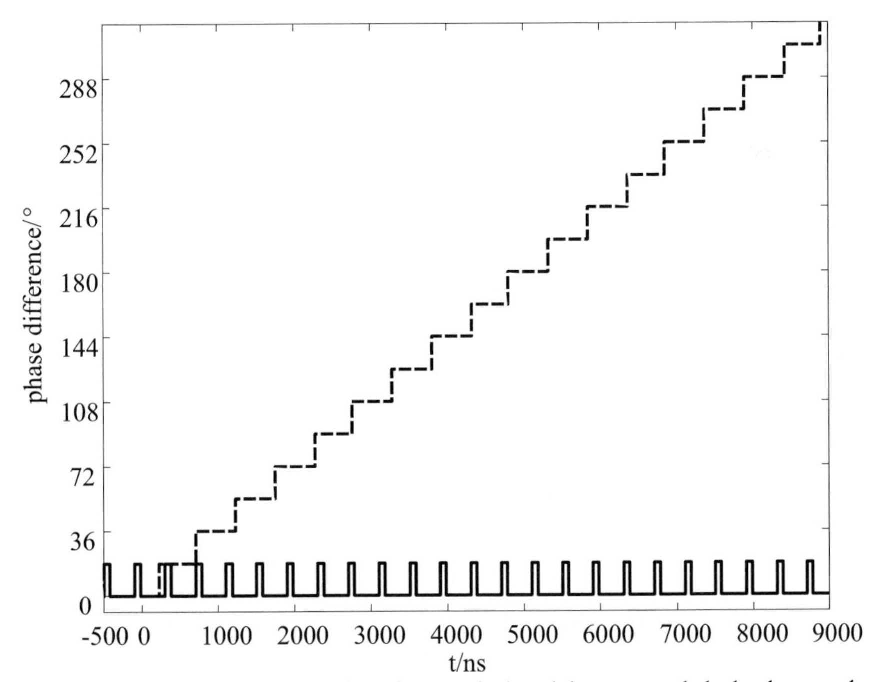

Figure 12-12. Phase difference of the reference clock and the generated clock when synchronizer is in use (solid line) and not in use (dashed line).

smaller one after the coarse synchronization is found. This kind of synchronization technique changes the phase drifting introduced by the inaccuracy of control word to a phase jitter, which can be tolerated without sampling errors if the overall interpolation ratio of the system is large enough.

The jitter is of the order of the output clock cycle (12.19), so the larger the interpolation ratio, the better the jitter can be handled, since the jitter is proportionally smaller compared to the input clock cycle.

The implementation of the phase compare block in Figure 12-10 is based on two digital saturating counters running at the output sampling frequency. A block diagram of the phase compare block is shown in Figure 12-11. The counters measure the phase difference of the reference clock and the generated clock in both directions. The result is given in output sampling periods. Because the counters have to be able to measure the worst case phase difference in output clock cycles, the number of bits needed in the counters is determined by the largest overall interpolation ratio for which the system is intended. The smaller of the measured differences is selected as the compensation word. The selection of the counter also defines the direction of the phase difference and therefore the sign of the compensation word. When the

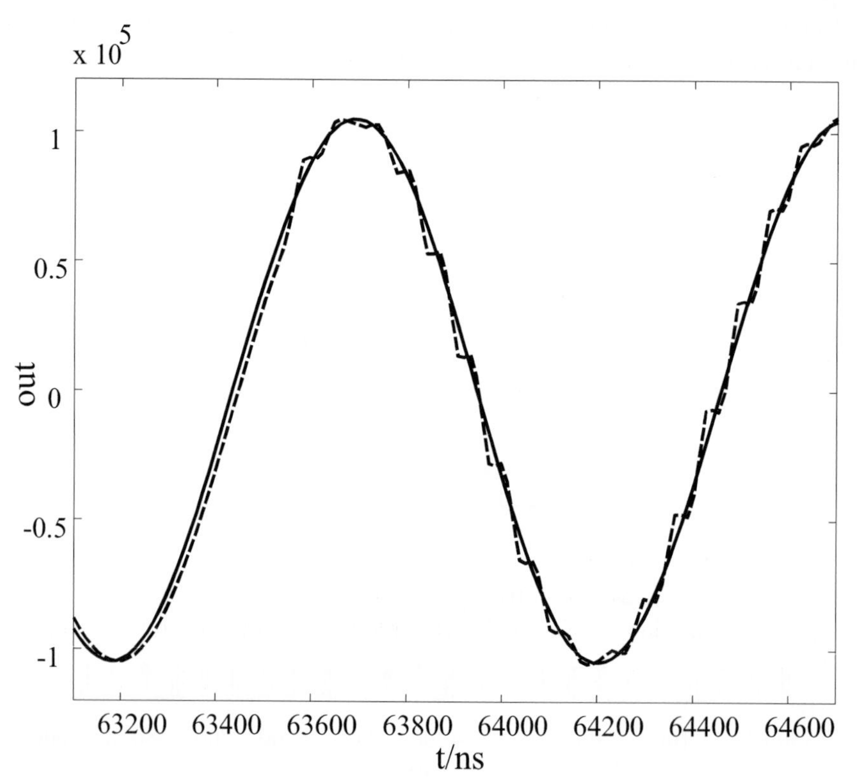

Figure 12-13. Output of 3rd-order Lagrangian interpolator when synchronizer is in use (solid line) and not in use (dashed line).

circuit is reset, the synchronizer is initialized to a fast_sync-mode and the compensation word is multiplied much larger by shifting the binary compensation word. Once the synchronization is found, the synchronizer sets automatically to the normal operation mode.

12.5.2 Simulations

Figure 12-12 shows the phase difference of the reference clock and the generated clock when the synchronizer is used or not. As can be seen, the synchronizer prevents a sampling error to occur in the input of the modulator circuit after 2272 ns of operation, since at that moment the phase difference exceeds 90°. In the simulation, an interpolation ratio of $F_{out}/F_s = 2.5$ is used. The input sampling rate is 30.72 MHz and the output sampling rate is 76.8 MHz. The accuracy of the phase accumulator is reduced to 7 bits in order to see the effects more quickly.

Figure 12-13 shows the output of a third-order Lagrange interpolator using a proposed synchronizer (solid line) and the output with the synchronizer turned off after the synchronization is once achieved (dashed line). The input signal is a sine with a frequency of 0.0318 F_{sin}. The sampling rates are similar to the previous simulation and the accuracy of the phase accumulator is 14 bits. As can be seen from Figure 12-13, approximately 63500 ns after the synchronizer is switched off, the interpolator begins to make sampling faults, i.e., it doubles some samples and skips others. The synchronized output remains correct.

REFERENCES

[Ana01] Analog Devices, "Four Channel, 104 MSPS Digital Transmit Signal Processor (TSP)," Preliminary Technical Data AD6623, 2001.

[Che93] D. Chester, "A Generalized Rate Change Filter Architecture," IEEE International Conference on Acoustics, Speech, and Signal Processing 1993, Vol. 3, pp. 181-184.

[Che98] D. Chester, "Rate Change Filter and Method," U. S. Patent 5,717,617, Feb. 10, 1998.

[Cho01] K. H. Cho, and H. Samueli, "A Frequency-Agile Single-Chip QAM Modulator with Beamforming Diversity," IEEE Journal of Solid-State Circuits, Vol. 36, No. 3, March 2001, pp. 398-407.

[Egi96] K. Egiazarian, T. Saramäki, H. Chugurian, and J. Astola, "Modified B-Spline Interpolators and Filters: Synthesis and Efficient Implementation," IEEE International Conference on Acoustics, Speech and Signal Processing 1996, Vol. 3, pp. 1743 -1746.

[Eru93] L. Erup, F. M. Gardner, and R. A. Harris, "Interpolation in Digital Modems-Part II: Implementation and Performance," IEEE Transactions on Communications, Vol. 41, No. 6, June 1993, pp. 998-1008.

[Far98] C. W. Farrow, "A Continuously Variable Digital Delay Element," in Proc. IEEE International Symposium on Circuits and Systems 1988, pp. 2641-2645.

[Fu99] D. Fu, and A. Willson, Jr., "Design of an Improved Interpolation Filter Using a Trigonometric Polynomial," IEEE International Symposium on Circuits and Systems 1999, Vol. 4, pp. 363-366.

[Gar93] F. M. Gardner, "Interpolation in Digital Modems-Part I: Fundamentals," IEEE Transactions on Communications, Vol. 41, No. 3, March 1993, pp. 501-507.

[Got01] A. Gotchev, K. Egiazarian, J. Vesma, and T. Saramäki, "Edge-Preserving Image Resizing Using Modified B-Splines," IEEE International Conference on Acoustics, Speech, and Signal Processing 2001, Vol. 3, pp. 1865-1868.

[Gra01] Graychip, Inc., "Multi-standard Quad DUC Chip," GC4116 Data Sheet, Rev 1.0, April 2001.

[Hen99] M. Henker, T. Hentschel, and G. Fettweis, "Time-Variant CIC-Filters for Sample Rate Conversion with Arbitrary Rational Factors," IEEE International Conference on Electronics, Circuits and Systems 1999, Vol. 1, pp. 67-70.

[Hou78] H. Hou, and H. Andrews, "Cubic Splines for Image Interpolation and Digital Filtering," IEEE Transactions on Acoustics, Speech, and Signal Processing, Vol. ASSP-26, No. 6, Dec. 1978, pp. 508-517.

[Int01] Intersil Corporation, Quad Programmable UpConverter, ISL5217 Data Sheet, File Number 6004, April 2001.

[Laa96] T. Laakso, V. Välimäki, M. Karjalainen, and U. Laine, "Splitting the Unit Delay," IEEE Signal Processing Magazine, Vol. 13, Jan. 1996, pp. 30-60.

[Mit98] S. Mitra, "Digital Signal Processing, A Computer-Based Approach," McGraw-Hill, Singapore 1998.

[Ram84] T. A. Ramstad, "Digital Methods for Conversion Between Arbitrary Sampling Frequencies," IEEE Transactions on Acoustics, Speech, and Signal Processing, Vol. ASSP-32, No. 3, June 1984, pp. 577-591.

[Sam00] H. Samueli, and J. Laskowski, "Variable Rate Interpolator," U. S. Patent 6,144,712, Nov. 7, 2000.

[Sne99] J. Snell, "Interpolator using a Plurality of Polynomial Equations and Associated Methods," U. S. Patent 5,949,695, Sep. 7, 1999.

[Uns91] M. Unser, A. Aldroubi, and M. Eden, "Fast B-Spline Transforms for Continuous Image Representation and Interpolation," IEEE Transactions on Pattern Analysis and Machine Intelligence, Vol. 13, No. 3, March 1991, pp. 277- 285.

[Väl95] V. Välimäki, "Discrete-Time Modeling of Acoustic Tubes Using Fractional Delay Filters," Doctoral thesis, Helsinki University of Technology, Espoo, Finland, 1995.
[Ves00] J. Vesma, and T. Saramäki, "Design and Properties of Polynomial-Based Fractional Delay Filters," in Proc. IEEE International Symposium on Circuits and Systems 2000, Vol. 1, pp. 104-107.
[Ves95] J. Vesma, "Timing Adjustment in Digital Receivers Using Interpolation," Master of Science thesis, Tampere University of Technology, Tampere, Finland, 1995.
[Ves96a] J. Vesma, and T. Saramäki, "Interpolation Filters with Arbitrary Frequency Response for All-Digital Receivers," in Proc. IEEE International Symposium on Circuits and Systems 1996, pp. 568-571.
[Ves96b] J. Vesma, M. Renfors, and J. Rinne, "Comparison of Efficient Interpolation Techniques for Symbol Timing Recovery," in Proc. IEEE Globecom 96, London, UK, Nov. 1996, pp. 953-957.
[Ves98] J. Vesma, R. Hamila, T. Saramäki, and M. Renfors, "Design of Polynomial Interpolation Filters Based on Taylor Series," in Proc. IX European Signal Processing Conf., Rhodes, Greece, Sep. 1998, pp. 283-286.
[Ves99] J. Vesma, "Optimization and Applications of Polynomial-Based Interpolation Filters," Doctoral thesis, Tampere University of Technology, Tampere, Finland, 1999.

Chapter 13

13. FIR FILTERS FOR COMPENSATING D/A CONVERTER FREQUENCY RESPONSE DISTORTION

Three different designs are represented as compensating the sinc(x) frequency response distortion resulting from D/A converters, by using digital FIR filters. The filters are designed to compensate the signal's second image distortion. All filters are designed on a programmable logic device (PLD). The numbers of logic elements (LEs) needed in three different implementations are compared. The measured second image of the wideband code division multiple access (WCDMA) signal meets the adjacent channel leakage power specifications.

The output of the digital modulator could be heterodyned up to the desired frequency by employing a mixer/local oscillator (LO)/filter combination, as shown in Figure 13-1 (a). The alternative method is to use images to up-convert, as shown in Figure 13-1 (b). To obtain frequencies above Nyquist frequencies, the output of the D/A converter is band-pass filtered (BPF) by an 'image-selecting' band-pass filter, rather than a low-pass filter (LPF). This saves a number of analog components (LPF, mixer, LO).

In the digital to analog (D/A) converters, there is amplitude distortion in the spectrum of the signal being converted. The converters are usually implemented by using a Non-Return-to-Zero (NRZ) pulse shape (zero-order hold) so that the impulse response of the converter is a rectangle. The width of the rectangle is T_s, and the height is scaled to $1/T_s$. The frequency re-

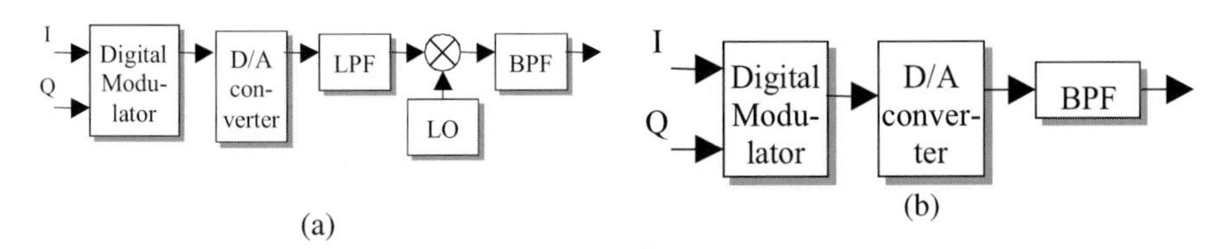

Figure 13-1. Up-conversion. (a) Conventional. (b) Using images.

sponse of the converter is calculated by taking a Fourier transform of the rectangle, which is given by

$$H(f)_{D/A} = \frac{\sin(\pi\ f/f_s)}{\pi\ f/f_s} \cdot e^{-j\pi\ f/f_S}, \tag{13.1}$$

where $\sin(\pi\ f/f_s)/(\pi\ f/f_s)$ can be written as $\text{sinc}(\pi\ f/f_s)$. Here $f_s = 1/T_s$ is the sampling rate of the D/A converter. The ideal pre-compensation filter has a linear phase response and the magnitude response is given by

$$|H(f)| = |1/\text{sinc}(\pi\ f/f_s)|. \tag{13.2}$$

The images exhibit a significantly poorer signal to noise and distortion

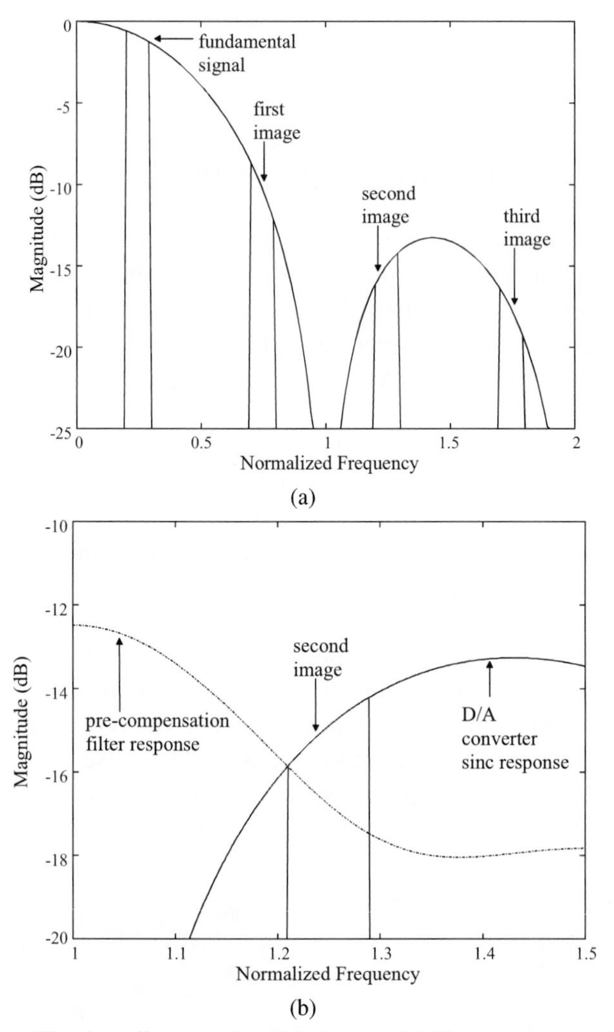

Figure 13-2. (a) The sinc-effect, signal, and its images. (b) The pre-compensation filter response, D/A converter sinc response and the second image.

ratio (SNDR) than does the fundamental signal because of the sinc attenuation. The amplitude of the image responses decreases according to sinc($\pi f/f_s$), while spurious responses due to the D/A-converter clock feedthrough noise and dynamic non-linearities generally roll off much more slowly with the frequency. Nevertheless, the first and second images may still meet the system SNDR requirements. The SNDR degradation can be compensated by using a more accurate D/A converter.

The sinc effect introduces a droop, which is not acceptable when the bandwidth of the signal is a wide (e.g. WCDMA signal [GPP00]). The signal, three images, and the D/A converter frequency response are represented in Figure 13-2 (a). If we take an odd numbered image, the fundamental signal frequency response has to be mirrored with respect to the center frequency. In this chapter, we compensate the second image, which is illus-

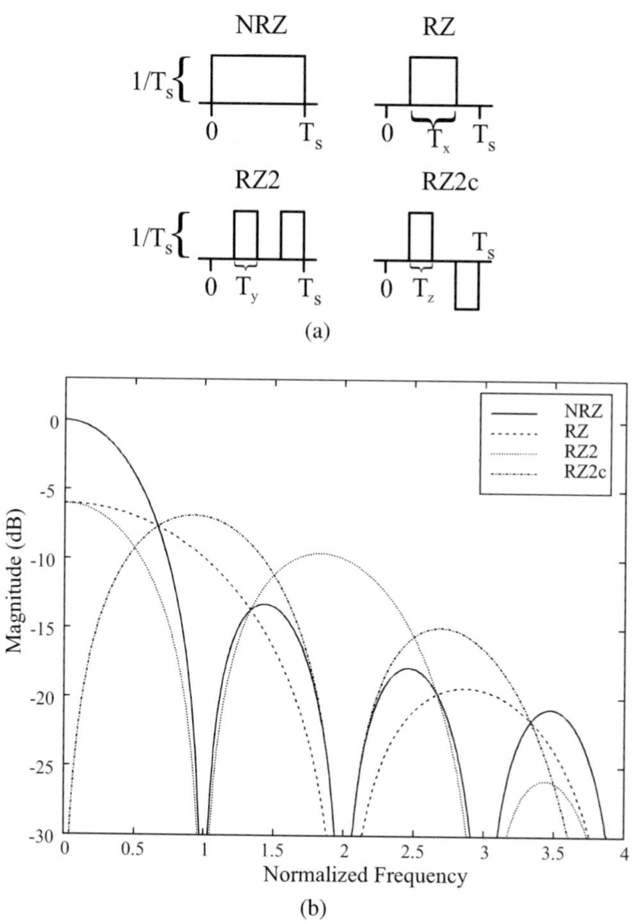

Figure 13-3. Four different D/A converter pulse shapes (a) in time domain (b) and their magnitude responses.

trated in Figure 13-2 (a). In the Figure 13-2 (b), there is the pre-compensation filter response, the second image, and the D/A converter sinc response. The sinc distortion is cancelled by the pre-compensation filter. This chapter shows three different implementations to compensate the sinc distortion by using digital FIR filters. In [Sam88], the real filter was designed to compensate the distortion of the fundamental signal (not image).

13.1 Four Different D/A Converter Pulse Shapes

Another solution for flattening the spectrum of the D/A converter is a change of the pulse shape [Bug99]. In Figure 13-3 (a), the four different pulse shapes are: a Non-Return-to-Zero (NRZ) pulse, a Return-to-Zero pulse (RZ)

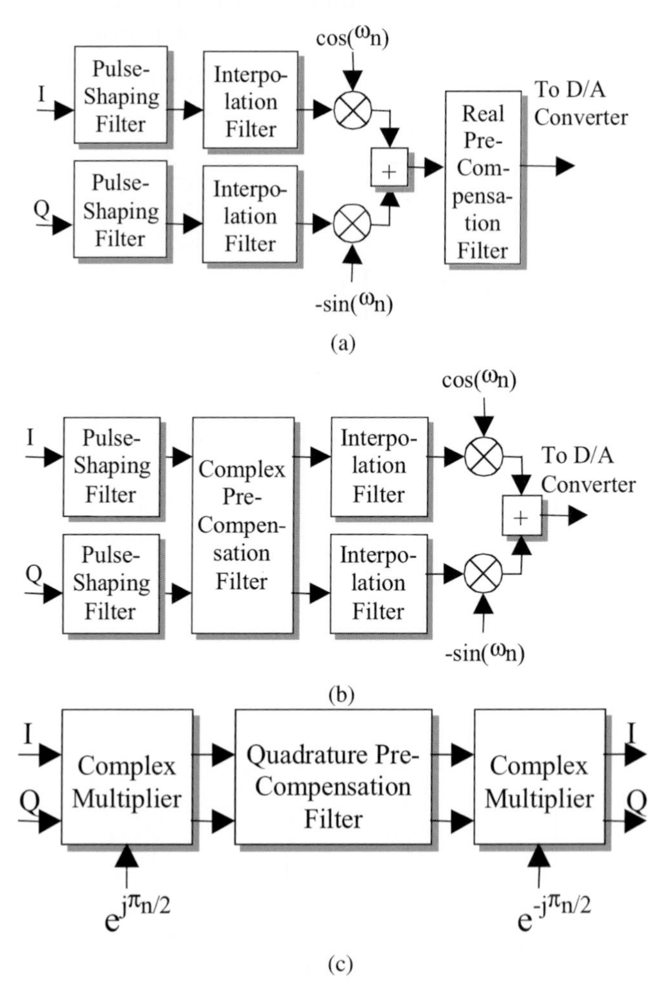

Figure 13-4. Pre-compensation filter in the digital modulator at (a) IF, (b) baseband (complex) and (c) baseband (quadrature).

[Bug99], a double RZ pulse (RZ2) and a double complementary pulse (RZ2c). The magnitude response of RZ pulse is

$$|\mathrm{RZ}(f)| = (T_x/T_s)|\mathrm{sinc}(\pi f / f_x)|, \tag{13.3}$$

where $f_x = 1/T_x$ and for RZ2 is

$$|\mathrm{RZ2}(f)| = (2T_y/T_s)\left|\,\mathrm{sinc}(\pi f / f_y)\cos(2\pi f / f_y)\right|, \tag{13.4}$$

where $f_y = 1/T_y$ and for RZ2c is

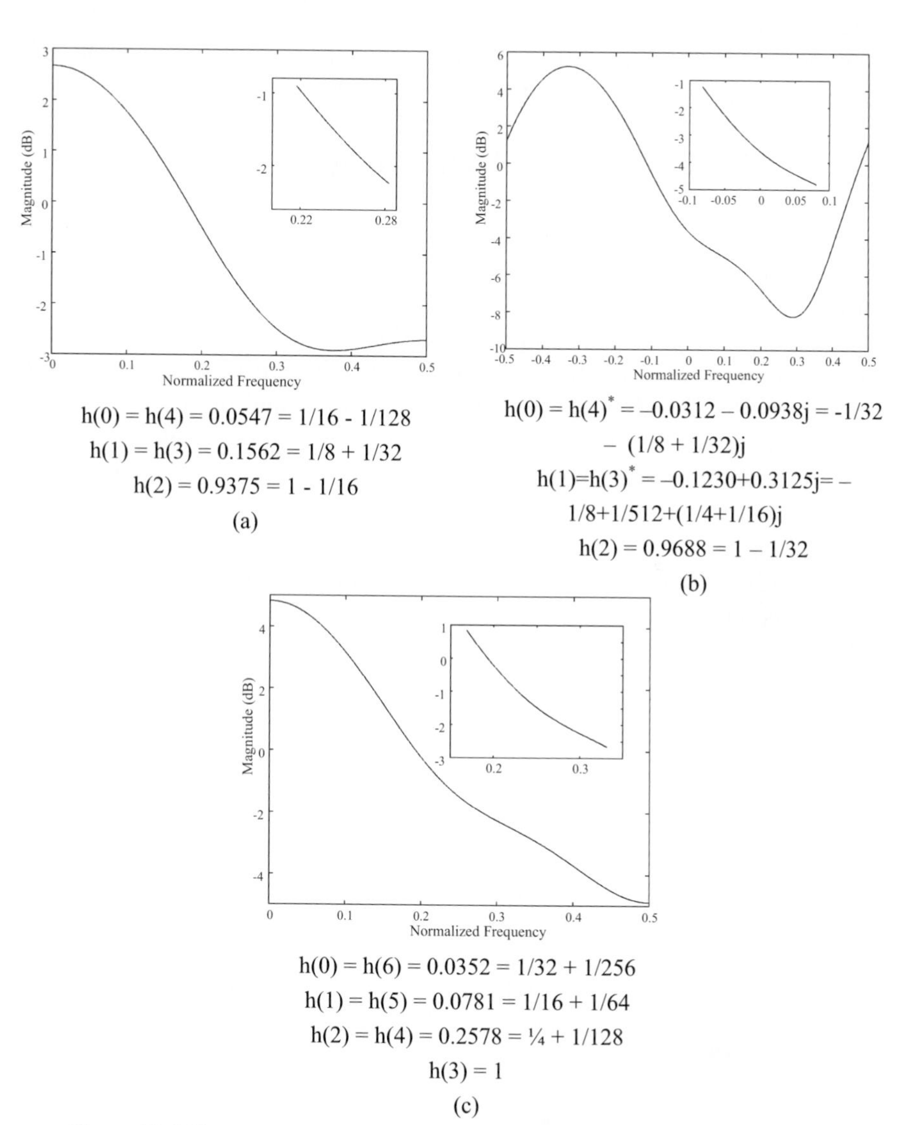

Magnitude (dB)
Normalized Frequency
h(0) = h(4) = 0.0547 = 1/16 - 1/128
h(1) = h(3) = 0.1562 = 1/8 + 1/32
h(2) = 0.9375 = 1 - 1/16
(a)
h(0) = h(4)* = –0.0312 – 0.0938j = -1/32 – (1/8 + 1/32)j
h(1)=h(3)* = –0.1230+0.3125j= – 1/8+1/512+(1/4+1/16)j
h(2) = 0.9688 = 1 – 1/32
(b)
h(0) = h(6) = 0.0352 = 1/32 + 1/256
h(1) = h(5) = 0.0781 = 1/16 + 1/64
h(2) = h(4) = 0.2578 = ¼ + 1/128
h(3) = 1
(c)

Figure 13-5. Frequency and impulse response of the (a) real FIR IF, (b) complex FIR and (c) quadrature FIR in the $f_s/4$.

$$|\mathrm{RZ2c}(f)| = (2T_z/T_s)\,|\mathrm{sinc}(\pi f/f_z)\sin(2\pi f/f_z)|, \tag{13.5}$$

where $f_z = 1/T_z$. The magnitude responses are shown in Figure 13-3 (b), when $T_x = T/2$ and $T_y = T_z = T/4$. The NRZ and RTZ pulses are optimized for signals near DC. For frequencies near f_s, the RZ2c pulse shows the largest gain. In the case when the frequencies are near $2f_s$, the RZ2 pulse results in the smallest loss of gain. Depending on the image frequency, different D/A converter pulse shapes give the largest gain. In order to be able to make the RZ, RZ2 and RZ2c pulses, a higher sampling frequency than f_s is required. Large voltage steps in these pulses cause clock jitter and waveform fall/rise asymmetry sensitivity. In this chapter, the filters were designed to pre-compensate the distortion from the NRZ pulse shape.

13.2 Different Implementation

In this chapter, the pre-compensation is carried out in three different ways. In the first method, it is performed in the IF by using a real filter that is shown in Figure 13-4 (a) [Sam88]. In the second case, we compensate the baseband signal by using a complex FIR filter. The block diagram of the pre-compensation filter in the baseband is shown in Figure 13-4 (b). In the third method, the baseband signal is upconverted to $f_s/4$ and the I and Q signals are both compensated by using real FIR filters. After pre-compensation, signals are down-converted back to the baseband. The block diagram of this method is shown in Figure 13-4 (c).

In all cases, the peak error must be below ± 0.045 dB over the 5 MHz signal bandwidth (WCDMA channel bandwidth [GPP00]). The center of the signal's second image being compensated is located at $5/4f_s$, which is shown in Figure 13-2.

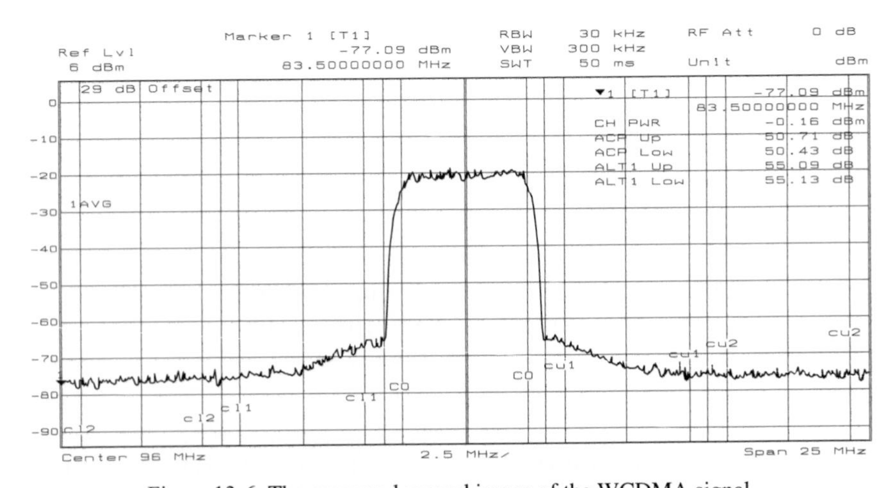

Figure 13-6. The measured second image of the WCDMA signal.

In the first case, the pre-compensation is carried out in IF by using a real linear phase FIR filter. The sampling frequency is 76.8 MHz, the signal frequency is 19.2 MHz and the center frequency of the image is located at 96 MHz. The output signal of the D/A converter is pre-compensated so that the droop in the second image is cancelled. The magnitude response of the pre-compensation filter at 19.2 MHz ($f_s/4$) is obtained from (13.2), where f_s is 76.8 MHz and f is 96 MHz ($5/4f_s$) ± 2.5 MHz. If a D/A converter pulse shape other than the NRZ is used, then the magnitude response of the pre-compensation filter is obtained from the inverse of (13.3), (13.4), or (13.5).

In the second case, the pre-compensation is performed in the baseband, where the sampling frequency is 30.72 MHz. Because the frequency response is not symmetric with respect to zero frequency, the pre-compensation FIR filter has to be complex. The pre-compensation located in the baseband allows the pre-compensation filter to run at a lower computational rate.

The third way to compensate discussed in this chapter is to up-convert the baseband signal to $f_s/4$, using the complex multiplier $e^{jn\pi/2}$. In this special case, when the IF center frequency is equal to a quarter of the sampling rate, considerable simplification is achieved, since the sine and cosine signals representing the complex phasor degenerate into two simple sequences [1 0 –1 0 …] and [0 1 0 –1 …], thus eliminating the need for high-speed digital multipliers and adders to implement the mixing functions. The only operations involved are sign inversions and the interchanging of real and imaginary parts. After the upconversion, two real filters compensate I- and Q-signals separately, and the compensated signal is down-converted back to the baseband with the $e^{-jn\pi/2}$ complex multiplier. In this case, the sampling frequency is the same as in the complex filter case.

Table 13-1. Three implementations

	Real FIR in the IF	Complex FIR	Quadrature in the $f_s/4$
Number of Taps	5	5	7
Sampling Frequency (MHz)	76.8	30.72	30.72
Number of Les	135	456	416
LEs utilized (%)	7	26	24
Max Clock Frequency (MHz)	131.57	125	75.18
Peak Error of the Compensation (dB)	±0.0278	±0.0325	±0.0374

13.3 Filter Design

The first step is to design the ideal coefficients. The real filter coefficients were designed by Matlab. In Matlab, there is a remez function, which uses the Parks-McClellan algorithm to calculate the coefficients; for the complex filter design there is function called cremez [Mat94].

A function was developed in which the user has to give five parameters: order of the desired filter, signal bandwidth, image center frequency, sampling frequency, and the weight of the pre-compensation error. There are also a small number of optional parameters. The function fits the filter response to match the sinc(x) compensation only on the signal band. The so-called 'don't care' regions cannot be left out of the signal band; if they were to be left out, then the frequency response in those regions could be tens of dBs higher than in the signal band. This would not be acceptable, so the gain in those regions is restricted.

The allowable filter coefficients were constrained to the set of numbers representable by a canonic signed-digit (CSD) code with 10 (ternary) bits of word length and only 2 nonzero digits by using an algorithm from [Sam89]. The ideal filter coefficients should be scaled so that the overflows cannot occur for baseband or IF WCDMA signal (narrow signal bandwidth respect to the sampling frequency). The second reason for scaling the coefficients is to minimize the error introduced by rounding the coefficients to the nearest representable CSD code [Sam88]. The word length of the input data is 14 bits. Inside the filter, the word length was increased to 15 bits so as to improve the accuracy of the filter. At the end of the filter, the signal is truncated back to 14 bits.

In Figure 13-5, the frequency and impulse responses of the designed filters are represented. The (*) in the impulse response of the complex filter stands for complex conjugate. The frequency responses in the signal bandwidth are shown inside the figures in the small boxes. The peak errors of the compensation over the signal bandwidth are shown in Table 13-1.

13.4 Implementations

All the filters were implemented with an Altera FLEX 10KA-1 series device [Fle98]. The filters were designed on an EP10K30ATC144-1 device. The number of logic elements (LEs) needed in three different implementations is shown in Table 13-1. The real filter in IF is the most efficient way to implement the pre-compensation. The quadrature filter in the IF includes two real filters, and complex multipliers $e^{jn\pi/2}$ and $e^{-jn\pi/2}$. That is why the number of logic elements (LEs) is so large.

13.5 Measurement Result

The simulation results in the real IF case were saved in the pattern generator memory. These data were transferred to the 14-bit D/A converter [Kos01]. The measured second image of the pre-compensated WCDMA signal is shown in Figure 13-6, where the ACP/ALT1 means the first/second adjacent channel leakage power ratio, respectively. The sampling frequency of the D/A converter is 76.8 MHz and the center frequency of the second image of the WCDMA signal is located at 96 MHz, as shown in Figure 13-6. The adjacent channel leakage power ratios are 50.43/55.09, as shown in Figure 13-6. The specifications are 45/50 dB, when the spectrum is measured at the base station RF port [GPP00]. The measurement result meets the specifications. There is some margin (5 dB) to take care of the other transmitter stages that might degrade the spectral purity of the signal.

13.6 Conclusion

Three different designs are represented to compensate the signal's second image distortion resulting from the D/A converter sinc(x) frequency response by using digital linear phase FIR filters. The measured WCDMA signal's second image meets adjacent channel leakage power specifications.

REFERENCES

[Bug99] A. R. Bugeja, B. S. Song, P. L. Rakers, and S. F. Gillig, "A 14-b, 100-MS/s CMOS DAC Designed for Spectral Performance," IEEE J. Solid-State Circuits, Vol. 34, No. 12, pp. 1719-1732, Dec. 1999.

[FLE98] FLEX 10K Embedded Programmable Logic Family Data Sheet, Altera Corp., San Jose, CA, Oct. 1998.

[GPP00] 3rd Generation Partnership Project; Technical Specification Group Radio Access Networks; UTRA (BS) FDD; Radio transmission and Reception, 3G TS 25.104, V3.3.0, June 2000.

[Kos01]M. Kosunen, J. Vankka, M. Waltari, and K. Halonen, "A Multicarrier QAM Modulator for WCDMA Base Station with on-chip D/A converter," Proceedings of CICC 2001 Conference, May 6-9 2001, San Diego, USA, pp. 301-304.

[Mat94]Matlab Signal Processing Toolbox User's Guide, The Math Works Inc., Natick, Mass., June 1994.

[Sam88] Henry Samueli, "The Design of Multiplierless FIR Filters for Compensating D/A Converter Frequency Response Distortion," IEEE Transactions on Circuits and Systems, Vol. 35, No. 8, pp. 1064-1066, August 1988.

[Sam89] Henry Samueli, "An Improved Search Algorithm for the Design of Multiplierless FIR Filters with Powers-of-Two Coefficients," IEEE Transactions on Circuits and Systems, Vol. 36, No. 7, pp. 1044-1047, July 1989.

Chapter 14

14. A DIRECT DIGITAL SYNTHESIZER WITH TUNABLE DELTA SIGMA MODULATOR

Table 14-1 shows spurious free dynamic ranges (SFDRs) to the Nyquist frequency in recently published direct digital synthesizers (DDSs) and D/A converters [Nic91], [Tan95], [Van98], [Ana02], [Sch03], [Jan02]. It is easy to achieve 100 dBc SFDR in the digital domain [Jan02]. In the wide output bandwidth DDSs, most spurs are generated less by quantization errors in the digital domain (see row 3 in Table 14-1) and more by analog non-idealities in the D/A-converter (see row 4 in Table 14-1), because the development of D/A converters is not keeping pace with the capabilities of the digital signal processing with faster technologies [Van01].

The multi-bit D/A converter is susceptible to glitches and spurious noise (as the output frequency increases), which is difficult to remove by filtering. The D/A converters with excellent amplitude resolution and frequency response tend to consume excess power and are expensive. As the 1-bit delta sigma ($\Delta\Sigma$) D/A converter has only two levels, any misplacement of the levels results only in gain error or offset. Neither of those are of great importance in many applications. The 1-bit $\Delta\Sigma$ D/A converter is an all digital circuit, which has numerous advantages over analog signal processing, such as flexibility, noise immunity, reliability and potential improvements in performance and power consumption because of the scaling of the technology. In addition, the design, synthesis, layout and testing of digital systems can be highly automated. The DDS with 1-bit $\Delta\Sigma$ D/A converter is attractive in digital transmitters, where it allows the output transistors to be operated in a power-efficient switching mode, and thus may result in an efficient power amplifier [Nor97], [Jay98], [Key01], [Iwa00].

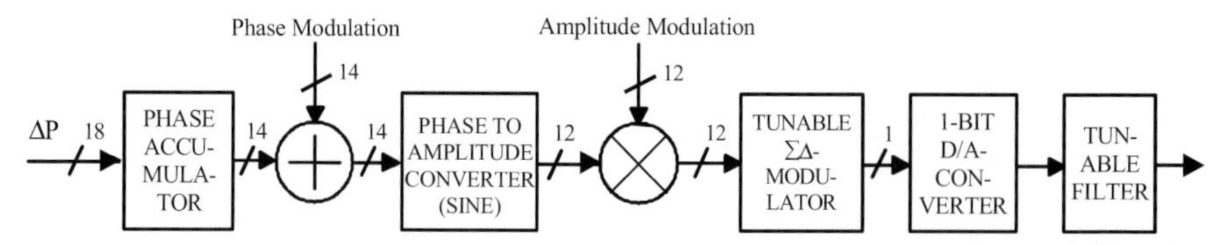

Figure 14-1. Direct digital synthesizer with tunable real ΔΣ modulator.

14.1 Direct Digital Synthesizer with Tunable ΔΣ Modulator

The DDS with a tunable real ΔΣ modulator is shown in a simplified form in Figure 14-1. It has the following basic blocks: a phase accumulator, a phase adder, a phase to amplitude converter (sine output), a multiplier, a real ΔΣ modulator, a 1 bit D/A converter and a filter. The DDS with a tunable complex ΔΣ modulator is shown in Figure 14-2. It has the following basic blocks: a phase accumulator, a phase adder, a phase to amplitude converter (quadrature output), two multipliers, a complex ΔΣ modulator, a 1 bit D/A converter and a filter. The DDSs are capable of frequency, phase, and quadrature amplitude modulation.

The phase value is generated using the modulo 2^j overflowing property of a j-bit phase accumulator ($j = 18$) in Figure 14-1 and Figure 14-2. The rate of the overflows is the output frequency (4.1). The phase to amplitude converter converts the digital phase information into the values of a sine (and cosine) wave(s) in Figure 14-1 and Figure 14-2. The ΔΣ modulator quantizes the phase to amplitude converter output to two levels such that most of the quantization noise power resides outside the signal band. The analog filter removes the out of band quantization noise along with any out-of-band noise and distortion introduced by the 1-bit D/A converter. The passband of the analog filter is tuned according to the DDS output frequency.

By setting the least significant bit of the phase accumulator to one the numerical period is maximal in (4.4). It has the effect of randomizing the errors introduced by the quantizied phase to amplitude converter samples, because, in a long output period, the error appears as "white noise" [Van01]. The disadvantage is that it introduces an offset into the output frequency of

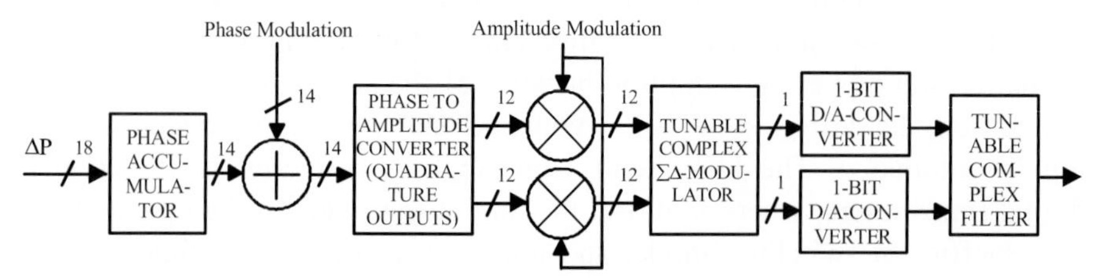

Figure 14-2. Direct digital synthesizer with tunable complex ΔΣ modulator.

Table 14-1.
Single-Tone SFDR

	[Nic91]	[Tan95]	[Van98]	[Ana02]	[Sch03]	[Jan02]
Sampling Frequency	150 MHz	200 MHz	150 MHz	165 MHz	400 MHz	200 MHz
Digital Data at D/A Converter Input †	90.3 dBc	84.3 dBc	72 dBc	-	-	100 dBc
Analog Signal at D/A Converter Output	51.48 dBc @ 11.1 MHz Output	58.50 dBc @ 12 MHz Output	52 dBc @ 75 MHz Output	57 dBc @ 50 MHz Output	63 dBc @ 120 MHz Output	-

† In the digital domain SFDR is achieved over tuning range (0 to Nyquist frequency)

the DDS. The offset will be small if the ratio of the sampling frequency to the power of two of the phase accumulator length is low in (4.2). Another method is to use a dither signal to randomize input quantization error to prevent idle tones [Nor97].

14.2 Quadrature Modulator

The frequency modulation could be superimposed on the hopping carrier by simply adding and subtracting a frequency offset to/from ΔP. The quadrature amplitude modulation (QAM) could be performed

$$I_{out} = I(n)\cos(\omega_{out}\, n) + Q(n)\sin(\omega_{out}\, n) = A(n)\cos(\omega_{out}\, n - P(n)),$$
$$Q_{out} = Q(n)\cos(\omega_{out}\, n) - I(n)\sin(\omega_{out}\, n) = A(n)\sin(\omega_{out}\, n - P(n)), \quad (14.1)$$

where $A(n) = \sqrt{I(n)^2 + Q(n)^2}$ and $P(n) = \arctan(\frac{Q(n)}{I(n)})$,

where arctan is the four quadrant arctangent of the in-phase ($I(n)$) and quad-

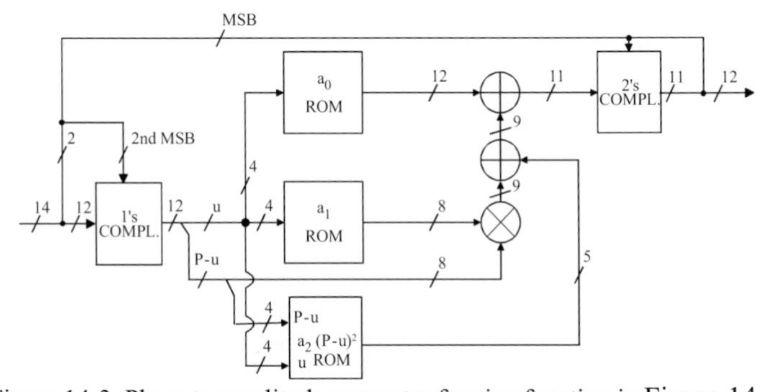

Figure 14-3. Phase to amplitude converter for sine function in Figure 14-1.

rature phase data ($Q(n)$). The in-phase output is only needed in Figure 14-1, which requires one adder for phase modulation ($P(n)$) before the phase to amplitude converter (sine) and a multiplier for amplitude modulation ($A(n)$) after the phase to amplitude converter according to (14.1). The quadrature output is needed in Figure 14-2, which requires two multipliers after the phase to amplitude converter (quadrature output) according to (14.1).

14.3 Phase to Amplitude Converters

The sine wave is separated into π/2 intervals in Table 9-1, and each interval is called a quadrant (q). The two most significant phase bits represent the quadrant number. The full sine wave can be represented by the first quadrant sine, using the symmetric and antisymmetric properties of sine, as shown in Table 9-1 [McC84]. Therefore, the two most significant phase bits are used to decode the quadrant, while the remaining 12 bits are used to address a one-quadrant sine look-up table in Figure 14-3. The quarter wave sine function is approximated by the sixteen equal length 2nd degree polynomial segments

$$\sin(\pi P/2 + 2\pi/2^{15}) = a_0(u) + a_1(u)\,(P-u) + a_2(u)\,(P-u)^2, \quad (14.2)$$

where $a_0(u)$, $a_1(u)$, $a_2(u)$ are polynomial coefficients. Please note that a 1/2 LSB offset was introduced into the phase in order to reduce hardware (see Figure 9-7). The phase address of the quarter of the sine wave "P" is divided into the upper phase address "u" and the lower phase address "P-u". While higher order polynomial can be employed, their contribution to the accuracy is very small and, therefore, of little weight in this application. The coeffi-

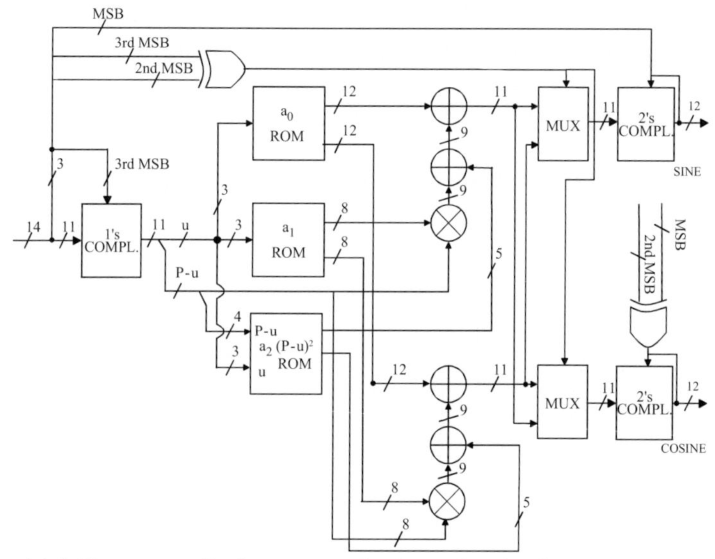

Figure 14-4. Phase to amplitude converter for sine and cosine function in Figure 14-2.

cients are chosen using least-squares polynomial fitting. These coefficients vary according to the upper phase address "u". The four most significant bits of the input phase are selected as the upper phase address "u", which is transferred simultaneously to two read-only memories (ROMs) as address signals as shown in Figure 14-3. The output of the "$a_0(u)$" ROM is the first term of the polynomial in (14.2) and is transferred to a first adder, where it will be summed with the remaining terms involved. The least significant bits "P-u" are multiplied by the output of the "$a_1(u)$" ROM to produce the second term of the polynomial. The third term is computed in the ROM by multiplying the coefficient "$a_2(u)$" and the square of the lower phase address "P-u". This is performed by selecting the upper bits of "P-u" and "u" values as a portion of the address for the ROM. This is possible since the last term only roughly contributes 4 LSBs to the output. The discrete Fourier transform (DFT) of the sine wave approximated by the sixteen equal length 2nd degree polynomial segments gives SFDR of 87.09 dBc.

The DDS architecture for both sine and cosine functions (quadrature output) is shown in Figure 14-4. The architecture is a simple extension of Figure 14-3, because the eight wave symmetry of a sine and cosine waveform would be advantageous [Tan95], [McC84]. The sine and cosine waves are separated into π/4 intervals in Table 9-2; each interval is called an octant (o). The three most significant phase bits represent the octant number. The sine and cosine curves for all octants can be represented by one octant of sine and one octant of cosine, using the symmetric and antisymmetric properties of sine and cosine. Hence, only the sine and cosine samples from 0 to π/4 in Figure 14-4 need be calculated, because the sine wave from π/2 to π/4 is the same as the cosine from 0 to π/4, and the cosine wave from π/2 to π/4 is the same as the sine from 0 to π/4. The Exclusive Ors (XORs), negators and multiplexers (MUXs) in Figure 14-4 are used to obtain full sine and cosine waves by changing polarity and/or exchanging the first octant sine and cosine values according to scheme defined by Table 9-2 [McC84].

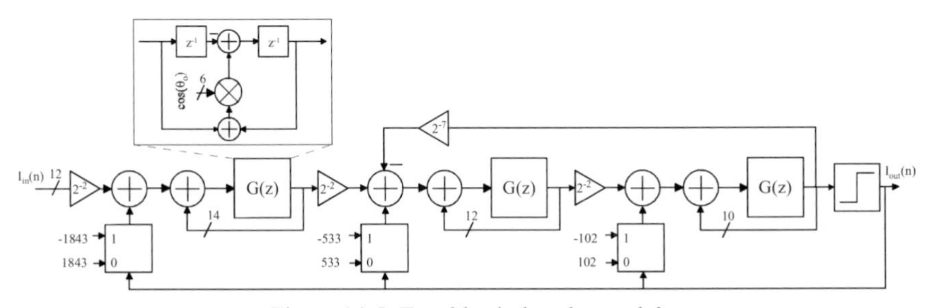

Figure 14-5. Tunable sixth order modulator.

14.4 Tunable ΔΣ Modulators

With a basic ΔΣ modulator, only a small fraction of the bandwidth can be occupied by the required signal. A larger fraction of the Nyquist bandwidth can be made available if a tunable ΔΣ modulator is employed. This is achieved by using discrete-time low-pass-to-bandpass transform

$$z^{-1} \rightarrow -z^{-1}\frac{z^{-1}-\cos(\theta_o)}{1-\cos(\theta_o)z^{-1}}, \tag{14.3}$$

which preserves both the realizability and maximum out of band gain constraint [Nor97]. The notch frequency of the all-pass transfer function is

$$f_o = f_s\,\theta_0 / 2\pi, \tag{14.4}$$

which is set the same as the DDS output frequency from (4.1).

A third-order noise transfer function (NTF) was designed using the delta sigma modulator design toolbox [Ftp] for a maximum out-of-band gain of 1.52 (3.62 dB) with zeros optimized for minimum in-band noise at an oversampling ratio of 64. The number of bits required in each stage and the interstage scaling coefficients were optimized using simulations. The complete modulator, including word lengths and coefficients, is depicted in Figure 14-5. The order of the NTF is doubled by the substitution (14.3), since each delay is replaced by a second order sub filter. Errors in approximating the coefficients for $\cos(\theta_o)$ simply result in a frequency shift of the tuned notch of the filter, because the NTF zeros are shifted around the unit circle of the z plane. The block diagram of the all-pass transfer function G(z) from equation (14.3) is shown in Figure 14-5. Tuning is trivially accomplished by changing the $\cos(\theta_o)$ multiplier of the all-pass network.

For a given order, the complex ΔΣ loop can achieve better performance than a real ΔΣ loop because it is not constrained to realize the zeros as complex conjugate pairs [Jan02]. The noise attenuation performance of the real tunable ΔΣ loop degrades, while the notch frequency deviates from dc or $f_s/2$ and the complex ΔΣ loop remains constant with the notch frequency [Jan97], [Azi95]. The complex 2nd order modulator is shown in Figure 14-6

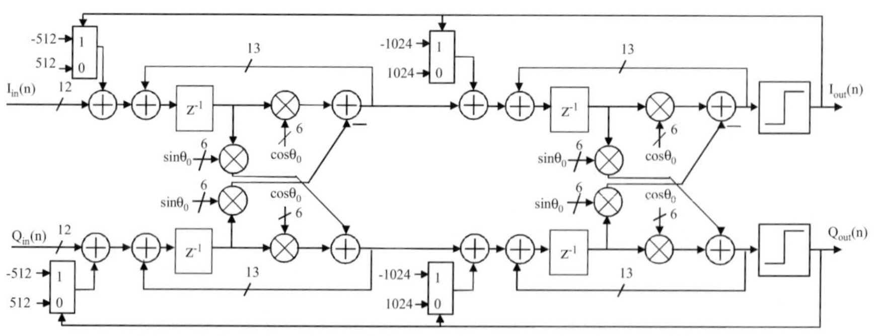

Figure 14-6. Structure of the complex second order modulator.

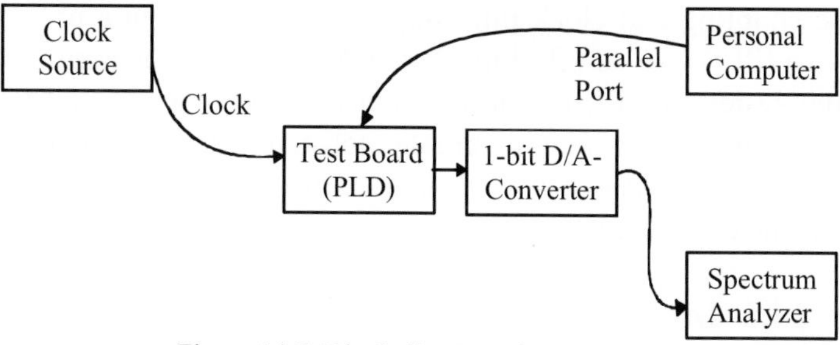

Figure 14-7. Block diagram of test system.

[Azi95]. Tuning is trivially accomplished by changing the sine and cosine coefficients in Figure 14-6. Errors in approximating the coefficients result in a degradation of the tuned notch of the filter if the angular frequency of the sine and cosine coefficients is different, because the NTF zeros are shifted away from the unit circle of the z plane. The complex modulator requires a complex analog post-filter as shown in Figure 14-2. If only the real analog output is needed, then an analog half-complex filter (Hilbert transformer) can be used.

14.5 1-bit D/A Converter

Since the output of the 1 bit D/A converter is an analog signal, it is amenable to clock jitter, waveform fall/rise asymmetry and noise coupling [Mag97].

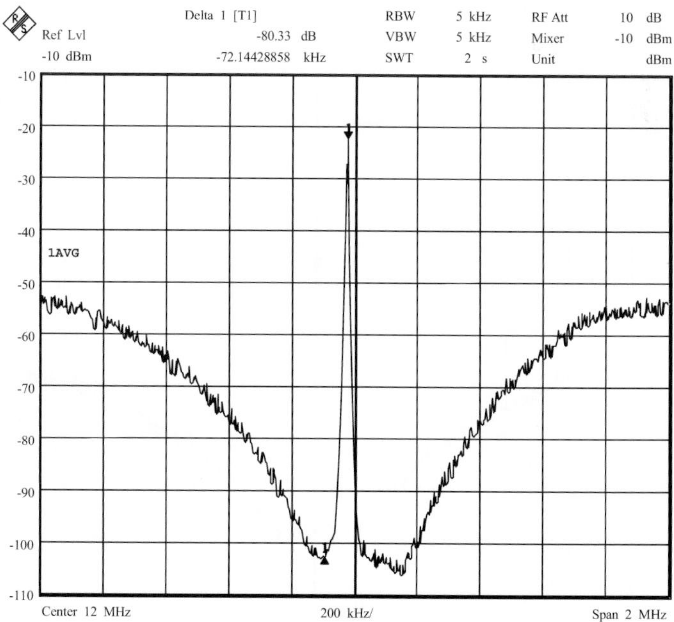

Figure 14-8. Spectrum of 12 MHz signal at 1-bit $\Delta\Sigma$ D/A converter output, where sampling frequency is 30 MHz.

The noise coupling and clock jitter might be a problem in a programmable logic device (PLD), so an off-chip 1 bit D/A converter was used. In the fully differential system used here, the D/A converter output rise and fall are inherently symmetric, so that a simple non-return-to-zero output suffices [Jen95]. The one bit D/A converter is a simple current-steering differential pair with a current source.

14.6 Implementations

The DDS with tunable $\Delta\Sigma$ modulator (real and complex) were implemented with the Altera FLEX 10KA-1 series devices [Fle98]. The DDS (in Figure 14-1) with the tunable real $\Delta\Sigma$ modulator requires 1089 (21 % of the total) logic elements (LEs) in the EPF10K100A device. The DDS (in Figure 14-2) with the tunable complex $\Delta\Sigma$ modulator requires 2098 (42% of the total) LEs in the EPF10K100A device. The maximum operating frequency of the real and complex realization is 30 and 40 MHz, respectively. The most significant bit of the 14-bit D/A converter was used as a 1 bit D/A converter [Kos01].

14.7 Measurement Results

To evaluate the DDS with tunable $\Delta\Sigma$ modulator (real and complex) a test

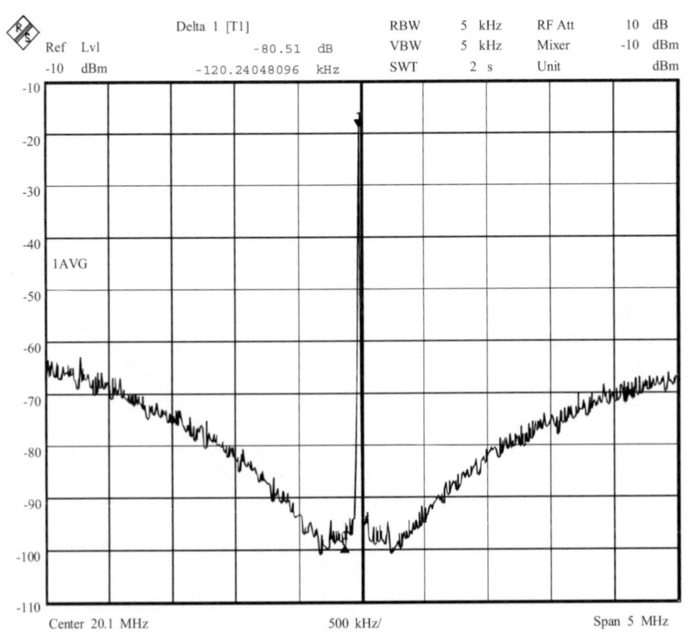

Figure 14-9. Spectrum of 20 MHz signal at 1-bit $\Delta\Sigma$ D/A converter output, where sampling frequency is 40 MHz.

board was built and a computer program was developed to control the measurement. The phase increment word and the other tuning signals were loaded into the test board via the parallel port of a personal computer. The block diagram of the test system is shown in Figure 14-7.

A spectrum plot of 12 MHz output from the DDS with real the tunable 1-bit $\Delta\Sigma$ D/A converter is illustrated in Figure 14-8. A spectrum plot of 20 MHz output from the DDS with the complex tunable 1-bit $\Delta\Sigma$ D/A converter is illustrated in Figure 14-9. The dynamic range is about 80 dB in the figures.

14.8 Conclusions

The DDS with the tunable (real or complex) 1-bit $\Delta\Sigma$ D/A converter were designed and implemented. Since the 1-bit $\Delta\Sigma$ D/A converter has only one bit, the glitch problems and spurious noise resulting from the use of the multi-bit D/A converter are avoided. The main drawback of the DDS with the tunable $\Delta\Sigma$ D/A converter is that the broadband frequency switching speed is determined by the tuning time of the passband of the analog filter. The inband of the DDS with the tunable $\Delta\Sigma$ D/A converter could be placed anywhere in the Nyquist interval; it therefore gives a high degree of flexibility in frequency planning when the passband of the analog filter is fixed.

REFERENCES

[Ana02] "AD9744 14-Bit, 165 MSPS TxDAC D/A Converter," Analog Devices Data Sheet, May 2002.

[Azi95] P. M. Aziz, H. V. Sorensen, and J. Van der Spiegel, "Performance of Complex Noise Transfer Functions in Bandpass and Multi Band Sigma Delta Systems," in Proc. IEEE International Symposium on Circuits and Systems (ISCAS), 1995,Vol. 1, pp. 641-644.

[Fle98] FLEX 10K Embedded Programmable Logic Family Data Sheet, Altera Corp., San Jose, CA, Oct. 1998.

[Ftp] ftp://ftp.mathworks.com/pub/contrib/v5/control/delsig.tar.

[Iwa00] M. Iwamoto, A. Jayaraman, G. Hanington, P. F. Chen, A. Bellora, W. Thornton, L. E. Larson, and P. M. Asbeck, "Bandpass Delta-Sigma Class-S Amplifier," Electronics Letters, Vol. 36, No. 12, pp. 1010-1012, June 2000.

[Jan02] I. Janiszewski, B. Hoppe, and H. Meuth, "Numerically Controlled Oscillators with Hybrid Function Generators," IEEE Transactions on Ultrasonics, Ferroelectrics and Frequency Control, Vol. 49, No. 7, pp. 995-1004, July 2002.

[Jan97] S. A. Jantzi, K. W. Martin, and A. S. Sedra, "Quadrature Bandpass ΣΔ Modulation for Digital Radio," IEEE J. of Solid State Circuits, Vol. 32, No. 12, pp. 1935-1950, Dec. 1997.
[Jay98] A. Jayaraman, P. F. Chen, G. Hanington, L. Larson, and P. Asbeck, "Linear High-Efficiency Microwave Power Amplifiers Using Bandpass Delta-Sigma Modulators," IEEE Microwave and Guided Wave Letters, Vol. 8, No. 3, pp. 121 –123, March 1998.
[Jen95] J. F. Jensen, G. Raghavan, A. E. Cosand, and R. H. Walden, "A 3.2-GHz Second-Order Delta-Sigma Modulator Implemented in InP HBT Technology," IEEE J. of Solid State Circuits, Vol. 30, No. 10, pp. 1119-1127, Oct. 1995.
[Key01] J. Keyzer, J. Hinrichs, A. Metzger, M. Iwamoto, I. Galton, and P. Asbeck, "Digital Generation of RF Signals for Wireless Communications with Band-Pass Delta-Sigma Modulation," IEEE Microwave Symposium Digest, Vol. 3, pp. 2127-2130, 2001.
[Kos01]M. Kosunen, J. Vankka, M. Waltari, and K. Halonen, "A Multicarrier QAM Modulator for WCDMA Base Station with on-chip D/A converter," In Proc. 2001 IEEE Custom Integrated Circuits Conference, May 2001, pp. 301-304.
[Mag97] A. J. Magrath, and M. B. Sandler, "Digital Power Amplification Using Sigma-Delta Modulation and Bit Flipping," J. Audio Eng. Soc., Vol. 45, No. 6, pp. 476-487, June 1997.
[McC84] R. D. McCallister, and D. Shearer, III, "Numerically Controlled Oscillator Using Quadrant Replication and Function Decomposition," U. S. Patent 4,486,846, Dec. 4, 1984.
[Nic91] H. T. Nicholas, and H. Samueli, "A 150-MHz Direct Digital Frequency Synthesiser in 1.25-μm CMOS with -90-dBc Spurious Performance," IEEE J. Solid-State Circuits, Vol. 26, pp. 1959-1969, Dec. 1991.
[Nor97] S. R. Norsworthy, R. Schreier, and G. C. Temes, "Delta-sigma Data Converters, Theory, Design, and Simulation," New York: IEEE Press, 1997.
[Sch03] W. Schofield, D. Mercer, and L. St. Onge, "A 16b 400MS/s DAC with <-80dBc IMD to 300 MHz and <-160dBm/Hz Noise Power Spectral Density," ISSCC Digest of Technical Papers, February 2003, San Francisco, USA, pp. 126-127.
[Tan95] L. K. Tan, and H. Samueli, "A 200 MHz Quadrature Digital Synthesizer/Mixer in 0.8 μm CMOS," IEEE J. of Solid State Circuits, Vol. 30, No. 3, pp. 193-200, Mar. 1995.
[Van01] J. Vankka, and K. Halonen, "Direct Digital Synthesizers: Theory, Design and Applications," Kluwer Academic Publishers, 2001.
[Van98] J. Vankka, M. Waltari, M. Kosunen, and K. Halonen, "Direct Digital Syntesizer with on-Chip D/A-converter," IEEE Journal of Solid-State Circuits, Vol. 33, No. 2, pp. 218-227, Feb. 1998.

Chapter 15

15. A DIGITAL QUADRATURE MODULATOR WITH ON-CHIP D/A CONVERTER

A logical progression to the previous multi-standard modulator [Van02] and multicarrier work [Kos01] is to move the digital to analog interface even closer to the antenna in the base station transmitter structure. In traditional transmitters, a complex baseband signal is digitally modulated to the first IF (intermediate frequency) and then mixed to the second IF frequency and to the RF frequency in the analog domain. The first analog IF mixer stage of the transmitter can be replaced with the digital quadrature modulator presented in this chapter as shown in Figure 15-1. The two complex modulators are in series with the quadrature modulator as shown in Figure 15-2. In the digital complex modulators, the baseband in-phase (*I*) and quadrature (*Q*) channels are modulated onto orthogonal carriers (*X*, *Y*) at the IF frequency at the lower sampling rate [Van02], [Kos01]. The tunable complex modulator, steered by the carrier NCO, enables the fine tuning of the transmitted carrier frequency with sub-Hz resolution [Van02], [Kos01], whereas this digital quadrature modulator is used for the coarse tuning. It is beneficial to implement the fine tuning at lower sampling frequencies and the coarse tuning at the higher frequencies, because of the smaller amount of hardware associated with the coarse tuning implementation. The quadrature modulator interpolates orthogonal input carriers by 16 and performs a digital quadrature modulation at carrier frequencies $f_s/4$, $-f_s/4$, $f_s/2$ (f_s is the sampling frequency). It allows the quadrature modulation to be performed in the digital domain with high precision and perfect I/Q-channel matching. The major limiting factor of digital IF modulator performance at base station applications is the D/A converter, because the development of D/A converters does not keep up with the capabilities of digital signal processing with faster technologies [Van01].

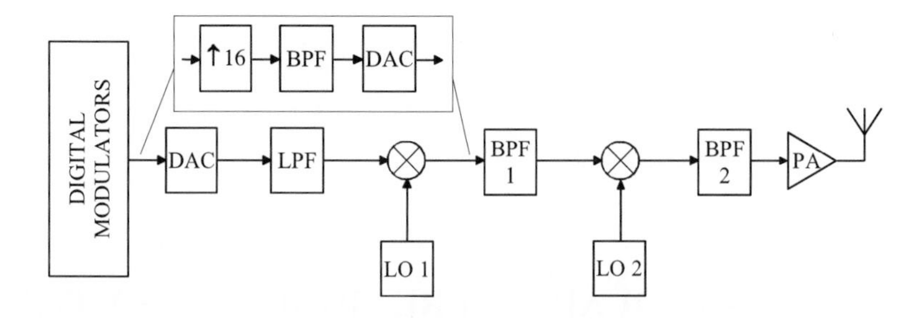

Figure 15-1. The digital quadrature modulator replacing the first analog IF mixer stage.

15.1 Multiplier Free Quadrature Modulation

The output of the complex modulator in Figure 15-2 is

$$\begin{aligned} X_i(n) &= I_i(n)\cos(\omega_{Ci}\, n) + Q_i(n)\sin(\omega_{Ci}\, n) \\ Y_i(n) &= Q_i(n)\cos(\omega_{Ci}\, n) - I_i(n)\sin(\omega_{Ci}\, n), \end{aligned} \tag{15.1}$$

where $I_i(n)$ and $Q_i(n)$ are pulse shaped and interpolated in-phase and quadrature-phase data symbols, which are upconverted to the orthogonal (in-phase ($X_i(n)$) and quadrature-phase ($Y_i(n)$)) carriers at frequency f_{Ci} [Van02], [Kos01]. The output of the quadrature modulator in Figure 15-2 is

$$O(n) = X(n)\cos(2\pi n f_{out} / f_s) + Y(n)\sin(2\pi n f_{out} / f_s), \tag{15.2}$$

where f_s and f_{out} are the sampling and output frequency, and $X(n)$, $Y(n)$ are interpolated in-phase and quadrature-phase carriers, respectively. Furthermore, if we require either the sine or cosine term to be zero in (15.2) at every n, then for each output value only one of the in-phase or quadrature-phase part needs be processed, which reduces the amount of hardware needed [Won91], [Dar70]. Solutions for f_{out}/f_s equal to 0, 1/2, 1/4, -1/4 in Table 4-2 correspond to our requirements. Using these values, high-speed digital multipliers and adders are not needed in the implementation of the quadrature

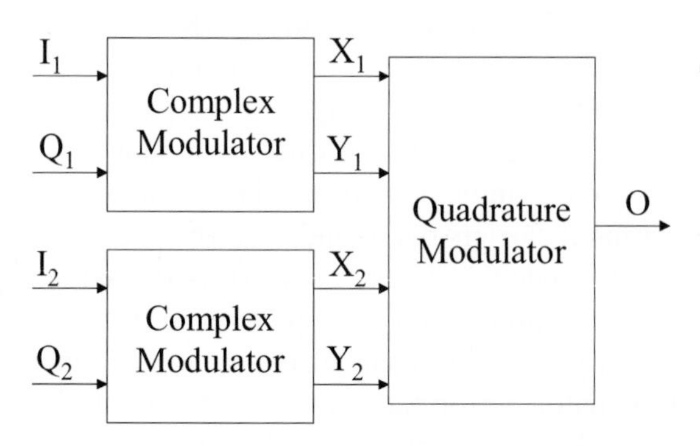

Figure 15-2. Two complex modulators in series with the quadrature modulator.

modulation in (15.2). Instead, multiplexers (MUXs) and negators (NEGs) can be used.

15.2 Interpolation Filters

The in-phase and quadrature-phase parts of the carriers from two complex modulators can be summed in the digital quadrature modulator (Figure 15-3), which allows the formation of the multi-carrier signal. The sampling rate of the input samples is first increased by four with two halfband interpolation filters as shown in Figure 15-3. The passband of the first half-band filter restricts the maximum output frequency of the complex modulator in Figure 15-2 to about two-fifths of the input sampling rate of the quadrature modulator. The last filter has an interpolation ratio of 4, therefore the polyphase decomposition has four polyphase filters. The quadrature modulation in (15.2) for f_{out}/f_s equal to 0, 1/2, 1/4, -1/4 (see Table 4-2) can be performed by negators in the polyphase filter outputs and 4:1 MUX as shown in Figure 15-3. In the lowpass mode ($f_{out}/f_s = 0$ in (2)), the output consists only of in-phase carriers (see Table 4-2). Therefore, only the X samples are driven to the polyphase filters. This is selected by the mode signal illustrated in Figure 15-3. Furthermore, the negators are disabled in the lowpass mode. In the bandpass mode ($f_{out}/f_s = 1/4$ or $f_{out}/f_s = -1/4$ in (15.2)), the output consists of the interleaved X and Y samples (see Table 4-2). Furthermore, every other second X and Y sample is negated. Therefore, the X samples are driven to the first and third polyphase filter and Y samples to the second and fourth polyphase filter. The third or fourth polyphase filter outputs are negated. The quadrature modulation by $f_{out}/f_s = 1/4$ or $f_{out}/f_s = -1/4$ can be performed by selecting whether to negate the Y-branches or not (control signals CTR4 and CTR5 in Figure 15-3). In the highpass mode ($f_{out}/f_s = 1/2$ in (15.2)), the output consists only of in-phase carriers and every second sample is negated (see Table 4-2). Therefore, the X samples are driven to the polyphase filters, and the second and fourth of the polyphase filter outputs are negated.

The digital quadrature modulator interpolates input carriers by 16. Therefore, the transition band for the analog reconstruction filter is increased and

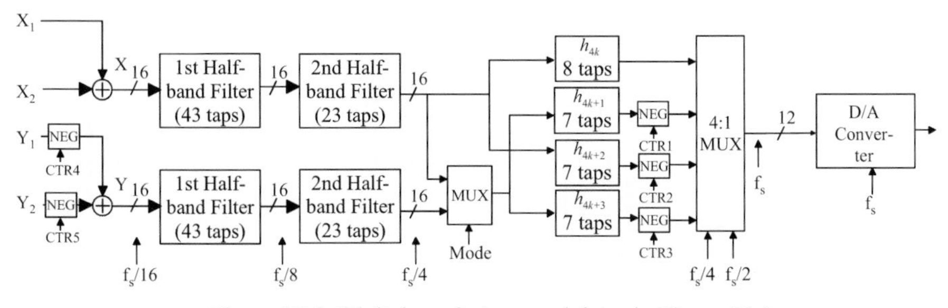

Figure 15-3. Digital quadrature modulator in Figure 15-2.

the images are suppressed more in the lowpass mode, thus reducing the complexity of the analog reconstruction filter. Furthermore, the sin(x)/x roll-off over the effective passband is significantly reduced. In the low-pass/highpass mode, the information in the quadrature-phase carrier is lost, because output consists only of in-phase carriers in those modes (Table 4-2). The output frequency of the complex modulator (f_{Ci}) should be higher than dc in those modes, because then the quadrature-phase data $Q_i(n)$ is modulated to the in-phase carrier in (15.1). Furthermore, the frequency should be considerably higher than dc in the lowpass/highpass mode, so that the images generated by the analog upconversion could be filtered by the bandpass filter (BPF 2 in Figure 15-1) (lowpass mode), and the images generated by the D/A conversion could be filtered by the analog reconstruction filter (BPF 1 in Figure 15-1) (highpass mode). In the bandpass mode there are no such restrictions.

The quadrature modulation in (15.2) is performed by the negators and 4:1 MUX shown in Figure 15-3. This may be considered to represent a frequency up-shift of the low-pass passband of the interpolation filters to the desired frequency. The digital modulator magnitude responses in the low-pass, bandpass and highpass modes are shown in Figure 15-4. The combination of the filters provides more than 74dB image rejection (passband ripple of 0.004dB). The transition band in the lowpass mode (Figure 15-4) is from 0.023 - 0.04 f_s. The images related to the last interpolation stage are shown in Figure 15-5. The image selection is achieved by changing the filter passband.

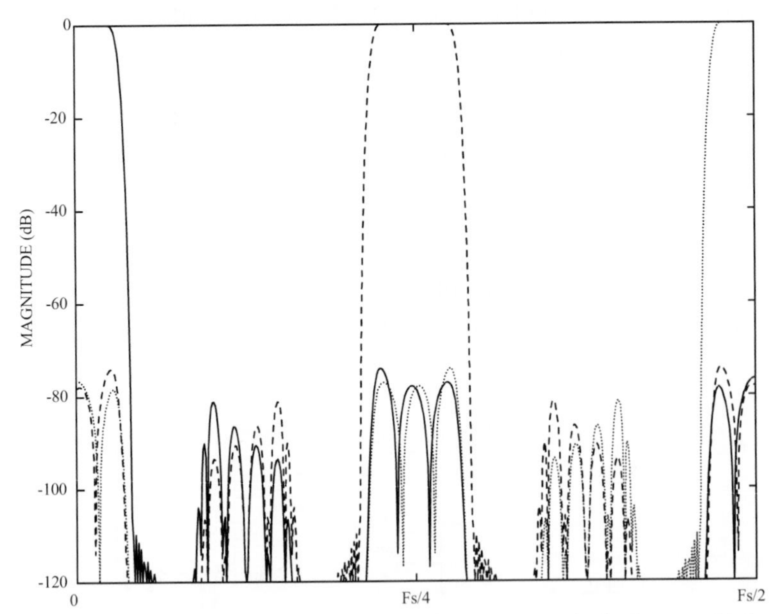

Figure 15-4. Digital quadrature modulator magnitude response in the lowpass (-), bandpass (--) and highpass mode (··) in Figure 15-3.

In the bandpass mode, the first or second image in Figure 15-5 can be chosen by the negation of Y branches (control signals CTR4 and CTR5 in Figure 15-3). This selection can be made independently from each orthogonal carrier, thus almost doubling the output bandwidth in the bandpass mode, as shown in Figure 15-5. If the odd image is used, then the output spectrum might be inverted (mirrored about carrier frequency) in the complex modulator [Van02]. This is achieved by changing the sign of the complex modulator carrier frequency (the direction of the vector rotation) in (15.1).

The internal wordlengths of the modulator are shown in Figure 15-3. The wordlengths were chosen such that the 12-bit D/A converter quantization noise dominates the digital output noise.

15.3 D/A Converter

From a systems perspective, integration of the D/A converter into the digital system is desirable to avoid high-speed multi-bit data crossing over the inter-chip boundary. The 12-bit on-chip D/A converter is based on a segmented current steering architecture. It consists of a 6b MSB matrix (2b binary and 4b thermometer coded) and a 6b binary coded LSB matrix, as shown in Figure 15-6. The dynamic linearity is important in this modulator because of the strongly varying signal, although the good static linearity is a prerequisite for obtaining a good dynamic linearity. For a current-steering D/A converter, the static linearity is mainly determined by the matching behavior of the current sources and the finite output impedance of the current source. The static linearity is achieved by sizing the current sources for intrinsic matching, using layout techniques and increasing current cell output impedance by biasing the switch transistors in the saturation region and adding cascode devices.

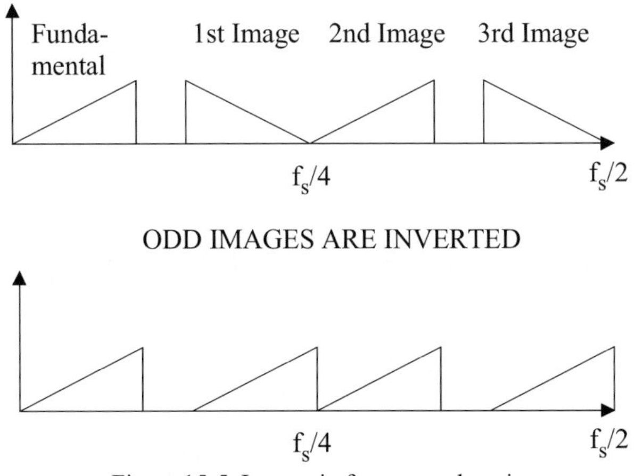

Figure 15-5. Images in frequency domain.

The integral non-linearity (INL) of a D/A converter is in this book defined to be the difference between the analog output value and the straight line drawn between the output values corresponding to the smallest and the largest input code, divided by the average LSB step. The INL yield is then given by the ratio of the number of functional D/A converters ($|INL| < \pm 1/2$ LSB) to the total number of try-outs. At the segmentation level of (4-8), the D/A converter suffers more severely from INL than differential non-linearity (DNL) according to the simulations, therefore the INL yield defines the requirements for standard deviation. In this case, for an INL yield of 99.99 %, a unit current source standard deviation of 0.26 % is required (Figure 10-3).

To minimize these three effects, a well designed and carefully laid out synchronized switch driver is used [Bos01]. A major function of the switch driver shown in Figure 15-7 is to adjust the crosspoint of the control voltages, and to limit their amplitude at the gates of the current switches, in such a way that these transistors are never simultaneously in the off state and that the feedthrough is minimized. The crossing point of the control signals is set by using different rise and fall times for the differential output of the driver [Bos01], [Wu95]. A buffer is inserted between the latch and the current cell to adjust further the crossing point of the differential outputs. The reduced voltage swing is achieved by lowering the power supply of the buffer. Dummy switch transistors were used to improve the synchronization of the switch transistors control signals. The sizing of the dummy transistors was optimized by simulations. The cascode transistor between the current sources and the switches is used to increase the output impedance of the current cell, which improves the linearity of the D/A-converter shown in Figure 15-7 [Tak91]. To minimize the feedthrough to the output lines and to increase the output impedance further in the current cell, the drain of the switching tran-

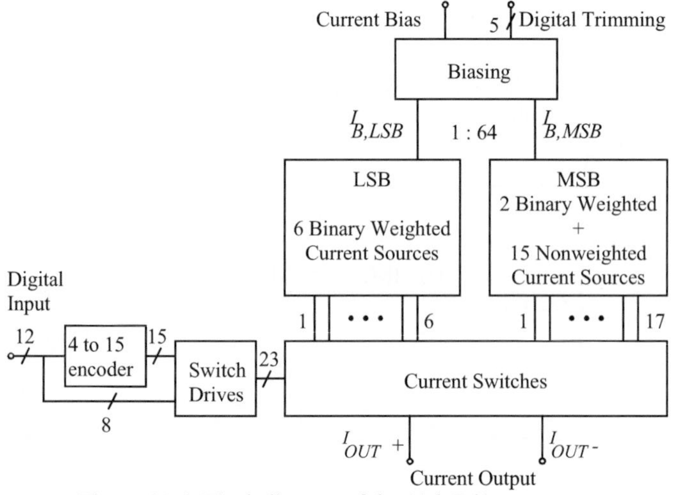

Figure 15-6. Block diagram of the 12-b D/A converter.

sistors is isolated from the output lines by adding cascode transistors [Tak91]. Unfortunately, this cascode increases the output settling time and causes different rise and fall times for single-ended outputs, although the latter is cancelled partly in the differential output.

By dividing the current source array into separately biased MSB- and LSB-arrays in such way that they both contain current sources for half the bits and that the MSB unit current source is 64 times wider than the LSB unit source, the number of unit current sources required falls from 1024 to 63+63 in a 12 bit D/A converter [Bas98]. This layout reduces the parasitic capacitances at the drain node of the current source transistors by reducing the amount of wiring needed to connect the transistors, as well as by reducing the parasitic capacitances caused by the transistors themselves [Bas98]. This improves the settling time of the converter [Chi94] and the output impedance at the higher signal frequencies [Bas98]. The scheme also reduces the area required by the current source matrices by allowing routing on top of the MSB transistors, as well as by reducing the amount of silicon area wasted in the process dependent spaces between the transistors [Bas98]. On the downside, the current generated by the MSB unit current sources has to be a very accurate 64 times the LSB unit current. This requires a trimmable biasing scheme for the MSB current sources.

15.4 Implementation and Layout

The digital part of the quadrature modulator, excluding the 4:1 MUX illustrated in Figure 15-3, was synthesized from the VHDL description using the standard 0.35 μm CMOS cell library. Multirate systems are efficiently implemented using the polyphase structure in which sampling rate conversion and filtering operations are combined (see section 11.8). The problem is that, for the series of polyphase filters in Figure 15-3, there is much more computation during some of the sampling instants than others. Thus, the switching noise is periodic; this tends to affect the timing on the D/A converter current

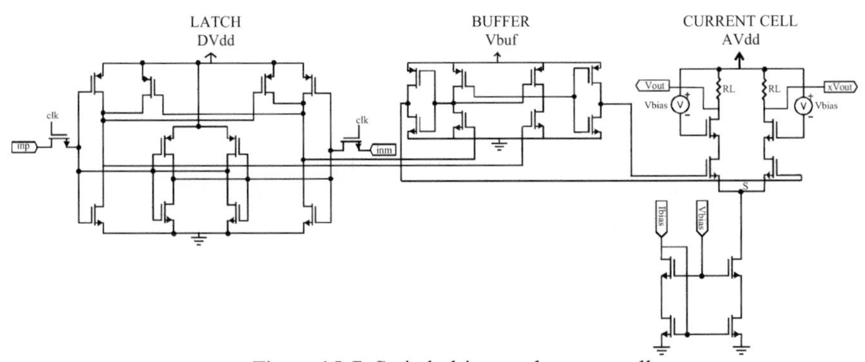

Figure 15-7. Switch driver and current cell.

switches, introducing a large phase modulation (PM) component. The use of the divided clocks in Figure 15-3 can cause more jitter at every other D/A converter sampling instant, which raises unwanted sideband in the bandpass mode. The PM component is reduced by shielding the D/A converter clock. Furthermore, the power supplies of the digital logic and the analog part are routed separately, while the digital and analog parts are placed to isolated wells to minimize the noise coupling to the analog output through the substrate. The last filter stage, shown in Figure 15-3, was implemented using a pipelined carry-save architecture due to the high speed requirements (see section 11.7). The taps of the folded transposed direct form FIR filters were realized with CSD coefficients. The static timing check and pre-layout timing simulations were performed for the netlist, and the chip layout was completed using Cadence place and route tools. Finally, based on the parasitic information extracted from the layout, the post-layout delays were back-annotated to ensure satisfactory chip timing. Since the output of the 4:1 MUX shown in Figure 15-3 operates at f_s, the multiplexer has been manually designed and laid out at the transistor level. To enhance the D/A converter dynamic performance, the layout of the switch drivers and current outputs were designed as symmetrical as possible. A clock delay line was used to ensure optimal settling of switch driver control signals at the sampling instants.

15.5 On-chip Capacitor

The implemented chip features an on-chip capacitor. This section explains

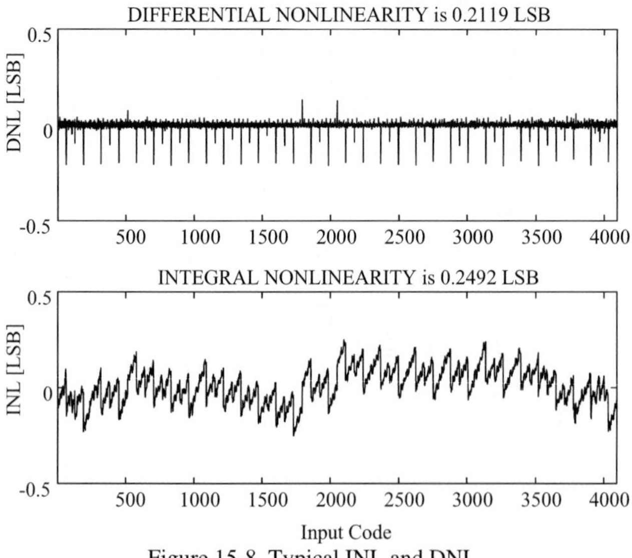

Figure 15-8. Typical INL and DNL.

why it is needed and how it is implemented. *di/dt* noise results from passing a non-constant current through an inductor. In today's digital IC design this often means the inductance of bonding wires. The noise voltage is given by

$$v_n(t) = L\frac{di(t)}{dt}. \tag{15.3}$$

The most important parameter with this noise is its peak value [Lar97].

15.5.1 Analytic First Order Model

Figure 15-9 presents the current paths in a typical CMOS circuit for an on-chip and off-chip driver. The MOS transistors are modeled as switches. A simple model of *di/dt* noise is the triangular approximation of the current pulse with a piecewise linear shape as in Figure 15-10. The peak value is I_p and the duration is t_f, the fall time of the discharged node. T is the rise time of the current pulse to the peak value [Lar97]. The approximation gives the maximum *di/dt* as I_p/T and [Lar97]

$$v_{n,\max} = L\frac{I_p}{T}. \tag{15.4}$$

This peak noise appears on V_{ss} in Figure 15-9 when the output node B is switched from high to low; the same would appear on V_{dd} when the opposite transition occurs.

When internal circuits are switched, the noise will simultaneously appear on both V_{dd} and V_{ss}, as shown by the current paths in Figure 15-9 [Lar97]. The capacitors C_{iv} and C_{ig} are capacitive loads connected to V_{dd} and V_{ss}, respectively, whereas C_{ev} and C_{eg} are capacitive loads connected to the off-chip supply voltage *pwr* and off-chip ground *gnd*, respectively.

15.5.2 Negative Feedback

(15.4) predicts that the peak noise can exceed the supply voltage, which clearly does not make sense. An improved model is obtained by writing the

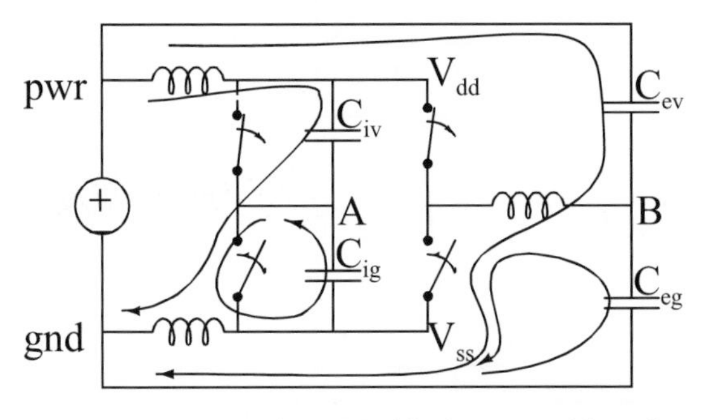

Figure 15-9. Current paths in a typical CMOS chip for an on-chip and an off-chip driver.

current as a function of the noise voltage. The increase in V_{ss} due to a falling transition in the output tends to reduce the current through the output N transistor (in Figure 15-9, the switch between V_{ss} and the bonding wire). If it is saturated, we have

$$i_{out}(t) = \frac{\beta}{2}\left(v_{in}(t) - V_t - v_n(t)\right)^2, \tag{15.5}$$

where v_{in} is the gate voltage of the output driver, V_t the threshold voltage and β the transconductance parameter [Lar97]. Combining this with the triangular current approximation yields for $v_{n,max}$

$$v_{n,\max} = v_{in}(T) - V_t + \frac{T}{L\beta}\left(1 - \sqrt{1 + 2(v_{in}(T) - V_t)\frac{L\beta}{T}}\right). \tag{15.6}$$

15.5.3 Reducing di/dt Noise

(15.4) predicts that the peak noise depends on the peak current I_p, therefore reducing the peak current also decreases the noise [Lar97]. [Lar97] presents several techniques for reducing the peak current.

Increasing the turn on time T of the output buffer also decreases the peak noise according to (15.4). Several techniques for this have been proposed [Lar97].

(15.3) shows that the noise can be decreased by decreasing the effective inductance. A commonly utilized method is to add more bonding wires for supplying V_{dd} and V_{ss}. In addition to the total inductance, the relative placement of the pins also has an impact on the reduction of *di*/*dt* noise.

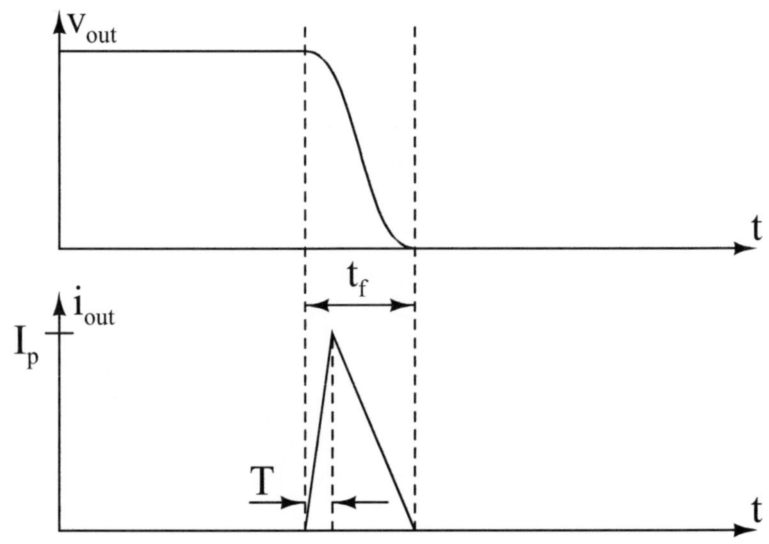

Figure 15-10. Triangular approximation of current.

Arbitrarily low levels of *di/dt* noise generated by internal circuits can be attained by adding decoupling capacitance between V_{ss} and V_{dd}. This is discussed in next section.

15.5.4 Decoupling Capacitance

Figure 15-11 (a) illustrates the effect of decoupling capacitance C_d on the di/dt noise generated by switching output buffers. Instead of entirely going through the V_{ss}, half the peak noise current goes through the V_{dd}, because the high frequency noise spike sees the decoupling capacitance as a short circuit [Lar97].

The effect of decoupling capacitance on the *di/dt* noise generated by switching internal circuits is illustrated in Figure 15-11 (b). The noise generated by the initial current spike can be made arbitrarily low by increasing C_d [Lar97]. The upper limit of the noise level, regarding the inductive bonding wires as open circuits, is

$$v_{n,\max} = \frac{\frac{C_{sw}}{2} V_{dd}}{2(C_d + \frac{C_{sw}}{2})}, \tag{15.7}$$

where C_{sw} is the switched capacitance [Lar97].

15.5.5 Resonance and Damping

The inductive bonding wires and the capacitance on the chip form an LC tank. Use of dedicated on-chip decoupling capacitors may cause resonance oscillations if the resonance and damping issue is not addressed [Lar98]. Figure 15-12 presents a simple second order model of a chip, it comprises the bonding wire inductances and a capacitance C_T that represents the aggregate capacitance of the chip and a series resistor R_T representing the aggregate resistance of the chip as defined in [Lar98]. This circuit has a resonance frequency given in (15.8) and a damping factor ζ given in (15.6) [Lar95]

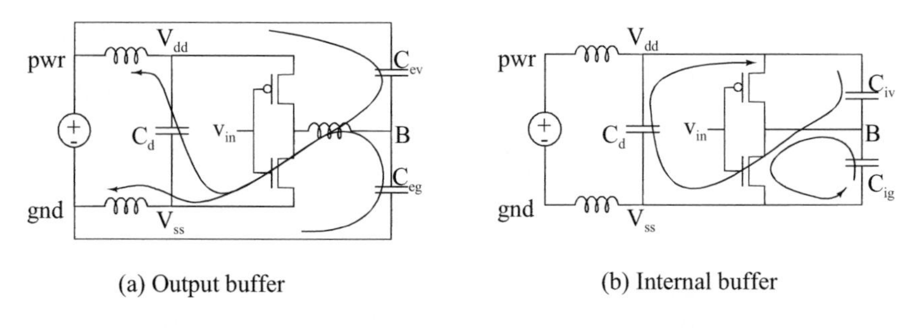

Figure 15-11. Effect of decoupling capacitance C_d.

$$\omega_{res} = \frac{1}{\sqrt{2LC_T}}, \tag{15.8}$$

$$\zeta = \frac{R_T}{2}\sqrt{\frac{C_T}{2L}}. \tag{15.9}$$

When a capacitance is switched during a clock cycle its charge is redistributed to the decoupling capacitance [Lar95]. The peak noise ('1' in the Figure 15-12) caused by this charge redistribution is given by (15.7). The peak to peak noise ('2' in the Figure 15-12) is given by [Lar95]

$$\Delta V_{pp} = v_{n,\max}\left(1 + e^{-\pi\varsigma/\sqrt{1-\varsigma^2}}\right). \tag{15.10}$$

Resonance can cause noise accumulation over several clock periods if the damping factor is low and the clock frequency is close to the resonance frequency. The noise accumulation can therefore be reduced by increasing damping or by using a clock frequency much lower than the resonance frequency [Lar95].

Damping can be increased by inserting a resistance in the resonance circuit. This can be accomplished in three ways: using the inherent resistance of the logic circuits, the parasitic resistance of the decoupling capacitance or the resistance of the power supply paths [Lar95]. Some other damping methods were reviewed in [Lar95] as well. Using the resistance of the logic circuits in the design of increased damping was deemed impractical in [Lar95].

The parasitic resistance of the decoupling capacitance can be designed such that a damping factor of about 0.3 to 0.4 can be attained [Lar95]. The optimal resistance of the decoupling capacitance is derived in [Lar98] to be

$$R_d = \frac{1}{\omega_{res} C_d}, \tag{15.11}$$

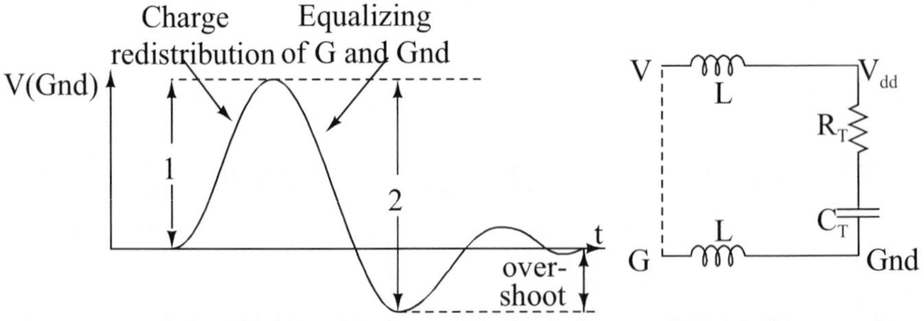

Figure 15-12. Noise waveform of the internal Gnd node for an underdamped system when internal circuits are switched and a second order model of the chip (V and G denote the off-chip supply voltage and ground respectively).

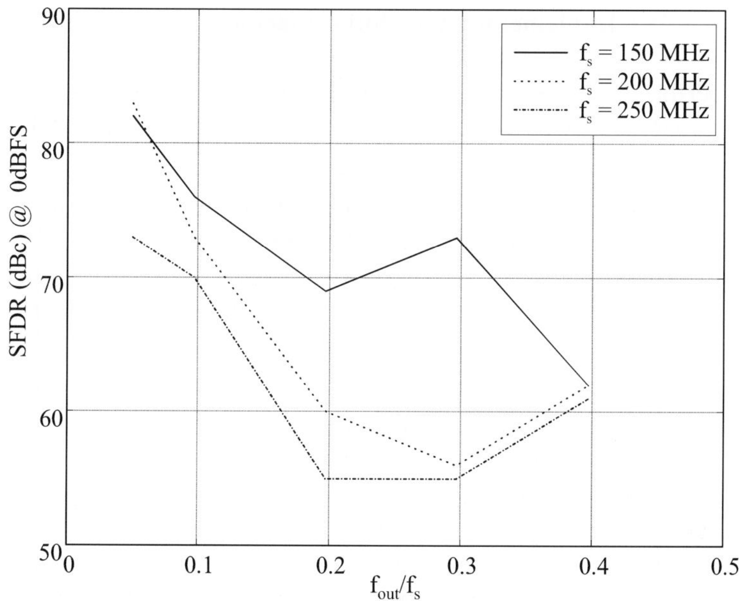

Figure 15-13. SFDR as function of relative output frequency at full-scale (0 dBFS).

where R_d is the resistance of the decoupling capacitor and C_d is the capacitance. Using a single decoupling capacitor with a large resistance can cause a resistive drop in the supply voltages, if there is a very large peak current such that the inactive circuits do not have low enough resistance to supply the peak current without a substantial resistive drop [Lar95]. In this case, two decoupling capacitors can be used. One with a low parasitic resistance is used to assure low resistive voltage drop and the other is used to optimize the damping [Lar95]. Resistance of power supply paths can be used to increase the damping factor above what is typically achieved by the parasitic resistance of the decoupling capacitance.

Table 15-1. Measured D/A Converter Performance

Resolution	12 bit
INL	0.25 LSB
DNL	0.21 LSB
Full-scale output current	17.2 mA
Sampling rate	Up to 500 MHz
Two tone SFDR (f_s = 150 MHz, f_{out1} = 39.75 MHz, f_{out2} = 45.75 MHz)	58.7 dBc
Power consumption (f_s = 500 MHz)	240 mW
Process	0.35μm CMOS
D/A converter core area	3.73 mm^2

15.5.6 Implemented On-chip Capacitor

In high performance systems, a fair share of the chip area is consumed by the on-chip capacitors. [Che95] reports that about 10% is needed for high per-

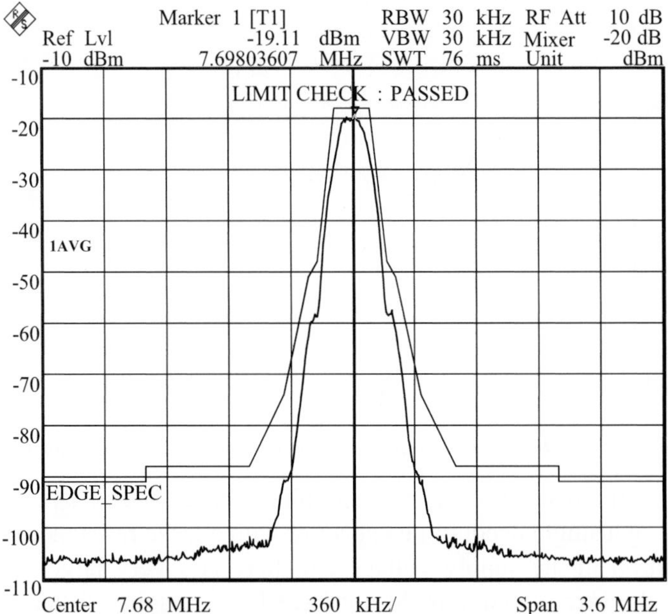

Figure 15-14. Power spectrum of the EDGE signal in the lowpass mode.

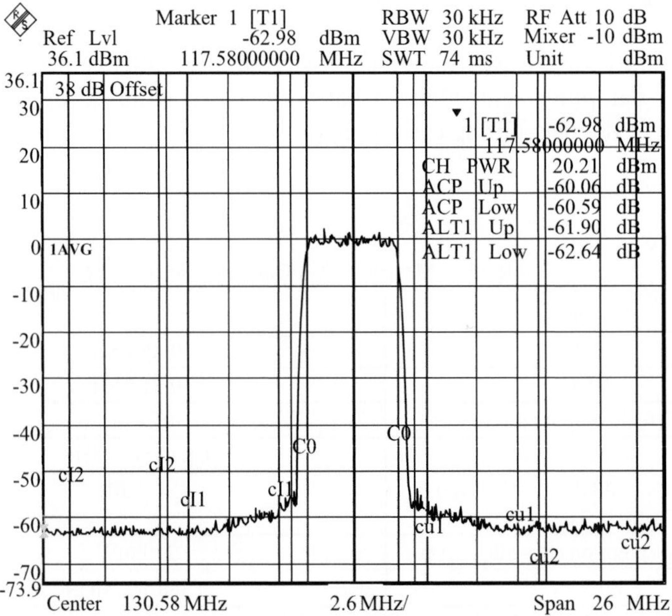

Figure 15-15. Power spectrum of the WCDMA signal in the bandpass mode.

formance circuits with a cycle time of 5 ns or less. Therefore, on-chip decoupling capacitors (total capacitance of 2 nF) are used in Figure 15-17 and Figure 16-22. The capacitor units consist of two overlapping metal layers. Instead of one big capacitor, it is better to insert smaller capacitors close to the hot spots of the chip, where the intense switching activity would otherwise bring about excessive power supply noise. In these circuits, the inductance of the bond wire is small, because the ground level is connected through several bonding wires and package pins.

15.6 Measurement Results

The orthogonal carriers to the digital quadrature modulator are generated by the multistandard modulator [Van02] in measurements. The on-chip D/A converter was used in measurements. Measurements are performed with a 50 Ω doubly terminated cable. Figure 15-8 shows that typical INL and DNL errors are 0.25/0.21 LSB, respectively. Figure 15-13 shows the spurious free dynamic range (SFDR) as a function of relative output frequency. The digital quadrature modulator was bypassed in the SFDR measurements (Fig. 10), because the output frequency response of the digital quadrature modulator

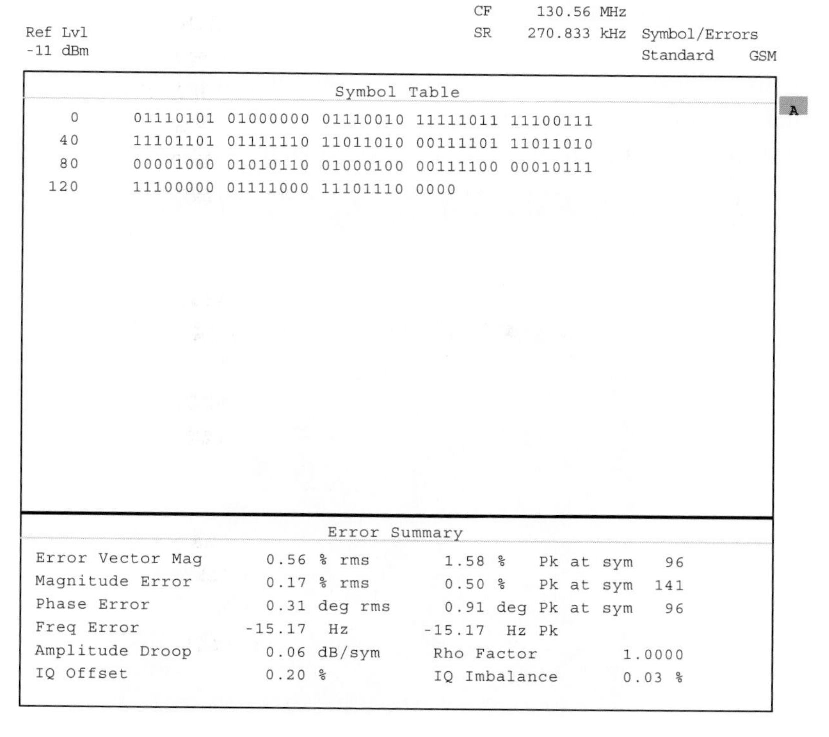

Figure 15-16. Measured phase and frequency errors.

has only lowpass, bandpass and highpass modes (see Figure 15-4). The data was driven directly to the D/A converter. The maximum operation frequency of the input pads was 250 MHz, which is the highest sampling frequency in Figure 15-13. The D/A converter performance is summarized in Table 15-1.

The digital quadrature modulator is designed to fulfill the derived spectral, phase, and EVM specifications of GSM, EDGE and WCDMA base stations. The EDGE output signal in the lowpass mode fulfills the EDGE spectrum mask requirements set out in Figure 15-14 [GSM99]. Figure 15-15 shows the WCDMA output with a crest factor of 8.48dB in bandpass mode, where the adjacent channel leakage powers (ACLR1/2) are 60.06 and 61.90dB, respectively, which fulfill the specifications (45/50 dB) [FDD00]. The ACLR1/2 are 77.07 and 82.12dB at the D/A converter input, so most of the errors are generated less by quantization errors in the digital domain and more by the D/A converter analog non-idealities. Figure 15-16 shows the phase error performance of the GSM signal, where the measured phase error is 0.31° rms with a maximum peak deviation of 0.91°. The phase error specifications are 5° with a peak of 20° [GSM99]. The phase error is low, because the quadrature modulation is performed digitally with high precision. The D/A converter sampling frequency was 491.52MHz in the figures.

15.7 Conclusion

The first analog IF mixer stage of a transmitter can be replaced with this digital quadrature modulator. The modulator interpolates orthogonal input carriers by 16 and performs $f_s/4$, $-f_s/4$, $f_s/2$ digital quadrature modulation. A

Figure 15-17. Chip microphotograph.

12b D/A-converter is integrated with the digital quadrature modulator. A segmented current source architecture is combined with a proper switching technique to reduce spurious components and enhance dynamic performance. The die area of the chip is 27.09 mm^2 (0.35μm CMOS technology). Total power consumption is 1.02W at 2.8V with 500MHz (0.78W digital modulator, 0.24W D/A converter). The IC is in a 160-pin CQFP package. Figure 15-17 displays the chip microphotograph.

REFERENCES

[Bas98] J. Bastos, A. M. Marques, M. S. J. Steyaert, and W. Sansen, "A 12-bit Intrinsic Accuracy High-Speed CMOS DAC," IEEE J. Solid-State Circuits, Vol. 33, No. 12, pp. 1959-1969, Dec. 1998.

[Bos01] A. Van den Bosch, M. A. F. Borremans, M. S. J. Steyaert, and W. Sansen, "A 10-bit 1-GSample/s Nyquist Current-Steering CMOS D/A Converter," IEEE J. Solid-State Circuits, Vol. 36, No. 3, pp. 315-324, Mar. 2001.

[Che95] H. H. Chen, and S. E. Schuster, "On-Chip Decoupling Capacitor Optimization for High-Performance VLSI Design," in Proc. International Symposium on VLSI Technology, Systems, and Applications, 1995, pp. 99-103.

[Chi94] S. Y. Chin, and C. Y. Wu, "A 10-bit 125-MHz CMOS Digital-to-Analog Converter (DAC) with Threshold-Voltage Compensated Current Sources," IEEE J. of Solid State Circuits, Vol. 29, No. 11, pp. 1374-1380, Nov. 1994.

[Dar70] S. Darlington, "On Digital Single-Sideband Modulators," IEEE Trans. Circuit Theory, Vol. 17, No. 3, pp. 409-414, Aug. 1970.

[FDD00] 3rd Generation Partnership Project; Technical Specification Group Radio Access Networks; UTRA (BS) FDD; Radio transmission and Reception, 3G TS 25.104, V3.3.0, June 2000.

[GSM99] GSM Recommendation 05.05: "Radio Transmission and Reception," Dec. 1999.

[Kos01] M. Kosunen, J. Vankka, M. Waltari, and K. Halonen, "A Multicarrier QAM Modulator for WCDMA Basestation with on-chip D/A Converter," Proceedings of CICC 2001 Conference, May 6-9 2001, San Diego, USA, pp. 301-304.

[Lar95] P. Larsson, "Analog Phenomena in Digital Circuits," Ph.D. dissertation, No. 376, Linköping University, Linköping, Sweden, 1995.

[Lar97] P. Larsson, "di/dt Noise in CMOS Integrated Circuits," Analog Integrated Circuits and Signal Processing, Vol. 14, pp. 113-129, Sept. 1997.

[Lar98] P. Larsson, "Resonance and Damping in CMOS Circuits with On-Chip Decoupling Capacitance," IEEE Transactions on Circuits and Systems I: Fundamental Theory and Applications, Vol. 45, No. 8, Aug. 1998, pp. 849-858.

[Tak91] H. Takakura, M. Yokoyama, and A. Yamaguchi, "A 10 bit 80MHz Glitchless CMOS D/A Converter," in Proc. IEEE Custom Integrated Circuits Conf., 1991, pp. 26.5.1-26.5.4.

[Van01] J. Vankka, and K. Halonen, "Direct Digital Synthesizers: Theory, Design and Applications," Kluwer Academic Publishers, 2001.

[Van02] J. Vankka, J. Ketola, O. Väänänen, J. Sommarek, M. Kosunen, and Kari Halonen, "A GSM/EDGE/WCDMA Modulator with on-chip D/A Converter for Base Station," ISSCC Digest of Technical Papers, February 3 - 7, 2002, San Francisco, USA, pp. 236-237.

[Won91] B. C. Wong, and H. Samueli, "A 200-MHz All-Digital QAM Modulator and Demodulator in 1.2-μm CMOS for Digital Radio Applications," IEEE J. Solid-State Circuits, Vol. 26, No. 12, pp. 1970-1979, Dec. 1991.

[Wu95] T. Y. Wu, C. T. Jih, J. C. Chen, and C. Y. Wu, "A Low Glitch 10-bit 75-MHz CMOS Video D/A Converter," IEEE J. Solid-State Circuits, Vol. 30, No. 1, pp. 68-72, Jan. 1995.

Chapter 16

16. A GSM/EDGE/WCDMA MODULATOR WITH ON-CHIP D/A CONVERTER FOR BASE STATIONS

16.1 Supported Communication Standards

The Global System for Mobile communication (GSM) is a second generation (2 G) system that has rapidly gained acceptance and a worldwide market share. As the mobile communications market develops, interest is building up in data applications and higher data rate operations. Short message services (SMS) were first added to the GSM system followed by high-speed circuit switched data (HSCSD) and the general packet radio service (GPRS). All of these services use the same modulation format as the original GSM network (0.3 Gaussian minimum shift keying (GMSK)), and change the allocation of the bits and/or packets to improve the basic GSM data rate. As a step towards 3G, enhanced data rates for GSM evolution (EDGE) provides a higher data-rate enhancement of GSM. It uses the GSM infrastructure with upgraded radio equipment to deliver significantly higher data rates. The primary objective of the EDGE signal is to triple the on-air data rate while taking up essentially the same bandwidth as the original 0.3 GMSK signal. The wideband code division multiple access (WCDMA) was selected by the European Telecommunications Standards Institute (ETSI) for wideband wireless access to support 3G services because of its resistance to multi-path fading, and other advantages such as increased capacity. This technology has a wider bandwidth and different modulation format from GSM or EDGE.

The first generation of the 3G base station modulator should include support for GSM, EDGE and WCDMA. The digital IF modulator is designed using specifications related to those standards [GSM99c], [TDD00], [FDD00]. The main requirements of the modulator are shown in Table 16-1. By programming the GSM/EDGE/WCDMA modulator, different carrier

Table 16-1.

GSM/EDGE/WCDMA Modulator Specifications

Symbol rates/Chip rate	270.833 ksym/s (GSM/EDGE) 3.84 Msym/s (WCDMA)
Modulations	GMSK with BT = 0.3 (GSM), linearized Gaussian 3π/8-8PSK (EDGE), M-QAM (WCDMA)
Carrier Spacing	200 kHz (GSM/EDGE), 5 MHz (WCDMA)
Frequency error	2 Hz
Spurious Free Dynamic Range	-80 dBc
D/A converter sampling frequency	65 – 110 MHz
Power ramp duration	5 - 15 μs
Power ramp curve type	Hanning, Hamming, Blackman
Power control range	0 - -32 dB
Power control fine tuning step	0.25 dB

spacings, modulation schemes, power ramping, frequency hopping and symbol rates can be achieved. By combining the outputs of multiple modulators, multicarrier signals can be formed, or the modulator chips can be used for steering a phased array antenna. The formation of multi-carrier signals in the modulator increases the base station capacity. The major limiting factor of digital IF modulator performance at base station applications is the D/A converter, because the development of D/A converters does not keep up with the capabilities of digital signal processing with faster technologies [Van01a].

16.1.1 GSM System

In the GSM system, a constant envelope partial response Gaussian Minimum Shift Keying (GMSK) modulation is used [Mur81].

Figure 16-1 presents a simple block diagram of the GMSK system. The symbol rate used in the GSM system is 270.833ksym/s. The data bits $d_i \in \{0,1\}$ are differentially encoded by the rule

$$\alpha_i = 1 - 2\hat{d}_i, \tag{16.1}$$

where

$$\hat{d}_i = \text{mod}[d_i + d_{i-1}, 2], \tag{16.2}$$

where mod[$d_i + d_{i-1}$, 2] denotes modulo 2 addition and $\alpha_i \in \{-1, +1\}$ is the

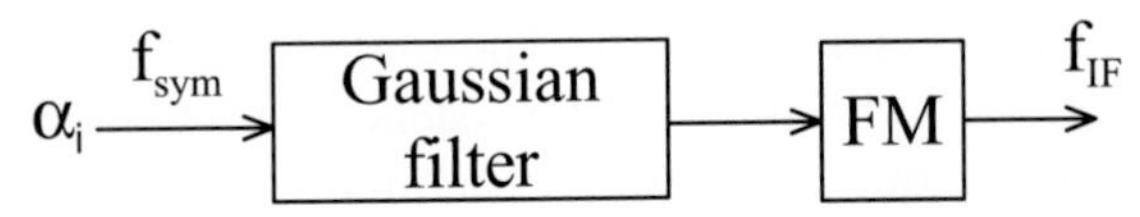

Figure 16-1. A simple block diagram of the GMSK system.

modulating data value [Dig99].

The differentially encoded data stream is filtered in a Gaussian pulse shaping filter. The impulse response of the pulse shaping filter is defined as

$$g(t) = h(t) * \mathrm{rect}\left(\frac{t}{T}\right), \tag{16.3}$$

where $*$ denotes convolution and the rectangular function rect(x) is defined by

$$\mathrm{rect}\left(\frac{t}{T}\right) = \begin{cases} \frac{1}{T} & |t| < \frac{T}{2} \\ 0 & \text{else} \end{cases}, \tag{16.4}$$

and $h(t)$ is defined by

$$h(t) = \frac{\exp\left(\frac{-t^2}{2\delta^2 T^2}\right)}{\sqrt{2\pi}\delta T}. \tag{16.5}$$

δ is

$$\delta = \frac{\sqrt{\ln(2)}}{2\pi BT} \tag{16.6}$$

where B is the 3dB bandwidth and T is the duration of one data bit. The GSM specification requires that $BT = 0.3$ [Dig99]. This definition is theoretical and the realization of (16.5) would be a filter of infinite length. This theoretical filter is associated with the tolerances defined in [GPP00c].

The phase of the modulated signal is

$$\varphi(t) = \sum_i \alpha_i \pi h \int_{-\infty}^{t-iT} g(u)\mathrm{d}u, \tag{16.7}$$

where the modulation index $h = 1/2$. This value implies that the maximum phase change during the data interval is $\pi/2$ radians [GPP00c].

The modulated intermediate frequency (IF) signal at the useful part of the burst can be expressed as

$$IF_{gsm}(t) = \sqrt{\frac{2E_c}{T}} \cos(2\pi f_c t + \varphi(t) + \varphi_0), \tag{16.8}$$

where E_c is the energy per modulating bit, f_c is the carrier frequency and ϕ_0 is a random phase, which is presumed to be constant during one burst [Dig99].

The modulation accuracy of the GSM is defined by the phase error, i.e. the difference between the phase error trajectory and its linear regression on the active part of the time slot. The spectral properties are defined by the spectrum mask and the timing of the burst is defined by the time mask. The required levels and the definitions of the performance metrics are specified in [GPP00c].

16.1.2 EDGE System

The Enhanced data rates for GSM evolution (EDGE) is a high-speed mobile data standard, intended to enable the second-generation GSM and TDMA networks to transmit data up to 384 kilobits per second. The EDGE provides the speed enhancements by changing the type of modulation. It triples the on-air data rate while meeting the same bandwidth occupancy (200 kHz) as the original GSM system. A linearized Gaussian 3π/8-8PSK modulation scheme [Lau86], [Jun94] is applied in the EDGE.

A block diagram of the EDGE system is shown in Figure 16-2. Although this section presents the whole signal generation chain in Figure 16-2, the blocks preceding the pulse shaping filters are left out from the implemented circuit presented in Figure 16-6.

In the EDGE system the data bits, arriving at a of 812.5 kbit/s, are Gray-coded from groups of three bits into an octal-valued symbol l according to Table 16-2. The Gray coding ensures that if a symbol is interpreted erroneously as an adjacent symbol, the error occurs only in one bit. It thus reduces the bit error probability. This way the symbol rate 812.5/3 ksym/s = 270.833 ksym/s will equal the symbol rate in the GSM system. After the Gray coding, an 8PSK modulation is performed. The 8PSK symbols are achieved by the rule

$$s_i = e^{j2\pi l/8}, \tag{16.9}$$

where l is given by Table 16-2. The 8PSK symbols are continuously rotated with 3π/8 radians before pulse shaping. The rotated symbols are defined as

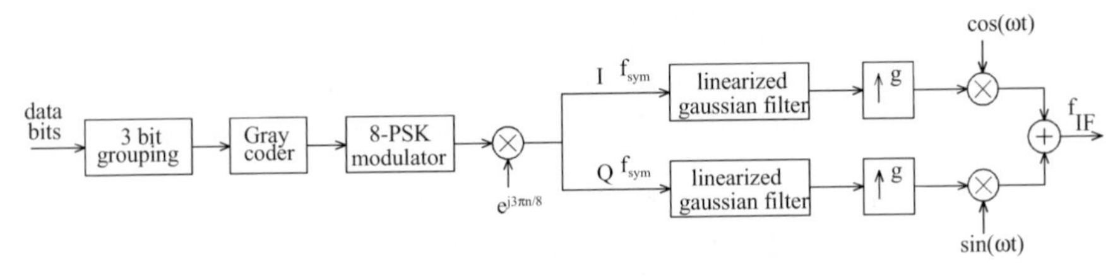

Figure 16-2. Generation of EDGE signal.

Table 16-2. Gray-coding of binary bit triplets into octal symbols

d_{3n},d_{3n+1},d_{3n+2}	0,0,0	0,0,1	0,1,0	0,1,1	1,0,0	1,0,1	1,1,0	1,1,1
l	3	4	2	1	6	5	7	0

$$\hat{s}_i = s_i \cdot e^{ji3\pi/8}, \tag{16.10}$$

The 3π/8-rotation of the symbols ensures that the modulating signal can never be near zero, hence the envelope of the transmitted signal can never be near zero either.

This symbol rotation and the so-called forbidden zone is illustrated in Figure 16-3. The original locations of the 8PSK symbols in the constellation diagram are marked with crosses and the symbols after rotation are marked with dots. The rotation of the symbols ensures a forbidden region around the origin. For example, an 8PSK transition from point 1 to point 2 turns to a transition from point 1 to point 3 after the rotation. Due to the rotation, a transition of two succeeding symbols always starts from the point marked with a cross and ends at the point without a cross, or vice versa. This means that the transition through zero can never occur. The transitions 1 - 3 and 3 - 4 in Figure 16-3 are examples of the transitions approaching the origin as nearly as possible. The transition 4 - 6 illustrates the original 8PSK zero crossing transition from point 4 to point 5 after the symbol rotation.

This zero evading behavior of the EDGE signal decreases the crest factor,

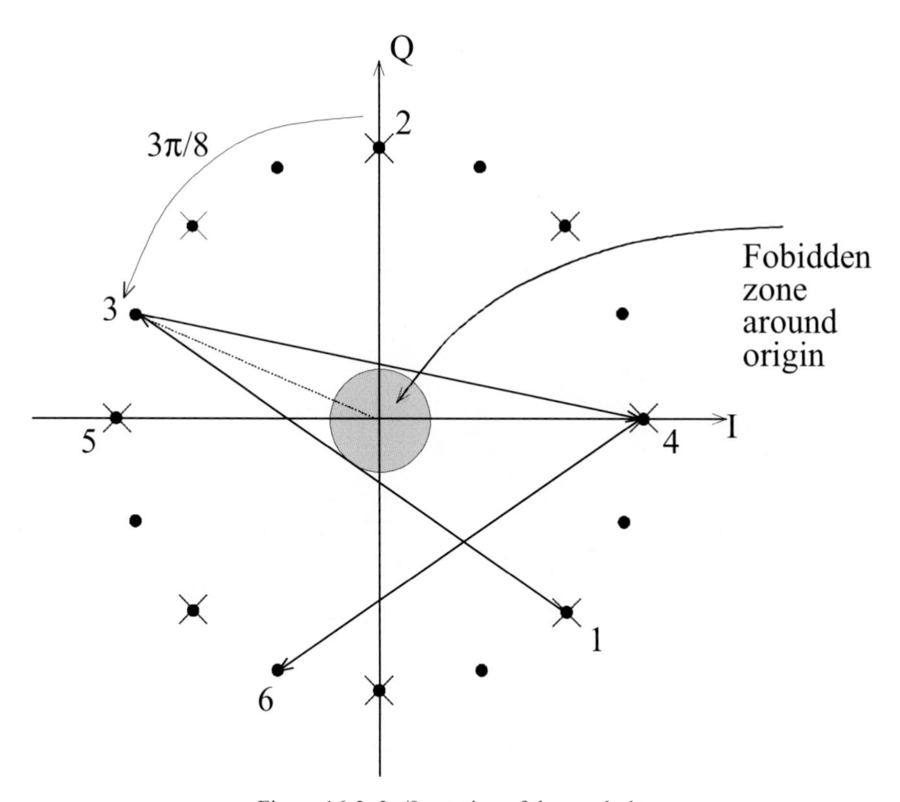

Figure 16-3. 3π/8-rotation of the symbols.

i.e. the ratio of the peak value and the average value of the signal. A low crest factor is advantageous in the performance of a digital-to-analog (D/A) converter and a power amplifier.

The modulating symbols are pulse shaped using a linearized GMSK pulse shaping filter [Lau86], [Jun94], allowing the 8PSK to fit into the GSM spectrum mask [Fur99]. The impulse response of the linearized GMSK filter is defined as [Dig99]

$$c_0(t) = \begin{cases} \prod_{i=0}^{3} S(t+iT) & 0 \le t \le 5T \\ 0 & \text{else} \end{cases}, \tag{16.11}$$

where T is the symbol period and

$$S(t) = \begin{cases} \sin\left(\pi \int_0^t g(u)\mathrm{d}u\right) & 0 \le t \le 4T \\ \sin\left(\frac{\pi}{2} - \pi \int_0^{t-4T} g(u)\mathrm{d}u\right) & 4T \le t \le 8T, \\ 0 & \text{else} \end{cases} \tag{16.12}$$

Figure 16-4. Constellation diagram of baseband EDGE signal.

$$g(t) = \frac{1}{2T}\left(Q\left(2\pi \cdot 0.3\frac{t - 5T/2}{T\sqrt{\log(2)}}\right) - Q\left(2\pi \cdot 0.3\frac{t - 3T/2}{T\sqrt{\log(2)}}\right)\right). \tag{16.13}$$

The error function $Q(t)$ is

$$Q(t) = \frac{1}{\sqrt{2\pi}}\int_t^\infty e^{-\frac{\tau^2}{2}}\,\mathrm{d}\tau. \tag{16.14}$$

The baseband signal is

$$y(t) = \sum_i \hat{s}_i \cdot c_0\left(t - iT + \frac{5}{2}T\right). \tag{16.15}$$

The in-phase and quadrature branches are obtained from the real and imaginary parts of $y(t)$, respectively:

$$I(t) = \Re\{y(t)\} = \sum_i \cos(\varphi_i) \cdot c_0\left(t - iT + \frac{5}{2}T\right), \tag{16.16}$$

$$Q(t) = \Im\{y(t)\} = \sum_i \sin(\varphi_i) \cdot c_0\left(t - iT + \frac{5}{2}T\right), \tag{16.17}$$

where ϕ_i is the angle of the rotated symbol $\hat{s}_i$. A constellation diagram of the baseband signal is presented in Figure 16-4. It can be clearly seen that the signal does not pass the region around zero.

The modulated IF signal is

$$IF_{edge}(t) = I(t)\cos(2\pi f_{out} t) - Q(t)\sin(2\pi f_{out} t), \tag{16.18}$$

where f_{out} is the carrier frequency.

The error vector magnitude (EVM), i.e. the magnitude of an error vector between the vector representing the actual transmitted signal and the vector representing the error-free modulated signal, is used to define the accuracy of the modulation in the EDGE system. The spectrum mask and the time mask define the spectral and timing properties of the signal, respectively. These performance metrics are defined and the required levels are specified in [GPP00c].

Equations (16.11) - (16.14) given by [Dig99] are a very complicated way to define a filter pulse. By transferring the pulse peak to the origin

$$c_0'(t) = c_0\left(t + \frac{5}{2}T\right), \tag{16.19}$$

it is possible to approximate it by an exponent function

$$c_0'(t) \approx e^{P(t)}, \tag{16.20}$$

where $P(t)$ is an M degree polynomial. Due to the even symmetry of the pulse c'_0, it can be assumed that the odd coefficients are zero. The coefficients of the polynomial $P(t)$ are achieved by calculating c'_0 using (16.11) and fitting an M degree polynomial to the natural logarithm of the result. For $M = 6$,

$$c_0'(t) = e^{0.007837(t/T)^6 - 0.2117(t/T)^4 - 1.0685(t/T)^2 - 0.0717}, \tag{16.21}$$

which is valid for all t and does not involve integrated error functions. The calculated peak and root mean square (rms) errors of the compared exact and approximate pulses over the interval $-5T/2 \leq t \leq 5T/2$ are 1.6% and 0.27%, respectively. This approximation is advantageous during the system level simulations when generating a reference signal for a device being designed. During this generation, the coefficients of the pulse shaping filter have to be recalculated for each symbol if the sampling rate is simultaneously converted with some rational number ratio, which is often the case with multimode systems. This can be efficiently computed using this approximation.

16.1.3 WCDMA System

The Wideband Code Division Multiple Access (WCDMA) system uses a Quadrature Amplitude Modulation (QAM). The modulating chip rate for WCDMA is 3.84 Mcp/s. From the modulator's point of view, the chip rate equals the symbol rate, and so the term symbol rate is used. Figure 16-5 presents the QAM modulation of the complex-valued chips sequence generated by the spreading process. The spreading process is described in [GPP00a], for example.

The I and Q branches, obtained from real and imaginary parts of the complex-valued chip sequence respectively, are pulse shaped using a Root-Raised-Cosine filter in order to reduce the signal bandwidth. The impulse

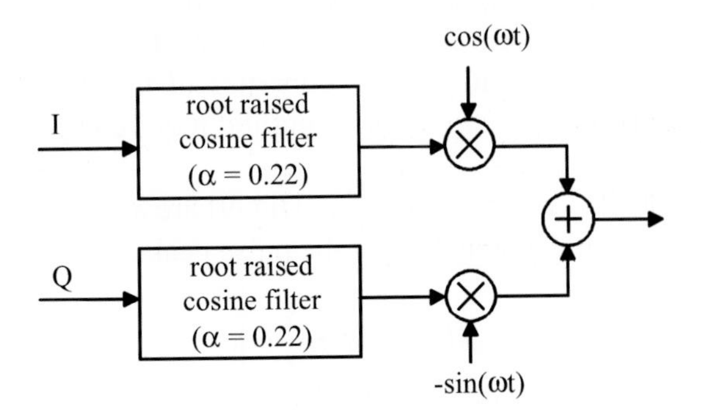

Figure 16-5. QAM modulation.

response of the pulse shaping filter is defined as

$$h(t) = \frac{\sin\left(\pi \frac{t}{T}(1-\alpha)\right) + 4\alpha \frac{t}{T}\cos\left(\pi \frac{t}{T}(1+\alpha)\right)}{\pi \frac{t}{T}\left(1-\left(4\alpha \frac{t}{T}\right)^2\right)}, \tag{16.22}$$

where T is the symbol (chip) duration and α is the roll-off factor defining the used transmission bandwidth. The roll-off factor is specified as $\alpha = 0.22$ in the WCDMA applications [GPP00b].

The signal is upconverted by multiplying it with a sinusoidal carrier according to the equation

$$IF_{wcdma}(t) = I(t)\cos(2\pi f_{out} t) - Q(t)\sin(2\pi f_{out} t), \tag{16.23}$$

where $I(t)$ and $Q(t)$ are filtered I and Q symbols and f_{out} is the carrier frequency [GPP00a].

The performance of the WCDMA signal is measured by the EVM, adjacent channel leakage power ratio (ACLR) and peak code domain error (PCDE). The required levels and definitions of the performance metrics are specified in [GPP00b].

16.2 GSM/EDGE/WCDMA Modulator

The block diagram of the modulator chip is shown in Figure 16-6. The use of different modulation formats requires programmable pulse shaping filter coefficients. The reconfiguration of new modulation formats can be achieved between bursts (e.g., GSM/EDGE). The two half-band filters increase pre-oversampling ratios, which reduces the complexity of the re-sampler (the order of the polynomial interpolator). The re-sampler circuit allows the sam-

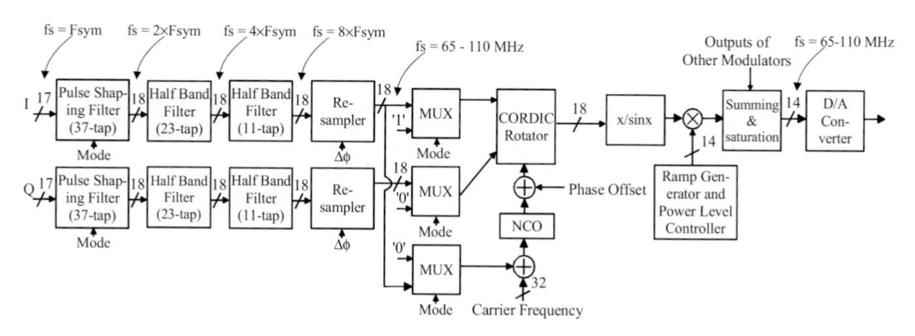

Figure 16-6. GSM/EDGE/WCDMA modulator chip. The symbol rates (Fsym) are shown in Table 16-1.

pling rate of the on-chip D/A converter to have a variable non-integer relationship with the input symbol rates [Eru93]. This block is needed, because the specified D/A converter sampling rates and input symbol rates shown in Table 16-1 do not have an integer frequency relationship. The coordinate rotation digital computer (CORDIC) rotator translates the baseband-centered spectrum to a programmable carrier center frequency [Vol59]. The IF signal is filtered by an x/sinx filter for compensating the sample and hold response of the on-chip 14 bit D/A converter [Sam88]. The internal wordlengths of the modulator are shown in Figure 16-6. The wordlengths were chosen so that the 14-bit D/A converter quantization noise dominates the digital output noise.

16.3 Pulse Shaping and Half-band Filters

The input symbols are filtered using Gaussian (BT = 0.3)/linearized Gaussian/square root raised cosine (α = 0.22) pulse shaping filter in GSM/EDGE/WCDMA mode, respectively [GSM99a], [FDD00]. The square root raised cosine filter (excess bandwidth ratio α = 0.22) was designed to maximize the ratio of the main channel power to the adjacent channels' power under the constraint that the error vector magnitude (EVM) is below 2 % (-34 dB) (see section 0) [Van00]. In the GSM/EDGE systems, a quarter of a guard bit is inserted after each burst, resulting in a burst length of 156.25 symbols [GSM99b]. Therefore, the input symbols to the pulse shaping filter

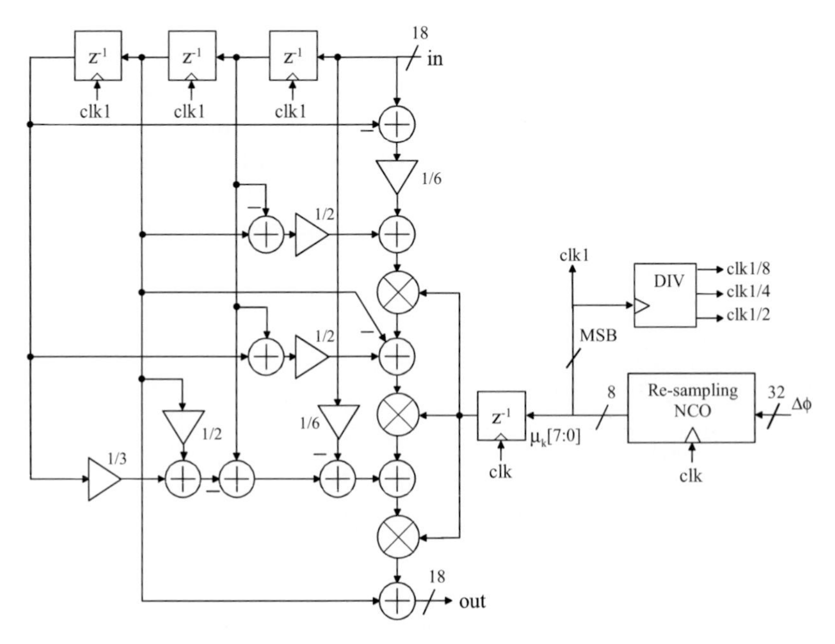

Figure 16-7. Re-sampler.

have to be oversampled by 4 in the GSM and EDGE modes. The pulse shaping filter is implemented using programmable canonic signed digit (CSD) coefficients [Hau93].

The pulse shaping and the half-band filters are implemented using a polyphase structure (see section 11.8) [Fli94]. Taking advantage of the fact that in the modulator data streams in the I and Q paths are processed with the same functional blocks, a further hardware reduction can be achieved by pipeline interleaving techniques (see section 11.11) [Par99]. The pulse shaping and the half-band filters are modified to handle two channels by doubling the sampling rate and delay elements between taps. This results in a more efficient layout with a penalty in terms of increased power dissipation. The fixed coefficients of the half-band filters are implemented using CSD numbers (see section 11.6) [Sam89].

16.4 Re-Sampler

The re-sampler consists of a re-sampling NCO and a cubic Lagrange polynomial interpolator, as shown in Figure 16-7. The re-sampling NCO supplies sampling clocks for re-sampler (clk1), half-band and pulse shaping filters (clk1, clk1/2, clk1/4) as well as the trigger signal for the input data symbols (clk1/8). The frequency control word ($\Delta\phi$) is calculated from the ratio of the input sampling rate of the re-sampler (clk1) to output sampling rate (clk).

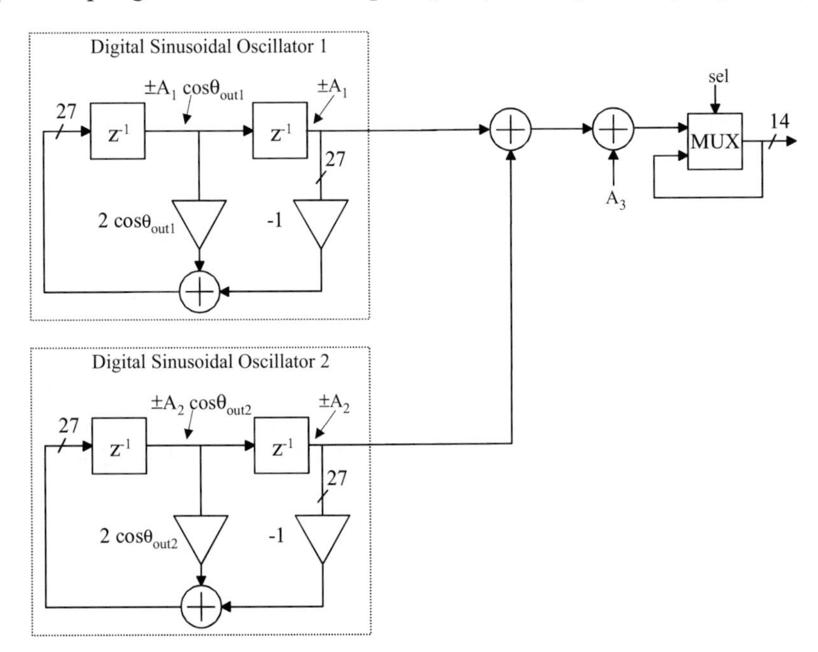

Figure 16-8. Ramp generator and output power level controller.

The trigger signals for different input symbol rates could be achieved by altering the frequency control word. The output of the re-sampling NCO (μ_k) may be considered to represent the phase offset between the input sampling rate of the re-sampler (clk1) to the output sampling rate (clk). Since the phase offset changes on every output sample clock cycle, the interpolation filter coefficients are time-varying. The time-varying filter coefficients with a long period can be easily implemented by polynomial-based interpolation filters using the so called a Farrow structure [Far98]. The cubic Lagrange polynomial interpolator was found suitable for our application because it fulfills spectral and time domain (phase error and EVM) specifications [GSM99c], [TDD00], [FDD00]. The cubic Lagrange polynomial interpolator is implemented with the Farrow structure [Eru93], where the number of unit delay elements is minimized as in Figure 16-7. Furthermore, the frequency error term generated by the external digital phase locked loop (DPLL) locking to an external symbol rate could be added to the frequency control word ($\Delta\phi$) (see section 12.5.1) [Ket03].

16.5 CORDIC Rotator

The in-phase output of the CORDIC rotator is

$$I_{out}(n) = I(n)\cos(\omega_{out}\, n) \pm Q(n)\sin(\omega_{out}\, n), \tag{16.24}$$

where $I(n)$, $Q(n)$ are pulse shaped and interpolated quadrature data symbols. The carrier frequency is

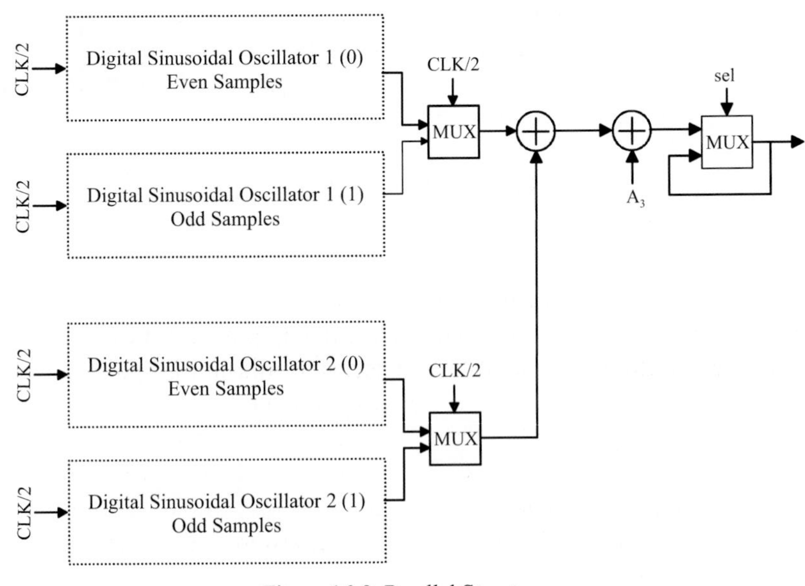

Figure 16-9. Parallel Structure.

$$f_{out} = \frac{\Delta P f_s}{2^j}, \; -\frac{f_s}{2} \leq f_{out} < \frac{f_s}{2}, \tag{16.25}$$

where ΔP is the j-bit phase increment word that can range between $-2^{j-1} \leq F_r < 2^{j-1}$, j is the numerically controlled oscillator (NCO) word length, and f_s is the sampling frequency.

The carrier frequency tuning resolution will be 0.0256 Hz by (4.2), when f_s is 110 MHz, and j is 32. The frequency resolution is better than the target frequency error specification in Table 16-1. The sign in (16.24) is controlled by the frequency control word. The spectral inversion (modulation spectrum is reversed) in the other up-conversion stages could be compensated by changing the sign of the carrier frequency control word (the direction of the vector rotation) in (16.25).

In the frequency modulation mode (GMSK modulation in GSM [GSM99a]), the pulse shaped and interpolated I frequency samples are added to the carrier frequency control word, and the inputs of the CORDIC rotator are set constant ($I(n)$ is 1 and $Q(n)$ is 0) as in Figure 16-6, then the output is from (16.24)

$$I_{out}(n) = \cos(\omega_{out} n + \theta(n)), \tag{16.26}$$

where $\theta(n)$ is the information-bearing component of the phase. In the EDGE/WCDMA mode (quadrature amplitude modulation [GSM99a], [FDD00]), the input to the adder before the NCO is set at zero and I/Q samples are filtered by the pulse shaping filter and interpolated prior to the CORDIC rotator.

The phase offset adds an offset to the NCO. This allows multiple modulators to be synchronized in order to produce carriers with a known phase relationship. The phase offset allows intelligent management of the relative phase of independent carriers. This capability supports the requirements of phased array antenna architectures [Stu81].

16.6 Ramp Generator and Output Power Level Controller

Summing the IF outputs from other devices allows the formation of a multicarrier signal in Figure 16-6. This necessitates power control to be implemented in the digital domain. Otherwise, it would not be possible to adjust the relative power of a single carrier with respect to the others. Therefore, a digital ramp generator and output power level controller is used as in Figure 16-6. The power control is realized by scaling the ramp curve [Van01b].

The programmable up/down unit allows power ramping on a time-slot basis as specified for GSM, EDGE and time division duplex WCDMA (TDD-

WCDMA) [GSM99c], [TDD00]. The ramp generator is based on a recursive digital sine wave oscillator [Van01b]. In previous work [Van01b], the ramp duration was fixed and the ramp generator could only generate cosine windows that have only one cosine term (e.g. a Hanning window). In Figure 16-8, two digital oscillators are used; these allow the Blackman window generation (two cosine terms). The Blackman window gives more attenuation to the switching transients than the Hanning window.

Another method for implementing the ramp generator is to use a look-up table LUT. Because the sampling frequency is high, the size of the LUT becomes large. The size of the LUT increases even more because of the high oversampling ratio required by the variable ramp duration (see Table 16-1). Alternatively, interpolation between the samples in the LUT can be used at the expense of increased complexity. Furthermore, the multiplier is needed to set the output power level. Therefore, the choice was made in favor of the recursive digital oscillators.

16.6.1 Ramp Generator

The ramp generator shown in Figure 16-8 has two recursive digital sine wave oscillators. The impulse response of the second-order system with complex-conjugate poles on the unit circle is a sinusoidal waveform (see Chapter 5). The amplitude and phase of the sinusoidal waveform is determined by the initial values $x(0)$ and $x(1)$ from (5.10) and (5.11). The output frequency of the digital oscillator (θ_{out}) can be altered by changing the coefficient α in (5.1) and the initial value in (5.10). The details and finite wordlength effects of the digital ramp generator and output power level controller are described in [Van01b].

16.6.2 Initial Values of Ramp Generator

The rising ramp of the Blackman window is given by

$$0.42\,A + 0.5\,A\cos(\pi t / T_r + \pi) + 0.08\,A\cos(2\pi t / T_r), \qquad (16.27)$$

where T_r is the ramp duration, t is $[0; T_r]$, and A is the amplitude of the ramp. The falling ramp of the Blackman ramp window is

$$0.42\,A + 0.5\,A\cos(\pi t / T_r) + 0.08\,A\cos(2\pi t / T_r). \qquad (16.28)$$

The cosine terms are implemented with the digital sine wave oscillators and the term 0.42 A is added to their output. The initial values for the falling Blackman ramp are

$$x1(0) = 0.5\,A = A_1, \qquad (16.29)$$

$$x1(1) = 0.5\,A\cos(\pi T_s / T_r) = A_1\cos(\theta_1), \qquad (16.30)$$

for the first oscillator. The initial values of the second oscillator are

$$x2(0) = 0.08A = A_2, \tag{16.31}$$

$$x2(1) = 0.08A\cos(2\pi T_s / T_r) = A_2 \cos(\theta_2). \tag{16.32}$$

The constant A_3 is $0.42A$. The initial values of the rising Blackman ramp for the first oscillator are the negatives of the falling ramp values. In the case of the Hanning window, the initial values are the same as in the Blackman case for the first oscillator, the values for the second oscillator are zero and the constant A_3 is $0.5A$. The ramp duration (T_r) can be altered by changing the output frequencies of the digital oscillators. The value (A) controls the amplitude of the ramp (output power level). During the ramp period the signal *sel* is low in Figure 16-8 and the multiplexer conducts the ramp signal to the multiplier (Figure 16-6). After the ramp duration (T_r) the signal *sel* becomes high; the output of the multiplexer is connected to the input of the multiplexer; and the output power level is constant.

16.6.3 Parallel Structure

The recursive digital oscillator shown in Figure 16-8 suffers from two major drawbacks: the quantization noise accumulates in the recursive structure and the maximum sampling rate of the digital oscillator is determined by its recursive parts. The oscillator produces only one cycle of sine wave in the

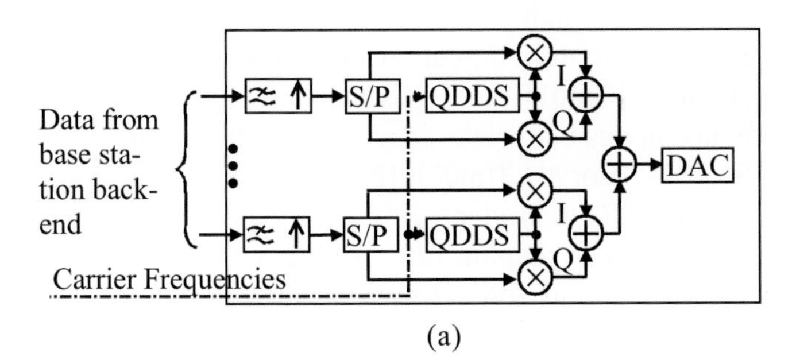

(a)

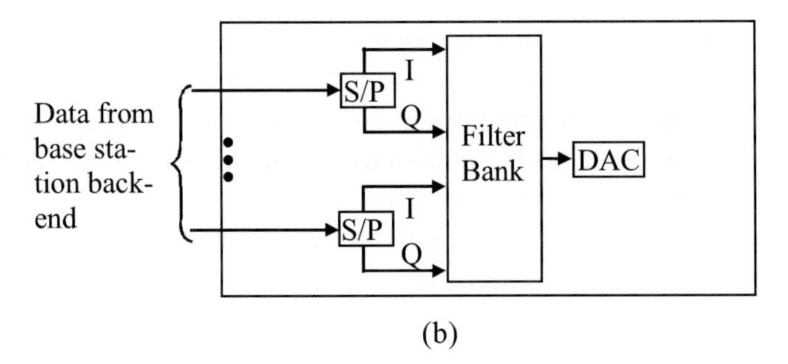

(b)

Figure 16-10. Multicarrier options: a) Parallel upconverters. b) Synthesis filter bank based.

Blackman ramp; after that new initial values are updated, so the problem of the accumulated noise is alleviated. The parallel structure is used to reduce the sampling rate of the digital oscillators. Then the multipliers can be implemented with higher wordlengths but with a reasonable area, as the speed requirement is not so stringent. The idea of the implemented parallel structure is to generate the desired sinusoidal oscillating signal with two oscillators, one of which generates the odd samples, and the other the even samples. This means that four oscillators are needed to generate Blackman ramps. The parallel structure is presented in Figure 16-9, where the sampling rate of the digital oscillator is halved. The odd and even oscillator outputs are alternately selected with a 2-to-1 multiplexer (MUX), the select signal of which is the divided clock. In order for the oscillators to generate correct samples to the multiplexer output, each oscillator must operate at a double output frequency (θ_{out}). A phase offset must be added to the odd oscillator so that the outputs are not duplicated. The initial values of the parallel ramp generator are calculated by choosing first the same initial values as in the normal case and calculating the next two values using the difference equation (5.1) and choosing the odd samples for the odd oscillators and the even samples for the even oscillators.

16.7 Multicarrier Modulator Architectures

In Figure 16-10 two multicarrier modulator architectures are presented: a bank of parallel quadrature digital upconverters [Van00], [Kos01], [Van02], [Int01], [Ana01], [Gra01], and pulse shaping filtering, interpolation and quadrature upconversion using a synthesis filter bank [Bel74], [Tsu78], [Sch81], [Kur82], [Cor90], [Im00], [Paš01]. In Figure 16-10(a), and Figure 16-10(b) it is possible to produce different modulation schemes by means of programming the pulse shaping filter. In Figure 16-10(a) the up-conversion to the IF frequency is performed by a quadrature direct digital synthesizer (QDDS), two multipliers and an adder [Cho01]. The upconversion can be also performed by a coordinate rotation digital computer (CORDIC) algorithm (see Chapter 6). In Figure 16-10(b) the pulse shaping, interpolation and upconversion to the IF frequencies are performed by the synthesis filter bank. The synthesis filter banks can be classified into three main types as described in [Sch81]; namely: per-channel approaches [Kur82], multistage techniques [Tsu78], [Paš01], and block-techniques which include the orthogonal transform of IDFT type [Bel74], [Cor90], [Im00]. The carrier frequency resolution depends on the number of channels in the synthesis filter bank; this approach therefore requires a considerable amount of hardware if a fine carrier frequency tuning and a small number of carriers are needed (see Table 16-1). In the synthesis filter bank, the carrier frequency resolution can also be improved by making fine frequency adjustments in the complex baseband [Paš01]. The fine carrier frequency resolution could be achieved with low hardware cost, as in Figure 16-10(a), by programming the QDDS [Van01]. The fine carrier frequency resolution gives a high degree of flexibility in IF and RF frequency planning. The choice was therefore made in favor of the parallel upconverter architecture (combining a number of parallel modulator outputs in Figure 16-6).

16.8 Design Flow

This chapter gives a brief overview of the design flow and the software used when the GSM/EDGE/WCDMA modulator circuit was implemented. A simplified diagram of the design flow is shown in Figure 16-11. Of course, some of the stages in the design flow were performed simultaneously supporting one another.

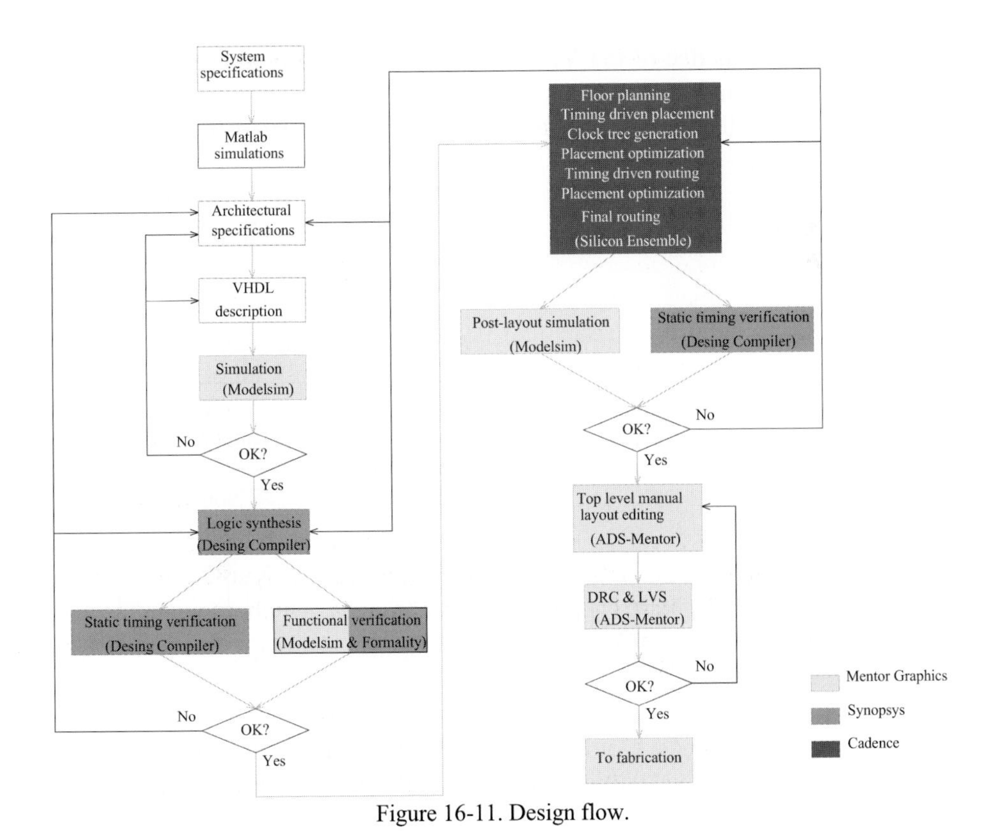

Figure 16-11. Design flow.

16.8.1 High Level Modeling

After and during the system specifications made in collaboration with Nokia Networks, high level modeling and simulation were performed. A simulation model with a graphical user interface was developed using the Matlab signal processing tool. The purpose of this model was mainly to model the effects of the finite precision in the digital circuit and the nonidealities of the D/A-converter on the precision of modulation and signal spectrum.

16.8.2 Hardware Description

After the high level modeling, the Register Transfer Level (RTL) description was written in VLSI Hardware Description Language (VHDL). Keeping in mind the future work, the VHDL descriptions were made modular and parameterizable. The test benches and the simulation scripts were developed to ease the simulations. The VHDL code was simulated using Modelsim and the results were viewed and verified using Matlab.

16.8.3 Logic Synthesis

When the performance of the VHDL description was found appropriate, the VHDL description was read into the Synopsys Design Compiler for logic synthesis. First, the timing information was provided from the technology

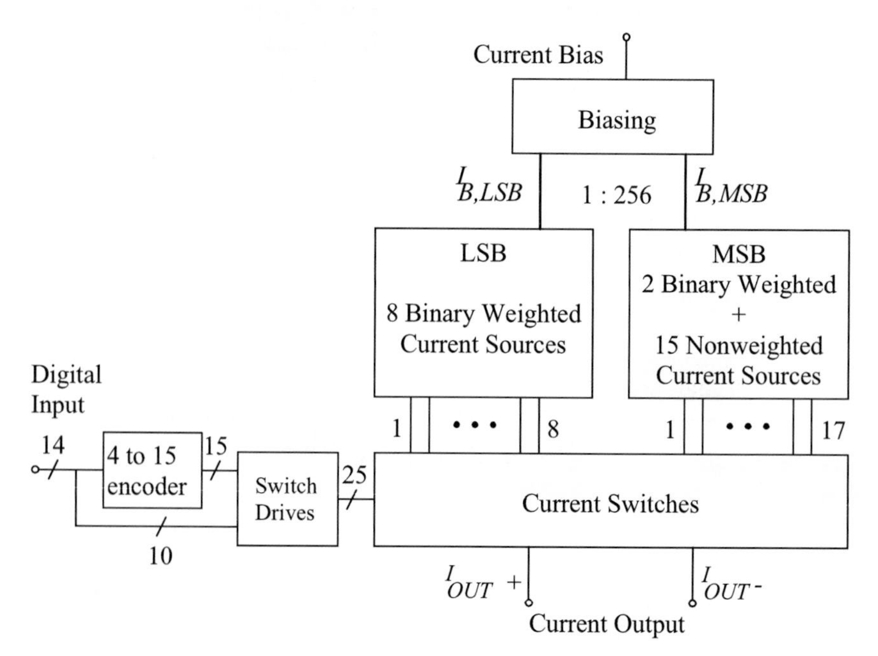

Figure 16-12. Block diagram of the 14-b D/A converter.

libraries. The VHDL description was synthesized to the logic elements meeting the given timing constraints. In some resource demanding blocks, for example, in re-samplers, an automatic pipelining feature of the Synopsys Behavioral Compiler was used [Syn99]. A top-down hierarchal compilation strategy described in [Bha99] with some modifications was used to compile the main blocks of the design. Thereafter, the design was grouped together and flattened. The static timing analysis was performed using the Design Compiler and the functionality of the synthesized design was verified by Formality and Modelsim.

16.8.4 Layout Synthesis

A floorplan with a so-called black-box reserving the area for the D/A-converter was generated in the Cadence Envisia Design Planner. The synthesized VERILOG netlist and the floorplan of the design were transferred to the Cadence Envisia Silicon Ensemble layout synthesis tool. In order to enable the timing driven place and route, the timing constraints were also provided for these tools. After the power planning and the initial placement of cells were performed, the clock tree was generated and the placement was optimized using the stand-alone programs CTGEN and PBOPT, respectively. Thereafter, the layout was routed. Exploiting the timing information available after the routing, another placement optimization was performed, followed by a new round of routing.

The timing and the functionality of the completed layout of the digital part was verified using Design Compiler and Modelsim, respectively.

16.8.5 Final Layout

After the digital part of the modulator was designed and verified, the layout was transferred into Mentor IC-station for the final manual editing. The ana-

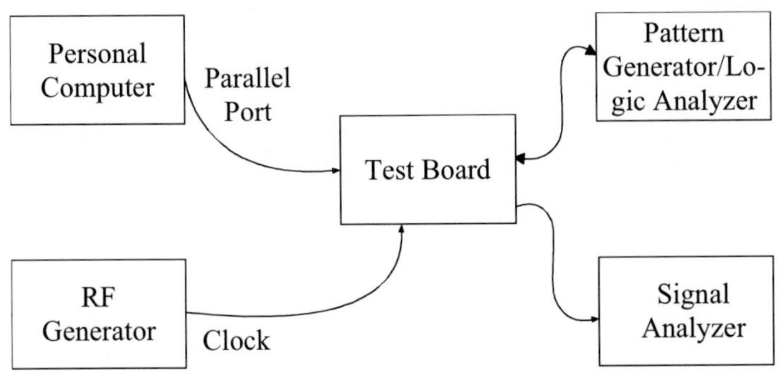

Figure 16-13. Block diagram of test system.

log part of the modulator i.e. the D/A-converter was inserted into the area reserved by the black-box. This manual editing also involved the addition of the bonding pads and the decoupling capacitors. A number of diodes were also added to protect the chip from the charges induced during the manufacturing process.

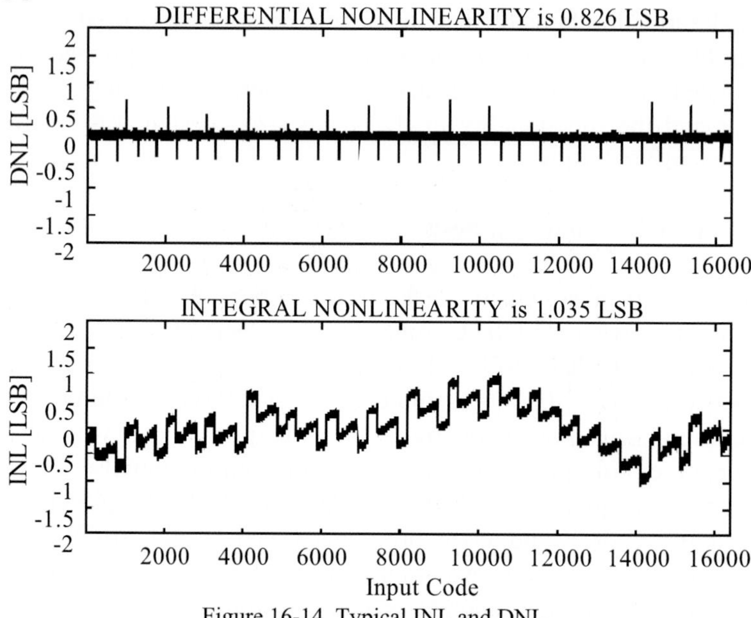

Figure 16-14. Typical INL and DNL.

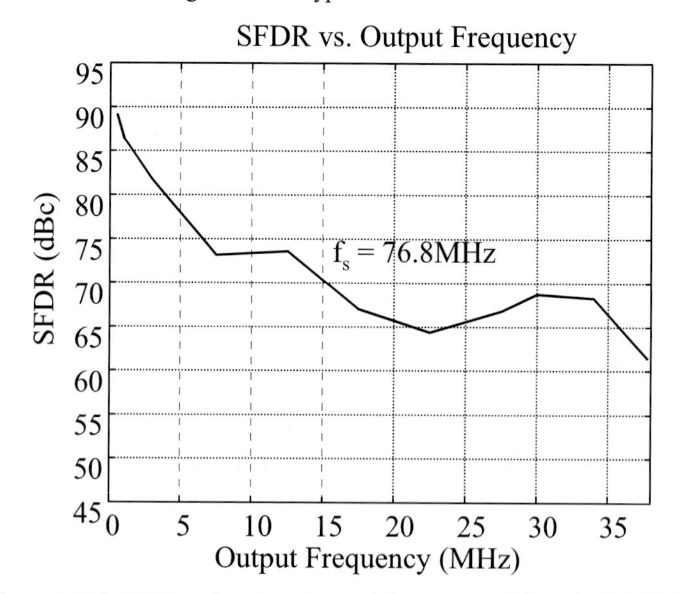

Figure 16-15. SFDR as function of output frequency at full-scale (0 dBFS).

The Design Rule Check (DRC) and the Layout Versus Schematic (LVS) were performed for the final layout using Mentor IC-station and Design Architect. After the successful DRC and LVS checks, the final layout was transferred to the fabrication in the GDSII format. The chip was fabricated using a 0.35 μm CMOS technology.

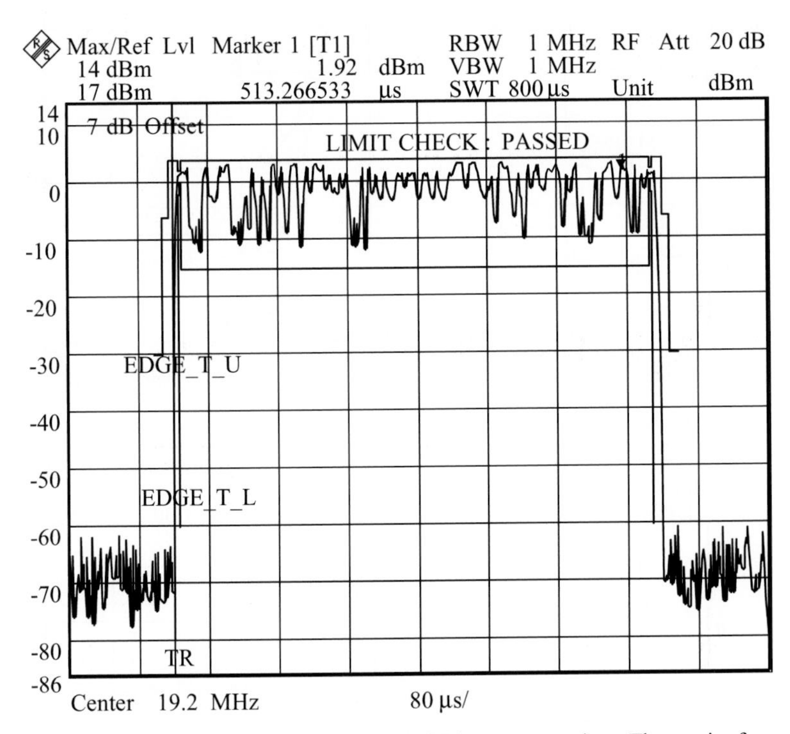

Figure 16-16. Transmitted power level of the EDGE burst versus time. The carrier frequency is 19.2 MHz.

Table 16-3 Spectrum due to Switching Transients (Peak-Hold Measurement, 30 kHz Filter Bandwidth, Reference ≥ 300 kHz with zero offset)

Offset (kHz)	Maximum Power Limit (dBc)		Measured Maximum Power (dBc) at Digital Output		Measured Maximum Power (dBc) at D/A converter Output	
	(GMSK)	(8-PSK)	(GMSK)	(8-PSK)	(GMSK)	(8-PSK)
400	- 60	-55	-77.3	-77.7	-65.64	-65.53
600	- 70	-65	-95.2	-94.2	-75.1	-75.36
1200	- 77	-77	-106.3	-106.8	-80.55	-78.57
1800	- 77	-77	-106.9	-107.7	-79.92	-79.23

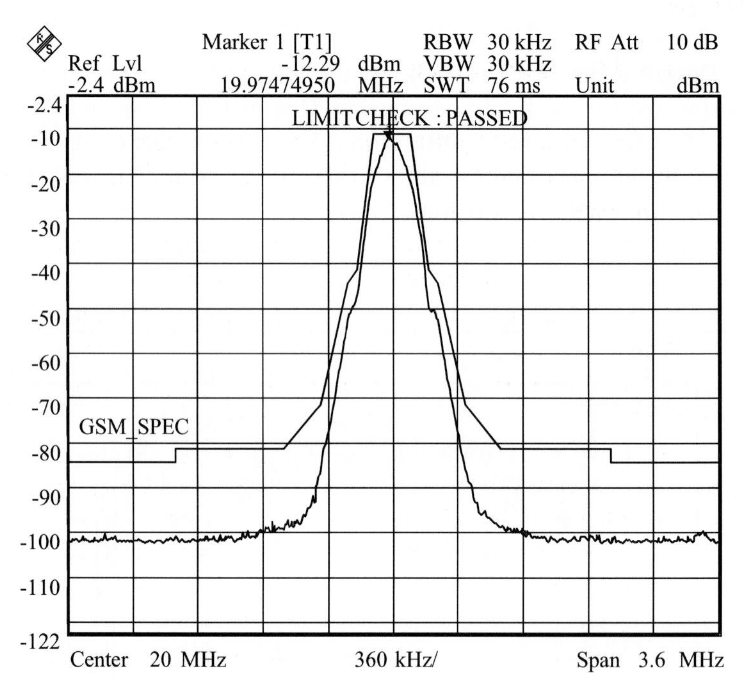

Figure 16-17. Power spectrum of GSM signal.

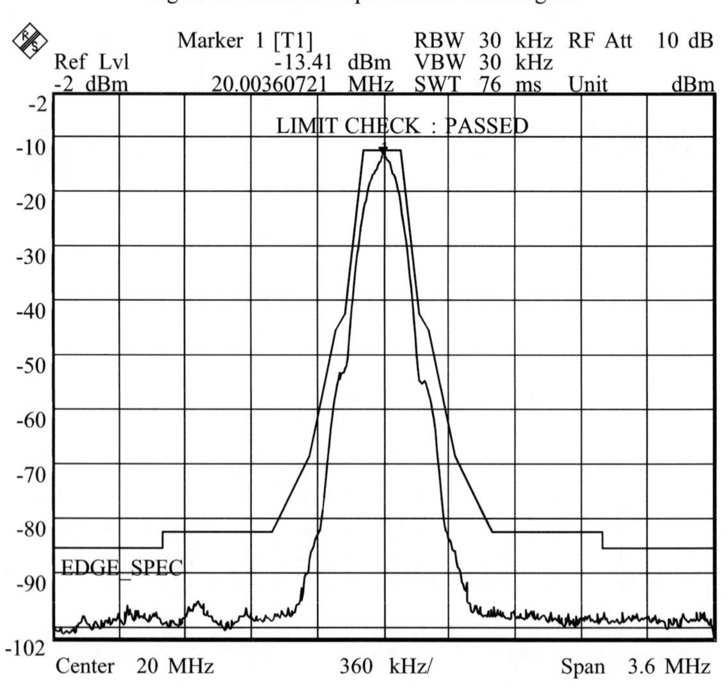

Figure 16-18. Power spectrum of EDGE signal.

16.9 D/A Converter

As the multicarrier feature requires high dynamic range requirements for the D/A converter, the wordlength was chosen to be 14 bit. The 14-bit on-chip D/A converter is based on a segmented current steering architecture [Bos01]. It consists of a 6b MSB matrix (2b binary and 4b thermometer coded), and an 8b binary coded LSB matrix (Figure 16-12). The static linearity is achieved by sizing the current sources for intrinsic matching [Bos01] and by using layout techniques; this is a prerequisite for obtaining good dynamic linearity. The cascode structure is used to increase the output impedance of the unit current source, which improves the linearity of the D/A-converter.

The dynamic linearity is important in this IF modulator because of the strongly varying signal. Therefore, well-designed and carefully laid out switch drivers and current switches are used. A major function of the switch driver in Figure 16-12 is to adjust the crosspoint of the control voltages, and to limit their amplitude at the gates of the current switches, in such a way that these transistors are never simultaneously in the off state and that the feedthrough is minimized. The crossing point of the control signals is set by delaying the falling edge of the signal [Tak91]. Dummy switch transistors are used to improve the synchronization of the switch transistors control sig-

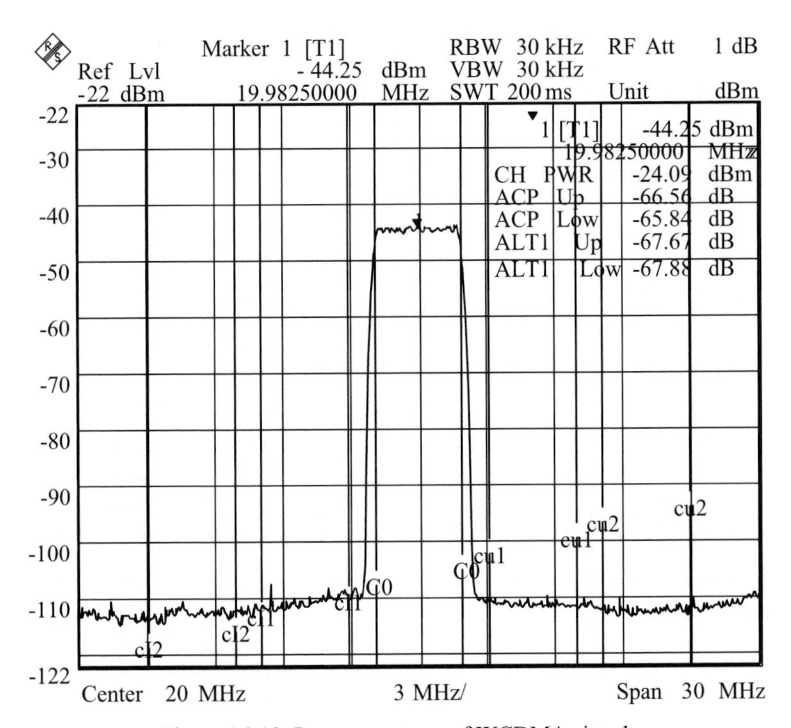

Figure 16-19. Power spectrum of WCDMA signal.

nals. Disturbances connected to the external bias current are filtered out on-chip with a simple one pole low-pass filter. The D/A-converter is implemented with a differential design, which results in reduced even-order distortions and provides a common-mode rejection of disturbances.

16.10 Measurement Results

To evaluate the multi-standard modulator, a test board was built and a computer program was developed to control the measurements. Figure 16-13 illustrates the block diagram of the multi-standard modulator test system. The on-chip D/A converter was used in measurements. Measurements are performed with a 50 Ω doubly terminated cable. The sampling rate of the D/A converter was 76.8 MHz in the measurement. Figure 16-14 shows that typical integral linearity (INL) and differential linearity (DNL) errors are

Table 16-4
Performance Summary

Signal Quality					
Measured at	GSM Phase Error (°)		EDGE EVM (%)		WCDMA EVM (%)
	peak	rms	peak	rms	rms
Digital Data at D/A Converter Input	0.75	0.29	1.263	0.27	1.11
Analog Signal at D/A Converter Output	1.71	0.74	1.55	0.37	1.18
Specifications at the base station RF port	20	5	22	7.0	17.5
Spectral Properties					
	GSM		EDGE	WCDMA	
	600 kHz offset		600 kHz offset	ACLR1	ACLR2
Digital Data at D/A converter input	-100		-90	72.9	73.3
Analog Signal at D/A Converter Output	-87.34		-84.58	65.84 From Figure 16-19	67.67 From Figure 16-19
Specifications at the base station RF port	-70		-70	45	50

1.04/0.83 LSB, respectively. The spurious free dynamic range (SFDR) is shown as a function of the output frequency in Figure 16-15. The SFDR to Nyquist frequency is better than 80 dBc at low synthesized frequencies, decreasing to 62 dBc at high synthesized frequencies in the output frequency band (single tone).

Figure 16-16 shows the measured ramp up and down profiles of the transmitted burst, which satisfy the EDGE base station masks. The allowed power of spurious responses originating from the power ramping before and after the bursts is specified by the switching transient limits. Some margin (3 dB) has been left between the values in [GSM99c] and the values specified for this implementation in Table 16-3 to take care of the other transmitter stages that might degrade the spectral purity of the signal. The power levels measured at the digital output and the D/A converter output meet the limits shown in Table 16-3.

The output signal in Figure 16-17 meets the GSM spectrum mask requirements [GSM99c]. The output signal in Figure 16-18 meets the EDGE spectrum mask requirements [GSM99c]. Figure 16-19 shows the WCDMA

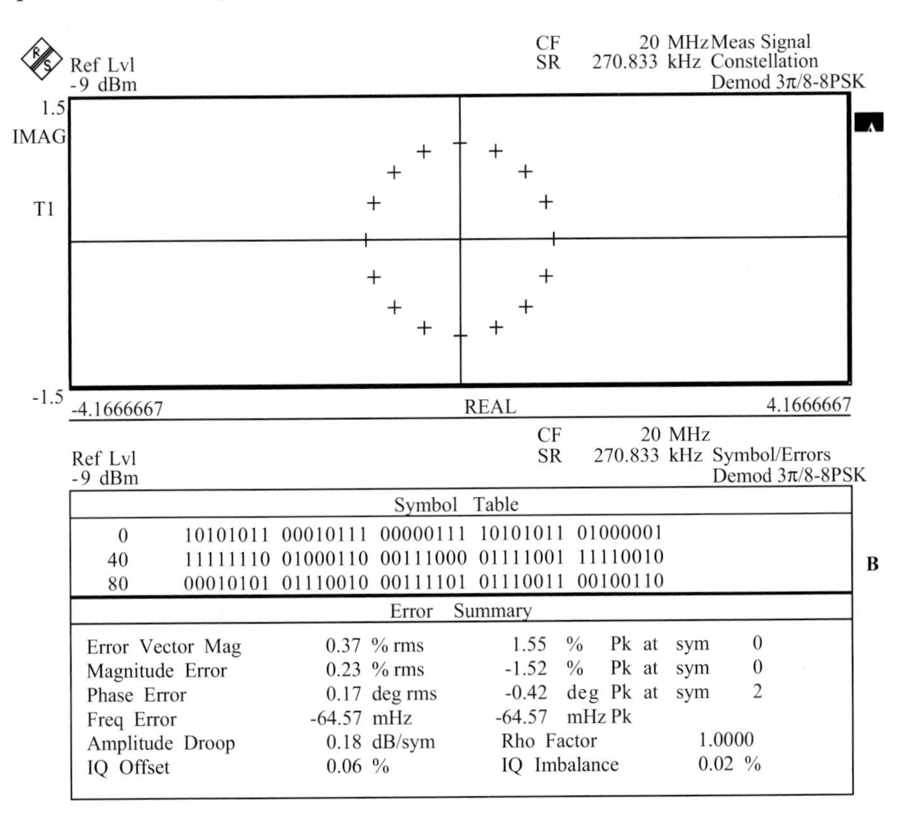

Figure 16-20. Measured EVM errors in EDGE mode.

output with a crest factor of 11.43 dB, where the adjacent channel leakage powers (ACLR1/2) are 65.84 and 67.67, respectively. Figure 16-20 shows the error vector magnitude (EVM) performance in EDGE mode, where the measured rms EVM is 0.37% with a maximum peak deviation of 1.55%. The signal performance is summarized in Table 16-4. The phase error, EVM and spectral performance [FDD00], measured at the digital output and the D/A converter output, meet the specifications shown in Table 16-4. In the multi-standard IF modulator, most of the errors are generated less by quantization errors in the digital domain and more by the D/A converter analog non-idealities, as shown in Table 16-4. There is some margin in the D/A converter output for taking care of the other transmitter stages that might degrade the signal quality. Combining a number of parallel modulator outputs allows the formation of multi-carrier signal in Figure 16-6. Figure 16-21 shows the multi-carrier signal at the D/A converter output. The first adjacent channel leakage power ratio is 57.19 dB, which meets the specification (45 dB) [FDD00].

16.11 Conclusions

A GSM/EDGE/WCDMA modulator with a 14-bit on-chip D/A converter was implemented. The pre-compensation filter, which compensates the sinc droop above the Nyquist frequency, makes it possible to use WCDMA signal images for up-conversion. The new programmable up/down unit allows power ramping on a time-slot basis as specified for GSM, EDGE and TDD-WCDMA. The multi-standard modulator meets the spectral, phase, and EVM specifications. The die area of the chip is 22.09 mm^2 in 0.35 μm CMOS technology. Power consumption is 1.7W at 3.3V with 110 MHz (maximum clock frequency). The IC is in a 160-pin CQFP package. Figure 16-22 displays the chip micrograph.

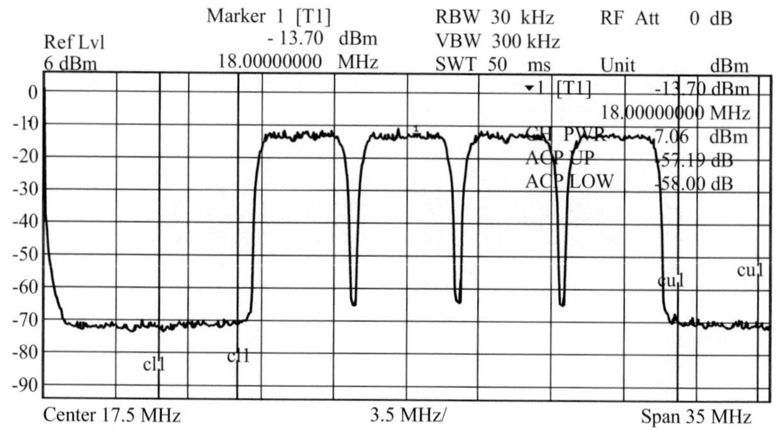

Figure 16-21. Power spectrum of multi-carrier WCDMA signal.

REFERENCES

[Ana01] "AD6623 Four Channel, 104 MSPS Digital Transmit Signal Processor (TSP)," Analog Devices Preliminary Technical Data Sheet, May 2001.
[Bel74] M. G. Bellanger, and J. L. Daguet, "TDM-FDM Transmultiplexer: Digital Polyphase and FFT," IEEE Trans. on Commun., Vol. COM-22, pp. 1199-1204, Sept. 1974.
[Bha99] Bhatnagar, H, Advanced ASIC Chip Synthesis Using Synopsys® Design Compiler™ and PrimeTime®, Kluwer Academic Publishers, USA 1999.
[Bos01] A. Van den Bosch, M. A. F. Borremans, M. S. J. Steyaert, and W. Sansen, "A 10-bit 1-GSample/s Nyquist Current-Steering CMOS D/A Converter," IEEE J. Solid-State Circuits, Vol. 36, No. 3, pp. 315-324, Mar. 2001.
[Cho01] K. H. Cho, and H. Samueli, "A Frequency-Agile Single-Chip QAM Modulator with Beamforming Diversity," IEEE J. of Solid State Circuits, Vol. 36, No. 3, pp. 398–407, Mar. 2001.
[Cor90] I. R. Corden, and R. A. Carrasco, "Fast Transform Based Complex Transmultiplexer Algorithm for Multiband Quadrature Digital Modulation Schemes," IEE Proceedings, Vol. 137, Pt. 1, No. 6, pp. 408-416, Dec. 1990.
[Dig99] Digital cellular telecommunication system (Phase 2+); Modulation (GSM 05.04) V8.1.0, European Telecommunications Standards Institute,

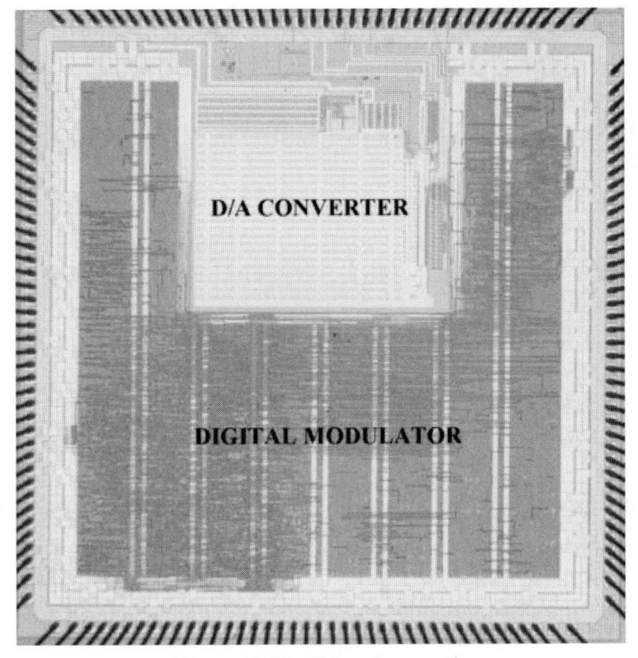

Figure 16-22. Chip micrograph.

December 1999.
[Eru93] L. Erup, F. M. Gardner, and R. A. Harris, "Interpolation in Digital Modems–Part II: Implementation and Performance," IEEE Trans. on Commun., Vol. COM-41, pp. 998-1008, June 1993.
[Far88] C. W. Farrow, "A Continuously Variable Digital Delay Element," in Proc. IEEE International Symposium on Circuits and Systems (ISCAS), June 1988, pp. 2641-2645.
[FDD00] 3rd Generation Partnership Project; Technical Specification Group Radio Access Networks; UTRA (BS) FDD; Radio transmission and Reception, 3G TS 25.104, V3.3.0, June 2000.
[Fli92] N. J. Fliege, and J. Wintermantel, "Complex Digital Oscillators and FSK Modulators," IEEE Trans. on Signal Processing, Vol. SP-40, No. 2, pp. 333-342, Feb. 1992.
[Fli94] N. J. Fliege, Multirate Digital Signal Processing, John Wiley & Sons, 1994.
[Fur99] A. Furuskär, S. Mazur, F. Muller, and H. Olofsson, "EDGE: Enhanced Data Rates for GSM and TDMA/136 Evolution," IEEE Personal Communications, June 1999, pp. 56-66.
[GPP00a] 3GPP Technical Specification Group Radio Access Network, Spreading and Modulation (FDD), TS 25.213 V3.2.0, March 2000.
[GPP00b] 3GPP Technical Specification Group Radio Access Network, UTRA (BS) FDD; Radio transmission and Reception, TS 25.104 V3.3.0, June 2000.
[GPP00c] 3GPP Technical Specification Group GERAN, Digital cellular telecommunication system (Phase 2+); Radio transmission and reception, TS 05.05 V8.6.0, September 2000.
[Gra01] "GC4116 Multi-standard Quad DUC Chip," Graychip Data Sheet, April 2001.
[GSM99a] GSM Recommendation 05.04: "Modulation," Dec. 1999.
[GSM99b] GSM Recommendation 05.02: "Digital Cellular Telecommunications System (Phase 2+); Multiplexing and Multiple Access on the Radio Path," July 1999.
[GSM99c] GSM Recommendation 05.05: "Radio Transmission and Reception," Dec. 1999.
[Hau93] J. C. Hausman, R. R. Harnden, E. G. Cohen, and H. G. Mills, "Programmable Canonic Signed Digit Filter Chip," U. S. Patent 5,262,974, Nov. 16, 1993.
[Im00] S. Im, W. Lee, C. Kim, Y. Shin, S. H. Lee, and J. Chung, "Implementation of SDR-Based Digital IF Channelizer/De-Channelizer for Multiple CDMA Signals," IEICE Trans. Commun., Vol. E83-B, No. 6, pp. 1282-1289, June 2000.

[Int01] "ISL5217 Quad Programmable UpConverter," Intersil Corporation Data Sheet, April 2001.
[Jun94] P. Jung, "Laurent's Representation of Binary Digital Continuous Phase Modulated Signals with Modulation Index 1/2 Revisited," IEEE Transactions on Communications, Vol. 42, No. 2/3/4, February/March/April 1994, pp. 221-224.
[Ket03] J. Ketola, J. Vankka, and K. Halonen, "Synchronization of Fractional Interval Counter in Non-Integer Ratio Sample Rate Converters," in Proc. ISCAS'03, Vol. II, May 2003, pp. 89-92.
[Kos01] M. Kosunen, J. Vankka, M. Waltari, and K. Halonen, "A Multicarrier QAM Modulator for WCDMA Basestation with on-chip D/A Converter," in Proc. Custom Integrated Circuits Conf., May 2001, San Diego, USA, pp. 301-304.
[Kur82] C. F. Kurth, K. J. Bures, P. R. Gagnon, and M. H. Etzel, "A Per-Channel, Memory-Oriented Transmultiplexer with Logarithmic Processing," IEEE Trans. on Commun., Vol. COM-30, pp. 1520-1527, July 1982.
[Lau86] P. Laurent, "Exact and Approximate Construction of Digital Phase Modulations by Superposition of Amplitude Modulated Pulses (AMP)," IEEE Transactions on Communications, Vol. COM-34, No. 2, February 1986, pp. 150-160.
[Mur81] K. Murota, "GMSK Modulation for Digital Mobile Radio Telephony," IEEE Transactions on Communications, Vol. 29, pp. 1044-1050, July 1981.
[Par99] K. K. Parhi, VLSI Digital Signal Processing Systems: Design and Implementation, John Wiley & Sons, 1999.
[Paš01] R. Paško, L. Rijnders, P. R. Schaumont, S. A. Vernalde, and D. Ďuračková, "High-Performance Flexible all-digital Quadrature up and down Converter Chip," IEEE J. of Solid State Circuits, Vol. 36, No. 3, pp. 408-416, Mar. 2001.
[Sam88] H. Samueli, "The Design of Multiplierless FIR Filters for Compensating D/A Converter Frequency Response Distortion," IEEE Trans. Circuits and Syst., Vol. 35, No. 8, pp. 1064-1066, Aug. 1988.
[Sam89] H. Samueli, "An Improved Search Algorithm for the Design of Multiplierless FIR Filters with Powers-of-Two Coefficients," IEEE Trans. Circuits and Syst., Vol. 36, No. 7, pp. 1044-1047, July 1989.
[Sam91] H. Samueli, "On the Design of FIR Digital Data Transmission Filters with Arbitrary Magnitude Specifications," IEEE Transactions on Circuits and Systems, Vol. 38, No. 12, Dec. 1991, pp. 1563-1567.
[Sch81] H. Scheuermann, and H. Göckler, "A Comprehensive Survey of Digital Transmultiplexing Methods," Proc. of IEEE, Vol. 69, No. 11, pp. 1419-1450, Nov. 1981.

[Stu81] W. L. Stutzman, and G. A. Thiele, "Antenna Theory and Design," John Wiley & Sons, Inc. 1981.

[Syn99] Synopsys, Behavioral Compiler Reference Manual, v1999.05.

[Tak91] H. Takakura, M. Yokoyama, and A. Yamaguchi, "A 10 bit 80MHz Glitchless CMOS D/A Converter," in Proc. IEEE Custom Integrated Circuits Conf., 1991, pp. 26.5.1-26.5.4.

[TDD00] 3rd Generation Partnership Project; Technical Specification Group Radio Access Networks; UTRA (BS) TDD; Radio transmission and Reception, 3G TS 25.105, V3.3.0, June 2000.

[Tsu78] T. Tsuda, S. Morita, and Y. Fujii, "Digital TDM-FDM Translator with Multistage Structure," IEEE Trans. on Commun., Vol. COM-26, pp. 734-741, May 1978.

[Van00] J. Vankka, M. Kosunen, I. Sanchis, and K. Halonen, "A Multicarrier QAM Modulator," IEEE Trans. on Circuits and Systems Part II, Vol. 47, No. 1, pp. 1-10, Jan. 2000.

[Van01a] J. Vankka, and K. Halonen, "Direct Digital Synthesizers: Theory, Design and Applications," Kluwer Academic Publishers, 2001.

[Van01b] J. Vankka, M. Honkanen, and K. Halonen, "A Multicarrier GMSK Modulator," IEEE Journal on Selected Areas in Communications, Vol. 19, No. 6, pp. 1070-1079, June 2001.

[Van02] J. Vankka, J. Ketola, J. Sommarek, Olli Väänänen, M. Kosunen, and K. Halonen, "A GSM/EDGE/WCDMA Modulator with on-chip D/A Converter for Base Stations," IEEE Trans. on Circuits and Systems Part II, Vol. 49, No. 10, pp. 645-655, Oct. 2002.

[Vol59] J. E. Volder, "The CORDIC Trigonometric Computing Technique," IRE Trans. on Electron. Comput., Vol. C-8, pp. 330–334, Sept. 1959.

Chapter 17

17. EFFECT OF CLIPPING IN WIDEBAND CDMA SYSTEM AND SIMPLE ALGORITHM FOR PEAK WINDOWING

In a WCDMA system, the downlink signal has typically a high Peak to Average Ratio (PAR). In order to achieve a good efficiency in the power amplifier, the PAR must be reduced, i.e. the signal must be clipped. In this chapter, the effects of several different clipping methods on Error Vector Magnitude (EVM), Peak Code Domain Error (PCDE) and Adjacent Channel Leakage power Ratio (ACLR) are derived through simulations. Also, a very straightforward algorithm for implementing a peak windowing clipping method is presented. The presented algorithm does not involve any iterative optimization algorithms, so it can be implemented as a FIR filter structure with feedback.

17.1 Introduction

In a Wideband Code Division Multiple Access (WCDMA) system, the downlink signal is a sum of signals intended for different users. The envelope of the composite signal is Gaussian distributed, which leads to a high Peak to Average Ratio (PAR) [Ozl95]. The PAR is measured by Crest Factor (CF). In order to achieve a good efficiency in the power amplifier, the PAR must be reduced, i.e. the signal must be clipped. Unfortunately, the signal quality is degraded in the clipping operation. The signal quality can be measured by Error Vector Magnitude (EVM), Peak Code Domain Error (PCDE) and Adjacent Channel Leakage power Ratio (ACLR) [GPP00]. In this chapter, the effects of several different clipping methods on the EVM, PCDE and ACLR are derived through simulations. Also, a very straightforward algorithm for implementing a peak windowing clipping method is presented. The presented algorithm does not involve any iterative optimization algorithms, so it can be implemented as a FIR filter structure with feedback.

17.2 Clipping Methods

17.2.1 Baseband Clipping

In the first case, the baseband signals *I* and *Q* are clipped independently, which leads to the situation where the constellation of the baseband symbols is circumscribed by a square. In the following text, Square refers to this method. In the second case, the amplitude of the complex symbol is clipped but the angle remains unchanged. In this case, the constellation is circumscribed by a circle. This method is referred to as Circle. It is obvious that baseband clipping has no effect on the ACLR because the clipping takes place before the pulse shaping filtering.

17.2.2 Adaptive Baseband Clipping

The problem of the baseband clipping, is that the pulse shaping filter tends to increase the Crest Factor and partially cancels the effect of clipping. An adaptive peak suppression algorithm intended to prevent the peak regrowth is presented in [Mil98]. In this chapter, the same idea is used in the following way. We assume that oversampling ratio of 2 is used in the filtering i.e. the input signal is zeropadded with a factor of 2 before the filter, while the filter ``fills in" the zero-valued samples with interpolated sample values. The clipping threshold after filtering is specified as A. The algorithm is as follows:

1. The unclipped signal is filtered and analyzed.
2. If the threshold is exceeded there are two options:

If the peak is not an interpolated sample, the corresponding sample of the unfiltered signal is scaled down by factor $k = A/A_s$, where A_s is the amplitude of the peak.

If the peak is an interpolated sample, the two adjacent samples of the corresponding sample of the unfiltered signal is scaled down by factor $k = A/A_s$.

3. After scaling operations the new version of the unfiltered signal is filtered.

As can be seen, this clipping algorithm requires at least one iteration loop. The optimal way to apply this algorithm is to use many iteration loops and decrease the clipping level step by step. In reality the complexity and the processing delay restrict the number of the iteration loops. In Section 17.4, this method is simulated in two cases. In the first case, referred to as Adaptive, the presented algorithm is used with one iteration loop. The block diagram of this is presented in Figure 17-1. In the second case, referred to as Iterative, the presented algorithm is used with 20 iteration loops.

17.2.3 IF Clipping

Another way to clip the signal is to operate with the IF signal. As a nonlinear operation, the clipping obviously distorts the signal and the ACLR is decreased. The ACLR can be increased by bandpass filtering after the clipping operation. The IF clipping with and without bandpass filter is simulated. Also, two clipping methods, Error Shaping [Yan00] and Windowing method [Nee98], [Pau96] are simulated. The windowing method is considered in details in the next section.

17.2.4 Windowing Algorithm

Conventional clipping causes sharp corners in a clipped signal. This leads to out of band radiation and reduces the ACLR. It is possible to increase the ACLR by smoothing those sharp corners. This is achieved by multiplying the signal to be clipped with a window function [Pau96], [Nee98]. The difference between conventional clipping and windowing is presented in Figure 17-2. Conventional clipping can be expressed as a multiplication

$$y(n) = c(n)\,x(n), \tag{17.1}$$

where

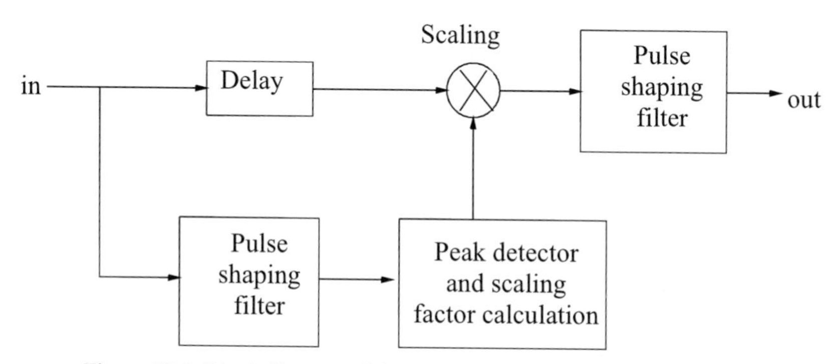

Figure 17-1. Block diagram of the adaptive baseband clipping structure.

$$c(n) = \begin{cases} 1 & ,|x(n)| \leq A \\ \dfrac{A}{|x(n)|} & ,|x(n)| > A, \end{cases} \tag{17.2}$$

where A is the maximum allowed amplitude for the clipped signal. The idea of this method is to replace the function $c(n)$ with the function

$$b(n) = 1 - \sum_{k=-\infty}^{\infty} a_k \, w(n-k), \tag{17.3}$$

where $w(n)$ is the window function and a_k a weighting coefficient.

To achieve the wanted clipping level, the function $b(n)$ must satisfy the inequality

$$1 - \sum_{k=-\infty}^{\infty} a_k w(n-k) \leq c(n), \tag{17.4}$$

for all n. To minimize the EVM, inequality (17.4) must be as near equality as possible. The difference between $c(n)$ and $b(n)$ depends on the window length W defined as a number of samples $w(n)$ which are not equal to zero, and on weighting coefficients a_k. The spectral properties of clipped signal depend on the window length W; choosing the W is a trade off between the EVM and ACLR. After the W is chosen, weighting coefficients a_k must be optimized. If it is assumed that clipping probability and window length are so low that windows do not overlap in the time domain, then the easiest way to form the function $b(n)$ is to find the part

$$\sum_{k=-\infty}^{\infty} a_k w(n-k), \tag{17.5}$$

by convolving the function 1 - $c(n)$ with the window $w(n)$, when $b(n)$ be-

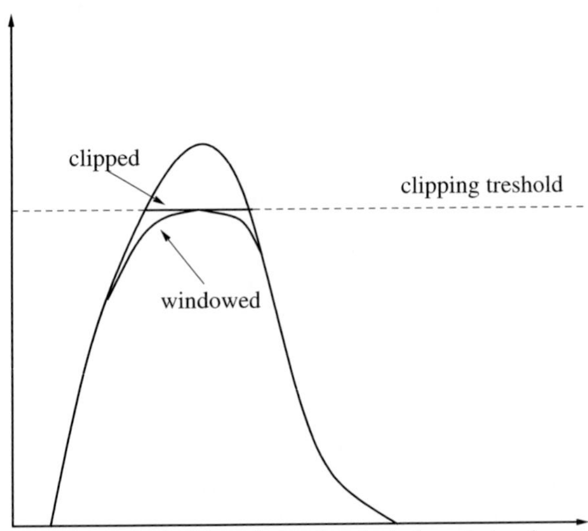

Figure 17-2. Clipped signal and windowed signal.

comes

$$b(n)=1-\sum_{k=-\infty}^{\infty}[1-c(k)]w(n-k). \qquad (17.6)$$

The convolution can be implemented as an FIR filter structure. The ideal convolution is not physically realizable, since it is noncausal and of infinite duration. In order to create a realizable filter, the impulse response, i.e. the window must be truncated and shifted to make the system causal. If the window function is symmetric then the FIR filter has symmetric impulse response $w(k) = w(W\text{-}1\text{-}k)$.

In a real system, windows unfortunately overlap and, as a result of convolution, the signal is clipped much more than needed. In the worst case,

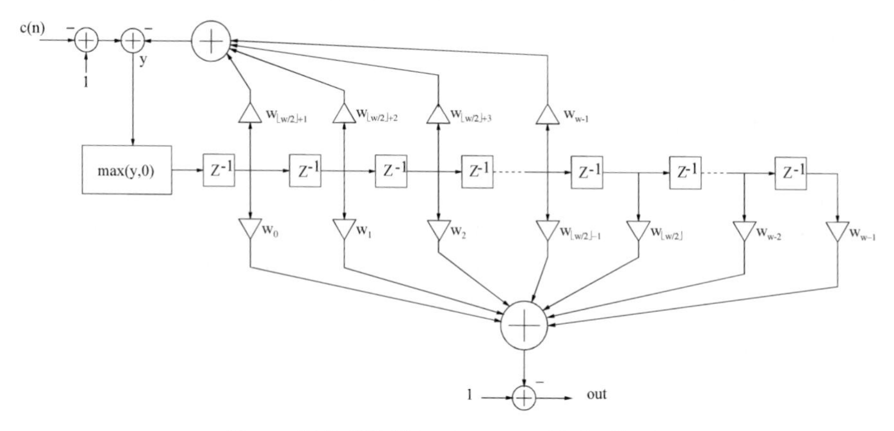

Figure 17-3. FIR filter structure with feedback.

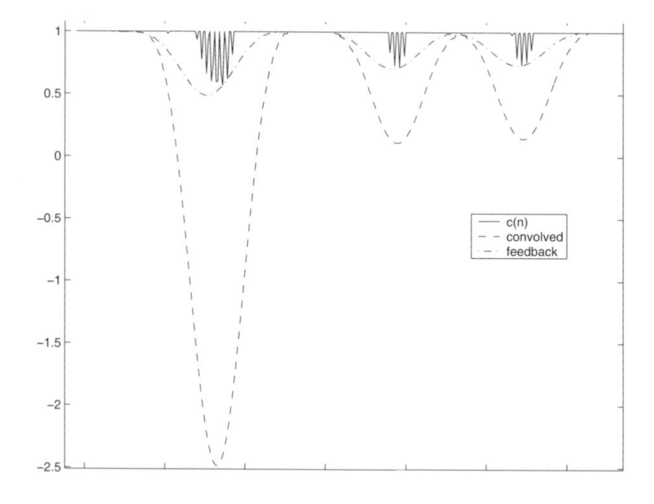

Figure 17-4. Function $b(n)$ formed by convolution (17.6) and by presented algorithm (Figure 17-3).

the sign of function $b(n)$ may become negative, which is fatal for the system. This can be seen in Figure 17-4. Hence another way to form the function $b(n)$ must be found. A simple solution to the problem mentioned above is to combine the conventional FIR structure with a feedback structure that scales down the incoming value if necessary. The proposed structure is presented in Figure 17-3. In Figure 17-3, '$\lfloor\ \rfloor$' denotes floor operation. The impulse response of the filter (coefficients w_n) is equal to the window function w. The previous values are used for calculating a correction term that can be subtracted from input while the output still satisfies (17.4). If the correction term is larger than the input value, signal y (Figure 17-3) becomes negative after the subtraction, which leads to an unwanted clipping result. This is prevented by adding a block that replaces negative values with zero.

Function $b(n)$ formed by convolution and by the presented algorithm is shown in Figure 17-4.

17.3 Simulation Model

All simulations were performed by using MATLAB 5.3. The test data was generated according to [GPP00]. The generation of the WCDMA signal is presented in Figure 17-5. In Figure 17-5, u_n is the complex data of the user n, c_n a spreading code, b_n a weighting factor, s a complex scrambling sequence and ω the angular frequency of the carrier.

The block diagram of the used modulator model is presented in Figure 17-6. The pulse shaping filter is Root Raised Cosine FIR filter with 1001 coefficients. The number of coefficients is chosen to be high so that the clipping is the only source of error. The interpolation ratio of the pulse shaping filter is 2. After the pulse shaping filter there are 3 half band filters. The function of the half band filters is to increase the sampling rate. Every half

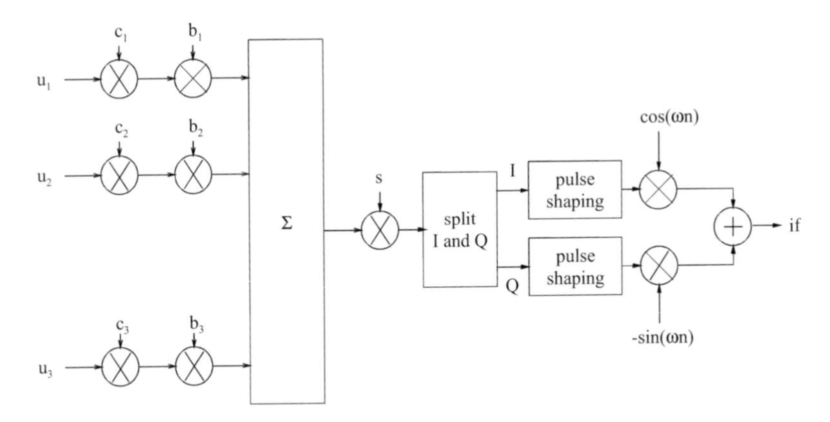

Figure 17-5. WCDMA modulator.

band filter interpolates by factor 2 so the oversampling ratio (OSR) at the IF becomes 16. The ACLR and CF are calculated by using signal IF in Figure 17-6.

Calculation of the EVM and PCDE is performed as it is presented in [GPP00]. In the multicarrier case, the EVM and PCDE are calculated for the carrier with the highest frequency. The ACLR is calculated by using the adjacent channel above the highest used frequency channel.

Simulations are performed in two cases. In the first case, the Crest Factor is minimized so that the ACLR, EVM and PCDE still fulfills the system specifications [GPP00]. In the second case, there is some margin left for the error caused by the following analogue parts. ACLR is specified to be more than 65 dB, EVM less than 3% and PCDE less than -49 dB.

17.4 Results

17.4.1 Single Carrier

The single carrier case is simulated by using 2 different test data. Simulation results for Test Model 3 [GPP00] with 32 active codes and 3 control channels are presented in Table 17-1 and Table 17-2. The Crest Factor before clipping is 15.418 dB. Results for Test Model 3 [GPP00] with 16 active codes and 3 control channels are presented in Table 17-3 and Table 17-4. The Crest Factor before clipping is 15.414 dB. In both cases, the spreading factor is 256.

The baseband clipping method Circle seems to be more efficient than method Square, even if the clipping ratio for Square is less than the clipping ratio for Circle. This can be explained by using the equation

$$IF = I\cos(\omega t) - Q\sin(\omega t) = \sqrt{I^2 + Q^2}\cos(\omega t + \phi), \tag{17.7}$$

which shows that the envelope of the signal IF is linearly dependent on the amplitude of the complex baseband signal. When the I and Q are clipped independently to value A, the critical maximum amplitude of the complex

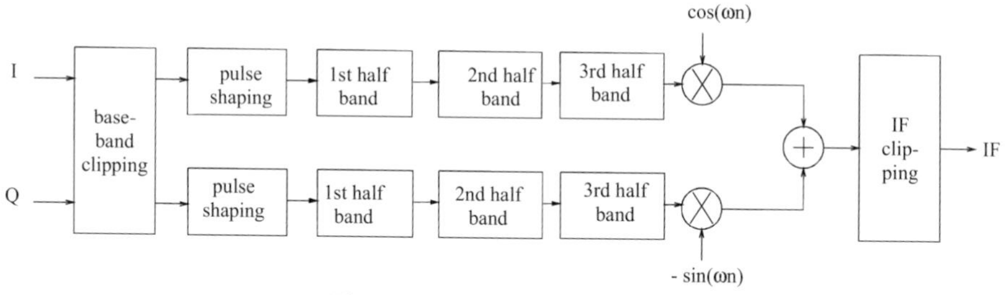

Figure 17-6. Used simulation model.

baseband signal becomes sqrt{2}A. If one of the signals (I or Q) is below the clipping level, and the other is clipped to value A, the amplitude of the complex symbol becomes less than sqrt{2}A. Because the limiting factor is the critical point (see Figure 17-7) from the point of view of the Crest Factor, it can be said that, in most cases, the amplitude of the complex signal is clipped too much without any advantage being gained, while the EVM is increased. The region clipped in vain is presented in Figure 17-7.

Taking into account the fact that, in the case of method Circle, the angle of the complex phasor does not change, which can be a useful property in the receiver, method Circle is more suitable for baseband clipping than method Square.

Adaptive clipping has better efficiency than the conventional baseband clipping methods. When a large amount of error is tolerated, the efficiency of the Adaptive method can be increased by using more iteration loops (method Iterative). In this case, the efficiency of the method Iterative becomes about same as the efficiency of the IF clipping with filtering. For high clipping levels, i.e. for low tolerated error levels, the efficiency of the adaptive clipping does not improve when the number of iterations is increased.

Clipping at IF is slightly less efficient than conventional baseband clipping. However, by combining the IF clipping to bandpass filtering, the efficiency becomes better than in the case of conventional baseband clipping. When clipping is performed at IF, the ACLR is the limiting parameter. By using the bandpass FIR filter with 50 coefficients, the ACLR can be increased so that the EVM and PCDE become the limiting parameters. The problem in this method is that filtering increases the Crest Factor after the clipping operation and it is difficult to find the optimal combination of the clipping ratio and the band pass filter.

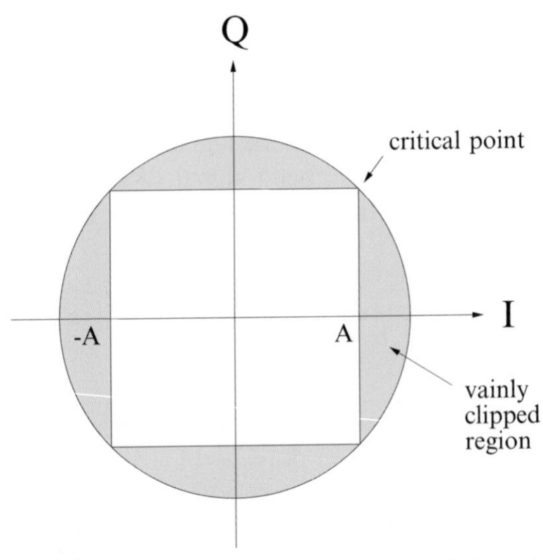

Figure 17-7. Vainly clipped region when I and Q clipped independently.

As the results in Table 17-1 and Table 17-2 show, no advantage is achieved by using error shaping. The idea of error shaping is to remove clipping noise from the signal band so that the clipping ratio can be decreased. Most of the clipping noise is in the signal band [Gro94], and filtering it out leads to a situation where the Crest Factor is increased. Even if the filter is a

Table 17-1. Simulation results for test data with 32 active codes and 3 control channels.

Method	ΔCF	EVM	PCDE	ACLR
Square	3.22 dB	16.2 %	-33.1 dB	92.0 dB
Circle	4.33 dB	17.5 %	-34.2 dB	92.0 dB
Adaptive	5.46 dB	17.5 %	-34.2 dB	92.0 dB
Iterative	6.15 dB	17.5 %	-33.6 dB	92.0 dB
IF	3.05 dB	0.73 %	-58.6 dB	50.1 dB
IF+filt	6.10 dB	16.4 %	-33.0 dB	52.6 dB
Error Shaping	5.48 dB	15.8 %	-33.0 dB	51.8 dB
Windowing	7.35 dB	17.5 %	-33.4 dB	56.1 dB

Table 17-2. Simulation results for test data with 32 active codes and 3 control channels. High margin.

Method	ΔCF	EVM	PCDE	ACLR
Square	1.80 dB	3.0 %	-50.8 dB	92.0 dB
Circle	2.05 dB	3.0 %	-50.8 dB	92.0 dB
Adaptive	2.33 dB	3.0 %	-51.0 dB	92.0 dB
Iterative	2.33 dB	3.0 %	-50.1 dB	92.0 dB
IF	1.00 dB	0.3 %	-58.9 dB	65.1 dB
IF+filt	2.57 dB	3.0 %	-51.5 dB	67.4 dB
Error Shaping	2.45 dB	3.0 %	-51.5 dB	66.5 dB
Windowing	3.65 dB	3.0 %	-49.9 dB	72.0 dB

Table 17-3. Simulations results for test data with 16 active codes and 3 control channels.

Method	ΔCF	EVM	PCDE	ACLR
Square	3.37 dB	9.9 %	-33.1 dB	92.0 dB
Circle	3.61 dB	10.1 %	-33.1 dB	92.0 dB
Adaptive	4.50 dB	11.4 %	-33.0 dB	92.0 dB
Iterative	4.71 dB	10.4 %	-33.0 dB	92.0 dB
IF	3.69 dB	0.7 %	-57.5 dB	50.0 dB
IF+filt	4.62 dB	8.4 %	-33.0 dB	57.3 dB
Windowing	6.17 dB	9.7 %	-33.0 dB	61.5 dB

Table 17-4. Simulations results for test data with 16 active codes and 3 control channels. High margin.

Method	ΔCF	EVM	PCDE	ACLR
Square	2.47 dB	3.0 %	-49.2 dB	92.1 dB
Circle	2.48 dB	3.0 %	-49.3 dB	92.1 dB
Adaptive	2.93 dB	3.0 %	-49.8 dB	92.0 dB
Iterative	2.93 dB	3.0 %	-49.7 dB	92.0 dB
IF	1.35 dB	0.3 %	-57.8 dB	65.1 dB
IF+filt	3.03 dB	3.0 %	-49.0 dB	65.9 dB
Windowing	4.09 dB	2.8 %	-50.0 dB	72.8 dB

FIR filter with only 10 coefficients, and the maximum attenuation in the stopband is 1.5 dB, the decrease of the clipping ratio cannot compensate the increase of the Crest Factor.

The windowing method is the most efficient of the presented clipping methods. The used window is Hamming window with the length of 75. In [Vää01], the properties of some common window functions are compared and it seems that Hamming window has the best relationship between the EVM and ACLR. Choosing a window length is a problematic issue. Results show that, in most cases, the EVM or PCDE is the limiting parameter and that there is some margin for ACLR. In theory, reducing the window length decreases the EVM, PCDE and ACLR which leads to a situation where the clipping ratio could be decreased. Simulations showed that, in this case, the Crest Factor increased, so no advantage was achieved. Another reason which makes the windowing method more interesting than IF clipping with filtering is that windowing method can be easily applied to a multicarrier system.

17.4.2 Multicarrier

The multicarrier signal is simulated by adding together four signals after the upconversion. All four signals are generated by using independent data and an equal number of codes. Baseband clipping is performed independently for each baseband signal with an equal clipping ratio. IF clipping is performed after carriers are combined. Results for data with 32 code channels are presented in Table 17-5 and Table 17-6. The Crest Factor before clipping is 13.745 dB.

In all cases, both baseband clipping methods are inefficient and no advantage is gained. When different signals are combined at IF, the high peaks of the individual signals can be cancelled out and, correspondingly, new peaks can be generated to the composite signal. For this reason, the adaptive baseband clipping is not considered in this context. An effective baseband clipping method would require a feedback structure which would,

Table 17-5. Multicarrier signal with 4 carriers, 32 code channels per carrier.

Method	ΔCF	EVM	PCDE	ACLR
Square	0.07 dB	17.5 %	-34.6 dB	90.1 dB
Circle	-0.03 dB	17.5 %	-34.4 dB	90.1 dB
IF	1.52 dB	0.41 %	-65.1 dB	50.0 dB
Windowing	5.57 dB	17.5 %	-34.7 dB	51.1 dB

Table 17-6. Multicarrier signal with 4 carriers, 32 code channels per carrier. High margin.

Method	ΔCF	EVM	PCDE	ACLR
Square	-0.06 dB	3.0 %	-50.0 dB	90.1 dB
Circle	-0.06 dB	3.0 %	-50.7 dB	90.1 dB
IF	0.48 dB	0.3 %	-66.5 dB	65.0 dB
Windowing	3.03 dB	3.0 %	-50.5 dB	65.0 dB

however, be very complex and hard to implement.

As in the single carrier case, the ACLR is a limiting parameter in IF clipping. Because it is difficult to apply filtering in the case of multicarrier signal, the windowing method becomes the most interesting. Results show that, by using the windowing method, the Crest Factor can be reduced significantly. The major problem, just as it was in the single carrier case, remains the method of choosing the window length. In simulations, efforts are made to minimize the Crest Factor.

17.5 Conclusions

The effects of several different clipping methods on the EVM, PCDE and ACLR are derived through simulations. In simulations, different types of test data are used and both single carrier and multicarrier cases are considered. Based on results, the clipping methods are compared. Also a peak windowing algorithm based on a FIR filter structure with feedback is presented.

REFERENCES

[GPP00] 3GPP Technical Specification Group Access Network Base station conformance testing, TS 25.141 V3.2.0. 2000.

[Gro94] R. Gross, and D. Veeneman, "SNR and Spectral Properties for a Clipped DMT ADSL Signal," IEEE International Conference on Communications, 1994, Vol. 2, pp. 843 -847.

[Mil98] S. L. Miller, and R. J. O'Dea, "Peak Power and Bandwidth Efficient Linear Modulation," IEEE Transactions on Communications, Vol. 46, No. 12, pp. 1639-1648, Dec. 1998.

[Nee98] R. van Nee, and A. de Wild, "Reducing the Peak-to-Average Power Ratio of OFDM," IEEE Vehicular Technology Conference, 1998, Vol. 3, pp. 2072-2076.

[Ozl95] F. M. Ozluturk, and G. Lomp, "Effect of Limiting the Downlink Power in CDMA Systems with or without Forward Power Control," IEEE Military Communications Conference, 1995, pp. 952-956.

[Pau96] M. Pauli, and H.-P, Kuchenbecker, "Minimization of the Intermodulation Distortion of a Nonlinearly Amplified OFDM Signal," Wireless Personal Communications 4: pp. 90-101, 1996.

[Vää01] O. Väänänen, "Clipping in Wideband CDMA Base Station Transmitter," Master's Thesis, Helsinki University of Technology, 2001.

[Yan00] Y. Hong-Kui, "Method & Apparatus for Reducing the Peak Power Probability of a Spread Spectrum Signal," European Patent Application EP1058400, Dec. 6, 2000.

Chapter 18

18. REDUCING PEAK TO AVERAGE RATIO OF MULTICARRIER GSM AND EDGE SIGNALS

In conventional base station solutions, the carriers transmitted are combined after the power amplifiers. An alternative to this is to combine the carriers in the digital domain. The major drawback of the digital carrier combining is a strongly varying envelope of the composite signal. The high PAR sets strict requirements for the linearity of the power amplifier. High linearity requirements for the power amplifier leads to low power efficiency and therefore to high power consumption. In this chapter, the possibility of reducing the PAR by clipping is investigated in two cases, GSM and EDGE.

18.1 Introduction

The Global System for Mobile communication (GSM) is a widespread second generation system that uses the 0.3 Gaussian Minimum Shift Keying (GMSK) modulation. The advantage of this modulation method is the constant envelope signal which makes it possible to use power efficient power amplifiers (PAs). Enhanced Data rates for GSM Evolution (EDGE) is an enhancement to the GSM system. The primary objective for the EDGE signal is to triple the on-air data rate while meeting essentially the same bandwidth occupancy as the original 0.3 GMSK signal. The EDGE system uses a $3\pi/8$-shifted 8-Phase Shift Keying (PSK) modulation, which is not a constant envelope modulation.

In conventional base station solutions, the carriers transmitted are combined after the power amplifiers. An alternative to this is to combine the carriers in the digital Intermediate Frequency (IF) domain (see section 16.7) [Van01], [Van02]. This saves a large number of analog components and, because there is no analog I/Q modulator, many problems such as dc offset can be avoided. The major drawback of the digital carrier combining is the strongly varying envelope of the composite signal. When a number of carriers are combined, according to the central limit theorem, the envelope of the composite signal becomes normally distributed with a high peak to average ratio (PAR). Typically, the PAR is measured by the Crest Factor (CF) defined as

$$CF = 10\log_{10}\left(\frac{\max(x(t)^2)}{E[x(t)^2]}\right), \tag{18.1}$$

where $x(t)$ is the composite signal. The high PAR sets strict requirements for the linearity of the power amplifier. In order to limit the adjacent channel leakage, it is desirable for the power amplifier to operate in its linear region. High linearity requirements for the power amplifier leads to low power efficiency and, therefore, to high power consumption (class-A amplifier). An alternative to the expense of a wide-dynamic-range power amplifier is the use of deliberate clipping to distort the signal digitally.

Several clipping techniques for other modulation methods that suffer from high PAR (OFDM, CDMA) are proposed. In this chapter, at first the properties of the GSM and EDGE multi-carrier signals are discussed and then conventional IF clipping and windowing method [Pau96], [Nee98] is applied in both cases.

18.2 Signal Model

In both cases, GSM and EDGE, a single carrier IF signal is generated using the burst format specified in [Dig99a]. The length of the test signal is 7 data bursts, which corresponds to 260000 samples at IF frequency. The oversampling ratio of 240 is used at the IF. This corresponds to a 65 MHz sampling frequency in the digital modulator because the symbol rate in the GSM/EDGE is 270.833 ksym/s. The multicarrier signal is generated by combin-ing several single carrier signals at IF using 600 kHz channel spacing. All combined signals are generated using independent random data; the initial phases of the carriers are chosen randomly.

The Crest Factor of the real valued single carrier GSM IF signal is approximately equal to the Crest Factor of the sinusoidal signal, 3.01 dB. Simulations have shown that the Crest Factor of the corresponding EDGE signal is about 6.18 dB.

If all the carriers are assumed to be statistically independent, the power of the composite signal is doubled when the number of carriers is doubled. In the worst case, all the carriers reach their maximum simultaneously, which means that when the number of carriers is doubled, the maximum of the composite signal is doubled and the peak power is multiplied by four. In this case, the PAR is doubled and the Crest Factor is increased about 3 dB. In reality, it is very unlikely that all the carriers would have their maximum simultaneously or that the Crest Factor would increase as much as predicted.

Simulated Crest Factors for composite signals with different numbers of carriers are presented in Table 18-1. The results show that the Crest Factor does not increase as much as in the worst possible case, but, in any case, for a large number of carriers it becomes very high in both cases, GSM and EDGE.

In the future, it will be possible to transmit the GSM and EDGE signals simultaneously using the same power amplifier. The Crest Factor of the signal with 16 carriers when the number of EDGE carriers is varied is presented in Table 18-2.

18.3 Clipping Methods

The conventional IF clipping can be expressed mathematically as

$$y = \begin{cases} A & x > A \\ x & |x| \le A \\ -A & x < -A, \end{cases}, \tag{18.2}$$

where x is the unclipped signal, A is the maximum amplitude allowed and y is the clipped version of signal x. It is obvious that, as a nonlinear operation, the clipping distorts the signal significantly and generates unwanted power to the adjacent channels. One possible solution to this problem is to smooth the sharp corners caused by clipping. This can be performed by multiplying the signal to be clipped with a window function [Pau96], [Nee98]. In a real system, windows unfortunately overlap, and, as a result of convolution, the signal is clipped much more than is needed. In the worst case, the sign of

Table 18-1. Simulated Crest Factors for signals with different number of carriers

Number of carriers	CF GSM	CF EDGE
1	3.010 dB	6.176 dB
2	6.020 dB	8.969 dB
4	9.012 dB	11.102 dB
8	11.397 dB	12.956 dB
16	14.258 dB	15.747 dB
32	17.395 dB	18.649 dB

Table 18-2. Crest Factor of the the signal with 16 carriers when the number of EDGE carriers is varied

Number of EDGE carriers	CF
1	14.372 dB
2	14.391 dB
4	14.266 dB
8	14.372 dB
12	14.783 dB

function $b(n)$ may become negative, which is fatal for the system. Hence, another way to form the function $b(n)$ must be found. A simple solution to the problem mentioned above is to combine the conventional FIR structure with a feedback structure that scales down the incoming value if necessary. The proposed structure is presented in Figure 17-3. In Figure 17-3, '$\lfloor\rfloor$' denotes floor operation. The impulse response of the filter (coefficients w_n) is equal to the window function w. The previous values are used for calculating a correction term that can be subtracted from input while the output still satisfies (18.4). If the correction term is larger than the input value, signal y (Figure 17-3) becomes negative after the subtraction, which leads to an unwanted clipping result. This is prevented by adding a block that replaces the negative values with zeros.

18.4 Results

The clipped signal must fulfil the system specifications [Dig99b]. In the case of GSM, the signal quality is measured by phase error, while in the case of EDGE, the signal quality is measured by Error Vector Magnitude (EVM). In both cases, the spectrum of the signal must fit in the spectrum mask.

18.4.1 GSM

Firstly, different window types are compared. The spectrum of the unclipped signal, the spectrum of the clipped signal and the spectrums of the windowed signals are presented in Figure 18-1. Results for Hanning and Blackman windows are presented. Other common window functions, i.e. Kaiser, Hamming and Gaussian, are investigated; the Hanning and Blackman windows are found to be better than the other windows.

The test signal consists of 16 carriers. The Crest Factor of the unclipped signal is 14.258 dB. In every case, the signal is clipped so that the Crest Factor becomes 10 dB. Figure 18-1 shows that the conventional clipping causes very high out of band radiation and is therefore not applicable in the case of GSM transmission. The Blackman window seems to give better spectral properties than the Hanning window.

The effect of the window length used is presented in Figure 18-2 and in Table 18-3. The Crest Factor of the test signal is clipped to 12 dB. In this example, the window length of 240 corresponds to the length of one symbol

Table 18-3. Phase error of the GSM signal as a function of the window length.

Window length	rms	peak
101	0.957	5.049
201	1.033	5.567
401	0.667	3.308
601	0.157	0.818

Table 18-4. Crest Factor reduction in the case of multicarrier GSM signal.

number of carriers	CF	ΔCF	rms	peak
8	9.131 dB	2.266 dB	0.199	0.866
16	10.207 dB	4.051 dB	0.230	1.212
32	10.658 dB	6.737 dB	0.360	2.805

(oversampling ratio of 240 is used). It is obvious that a long window gives better spectral properties than a short window, but the interesting result is that the long window gives better phase error performance than a short window. This is surprising because when the window length increases the difference between the transmitted and the ideal waveform increases, so therefore, intuitively, the phase error should increase.

The achieved Crest Factor reduction (ΔCF) in the case of 8, 16 and 32 carriers is presented in Table 18-4. The used window length is 601 and the clipping level is set so that the spectrum is the limiting element. The results show that the Crest Factor of the multicarrier GSM signal can be reduced significantly while the distortion is still kept in a tolerable level. The phase error specifications are 5° for rms and 20° for peak error [Dig99b]. In practice, implementing the windowing algorithm presented in [Vää02] with a window length of 601 might be very difficult and lead to a high area and power consumption in the circuit implementation.

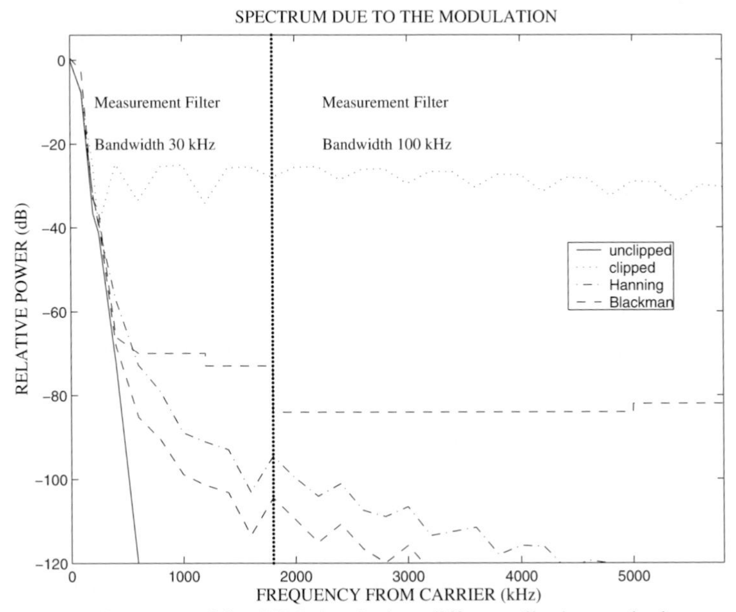

Figure 18-1. Spectrum of the GSM signal when different clipping methods are used.

18.4.2 EDGE

Again different window types are compared. The spectrum of the unclipped signal, the clipped signal and the spectrums of the windowed signals are presented in Figure 18-3. Two different windows, Hanning and Blackman are used. The test signal consists of 16 carriers and the Crest Factor of the unclipped signal is 15.747 dB. In every case, the signal is clipped so that the Crest Factor becomes 12 dB. Figure 18-3 shows that, as earlier, the conventional clipping causes very high out of band radiation and therefore it is not applicable in the case of EDGE transmission. The Blackman window seems to give better spectral properties than the Hanning window. The effect of the used window length is presented in Figure 18-4 and in Table 18-5. The Crest Factor of the test signal is clipped one decibel. The longer window gives a better spectral behavior but the EVM becomes high. A window long enough to meet the spectral specifications causes EVM that is very near, or above, the EVM specifications, 7% for rms and 24% for peak value [Dig99b]. In a real digital modulator, EVM this high cannot be tolerated because some margin must be left for the following analogue parts. In conclusion, it can be said that neither of the clipping methods discussed in this chapter can be used for EDGE clipping. Even one decibel reduction in the Crest Factor leads to an intolerable error. Generally, the EDGE signal seems to be very sensitive to clipping errors, which makes the Crest Factor reduction a very challenging problem.

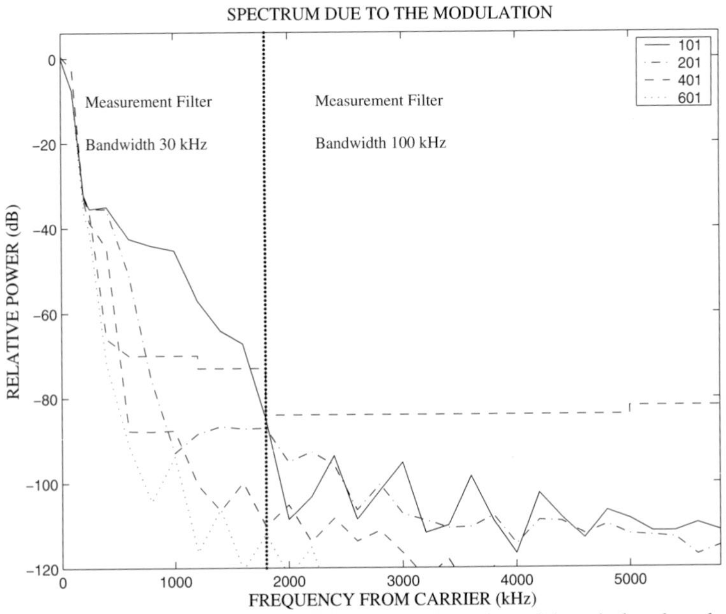

Figure 18-2. Spectrum of the GSM signal as a function of the window length.

Table 18-5. EVM of the clipped EDGE signal as a function of the window length.

Window length	rms EVM	peak EVM
101	2.763 %	11.526 %
201	3.683 %	15.156 %
401	5.690 %	21.920 %
601	6.944 %	24.783 %

The reason for the poor performance of the EDGE clipping is that the clipping seems to affect more the amplitude of the signal than the phase of the signal. Because the distortion in the case of the EDGE signal is measured by both amplitude error and phase error, the error metric EVM becomes high. In the case of GSM clipping, the error is measured by phase error only, so the signal can be clipped significantly. If we down-convert the clipped GSM signal, divide it into the in phase and quadrature branches and calculate the EVM as is done in the case of EDGE, it can be seen that, while the phase error remains low, the EVM can be high. For the signal with 0.19 degrees rms and 0.88 degrees peak phase error, the corresponding EVM values are 5.4% and 19.6%, respectively.

18.4.3 GSM/EDGE

As has been shown earlier, the EDGE clipping is much more complicated than the GSM clipping, so it can be assumed that when the GSM and EDGE carriers are transmitted simultaneously, performance of the EDGE signals restricts the clipping. When a signal with 15 GSM carriers and one EDGE

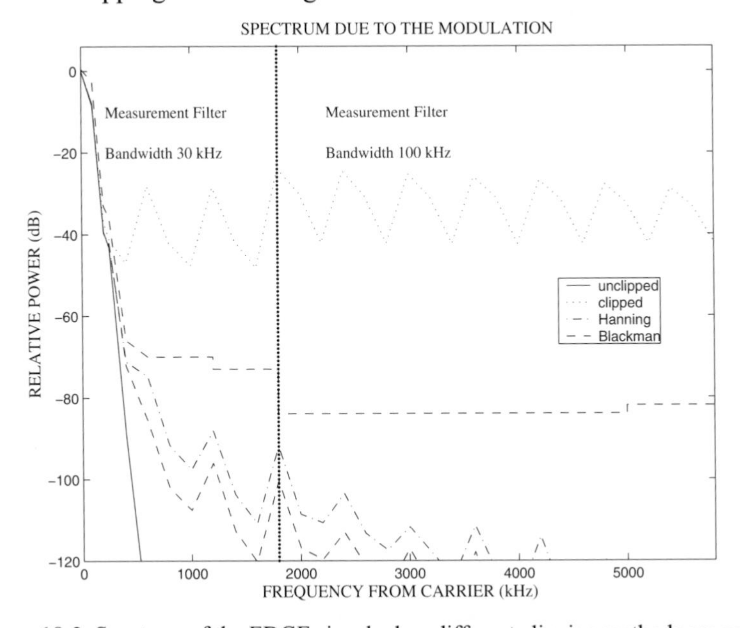

Figure 18-3. Spectrum of the EDGE signal when different clipping methods are used.

carrier is clipped by using the windowing method, the Crest Factor is reduced by about 1.5 dB from 14.37 dB to 12.88 dB. In this case, the rms EVM is 3.1 % and the peak EVM is 21.5 %, both of which fulfil the specifications but are intolerably high. When the number of the EDGE carriers is varied, the results are of the same kind; the peak EVM seems to remain especially problematic.

18.5 Conclusions

Two different clipping methods, conventional clipping and windowing method, are applied to GSM and EDGE multicarrier signals in order to reduce the Crest Factor. In the case of GSM, the windowing method is shown to be efficient and the Crest Factor is reduced significantly while the distortion is still kept at a tolerable level. In the case of EDGE, both clipping methods are proved to be inapplicable.

REFERENCES

[Dig99a] Digital cellular telecommunications system (Phase 2+); Modulation (GSM 05.04) V8.1.0 European Telecommunications Standards Institute 1999.

[Dig99b] Digital cellular telecommunications system (Phase 2+); Radio Transmission and reception (GSM 05.05) V8.3.0 European Telecommunica-

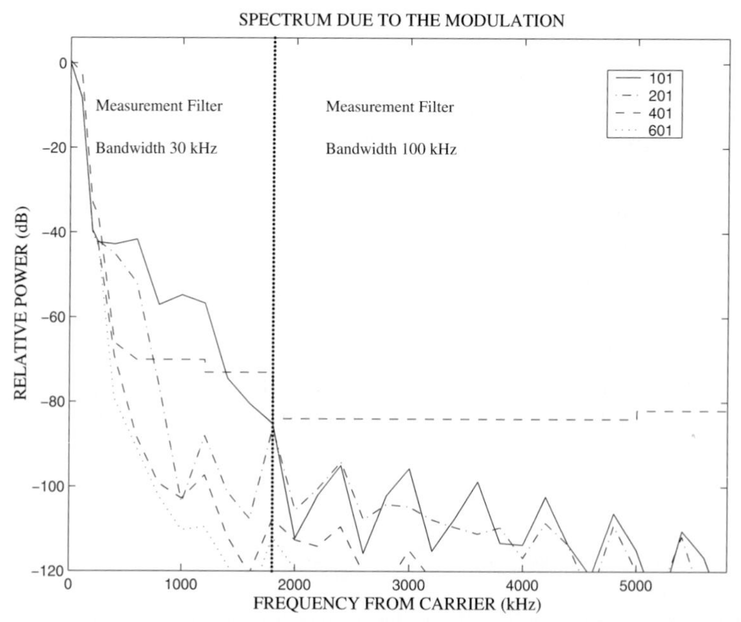

Figure 18-3. Spectrum of the clipped EDGE signal as a function of the window length.

tions Standards Institute 1999.
[Nee98] R. van Nee, and A. de Wild, "Reducing the Peak-to-Average Power Ratio of OFDM," IEEE Vehicular Technology Conference, 1998, Vol. 3, pp. 2072-2076.
[Pau96] M. Pauli, and H.-P, Kuchenbecker, "Minimization of the Intermodulation Distortion of a Nonlinearly Amplified OFDM Signal," Wireless Personal Communications 4, pp. 90-101, 1996.
[Vää02] O. Väänänen, J. Vankka, and K. Halonen, "Effect of Clipping in Wideband CDMA System and Simple Algorithm for Peak Windowing," World Wireless Congress, May 28-31, 2002, San Francisco, USA, pp.614-619.
[Van01] J. Vankka, J. Pyykönen, J. Sommarek, M. Honkanen, and Kari Halonen, "A Multicarrier GMSK Modulator for Base Station," ISSCC Digest of Technical Papers, February 5 - 7, 2001, San Francisco, USA, pp. 354-355.
[Van02] J. Vankka, J. Ketola, J. Sommarek, Olli Väänänen, M. Kosunen and K. Halonen, "A GSM/EDGE/WCDMA Modulator with on-chip D/A Converter for Base Stations," IEEE Trans. on Circuits and Systems Part II, Vol. 49, No. 10, pp. 645-655, Oct. 2002.

ADDITIONAL REFERENCES TO CLIPPING

ANALYSIS

R. Gross, and D. Veeneman, "SNR and Spectral Properties for a Clipped DMT ADSL Signal," IEEE International Conference on Communications, Vol. 2, Nov. 1994, pp. 843-847.
F. M. Ozluturk, and G. Lomp, "Effect of Limiting the Downlink Power in CDMA Systems with or without Forward Power Control," IEEE Military Communications Conference, Vol. 3, Nov. 1995, pp. 952-956.
J. H. van Vleck, and D. Middleton, "The Spectrum of Clipped Noise," Proceedings of the IEEE, Vol. 54, No. 1, Jan. 1966, pp. 2-19.
H. Pretl, L. Maurer, W. Schelmbauer, R. Weigel, B. Adler, and J. Fenk, "Linearity Considerations of W-CDMA Front-ends for UMTS," IEEE International Microwave Symposium Digest, Vol. 1, June 2000, pp. 433-436.
R. O'Neill, and L. B. Lopes, "Envelope Variations and Spectral Splatter in Clipped Multicarrier Signals," IEEE International Symposium on Personal, Indoor and Mobile Radio Communications, Vol. 1, Sept. 1995, pp. 71-75.
L. Xiaodong, and L. J. Cimini, "Effects of Clipping and Filtering on the Performance of OFDM," IEEE Communications Letters, Vol. 3, No. 5, pp. 131-133, May 1998.

V. K. N. Lau, "On the Analysis of Peak-to-Average Ratio (PAR) for IS95 and CDMA2000 Systems," IEEE Transactions on Vehicular Technology, Vol. 49, No. 6, pp. 2174-2188, Nov. 2000.
R. N. Braithwaite, "Exploiting Data and Code Interactions to Reduce the Power Variance for CDMA Sequences," IEEE Journal on Selected Areas in Communications, Vol. 19, No. 6, pp. 1061-1069, June 2001.
G. Jacovitti, and A. Neri, "Estimation of the Autocorrelation Function of Complex Gaussian Stationary Processes by Amplitude Clipped Signals," IEEE Transactions on Information Theory, Vol. 40, No. 1, pp. 239-245, Jan. 1994.
P. R. Pawlowski, "Performance of ODS-CDMA QPSK in a Soft-Limiting AM/AM Nonlinearity Channel," IEEE International Symposium on Personal, Indoor and Mobile Radio Communications, Vol. 2, Sept. 1998, pp. 789-794.
R. O'Neill, and L. B. Lopes, "Performance of Amplitude Limited Multitone Signals," IEEE Vehicular Technology Conference, Vol. 3, June 1994, pp. 1675-1679.
P. Banelli, S. Cacopardi, F. Frescura, and G. Reali, "Counteraction of Nonlinear Distortion in a Novel MCM-DS-SS Wireless LAN Radio Subsystem," IEEE Global Telecommunications Conference, Vol. 1, Nov. 1997, pp. 320-326.
Q. Shi, "OFDM in Bandpass Nonlinearity," IEEE Transactions on Consumer Electronics, Vol. 42, No. 3, pp. 253-258, Aug. 1996.
R. N. Braithwaite, "Nonlinear Amplification of CDMA Waveforms: An Analysis of Power Amplifier Gain Errors and Spectral Regrowth," IEEE Vehicular Technology Conference, Vol. 3, May 1998, pp. 2160-2166.
Y. Luan, J. Yang, and J. Li, "Effects of Amplitude Clipping on Signal-to-Noise Ratio and Spectral Splatter of Multicarrier DS-CDMA," International Conference on Communication Technology, Vol. 1, Aug. 2000, pp. 785-789.
B.-J. Choi, E.-L. Kuan and L. Hanzo, "Crest-Factor Study of MC-CDMA and OFDM," IEEE Vehicular Technology Conference, Vol. 1, Sept. 1999, pp. 233 -237.

BASIC METHODS

R. Dinis, and A. Gusmao, "On the Performance Evaluation of OFDM Transmission Using Clipping Techniques," IEEE Vehicular Technology Conference, Vol. 5, Sept. 1999, pp. 2923-2928.
P. Stadnik, "Baseband Clipping Can Lead To Improved WCDMA Signal Quality," Wireless System Design, pp. 40-44, Sept. 2000.

R. Greighton, "System and Method to Reduce the Peak-to-Average Power Ratio in a DS-CDMA Transmitter," U. S. Patent 6.529.560, Lucent Technologies, Inc., Mar. 4, 2003.
N. McGowan, and X. Jin, "CDMA Transmit Peak Power Reduction," U. S. Patent 6.236.864, Nortel Networks Limited, May 22, 2001.
S. Toshifumi, "Code Division Multiple Access Base Station Transmitter," U. S. Patent 5751705, NEC Corporation, May 12, 1998.
H. Muto, "Peak Clipping Circuit Provided in a Quadrature Modulator and a Method of Clipping Peak Levels of In-Phase and Quadrature Signals," U. S. Patent 6.044.117, NEC Corporation, Mar. 28, 2000.
P. Lundh, and G. Skoog, "A Method and Apparatus for Clipping Signals in a CDMA System," PCT WO 0045538, LM Ericsson, Jan. 25, 2000.
S. Bhagalia, "Processing CDMA Signals," U. S. Patent 5742595, DSC Communications Corporation, Apr. 21, 1998.

WINDOWING

M. Pauli, and H.-P. Kuchenbecker, "On the Reduction of the Out-of-Band Radiation of OFDM-Signals," IEEE International Conference on Communications, Vol. 3, June 1998, pp. 1304-1308.
R. van Nee and A. de Wild, "Reducing the Peak-to-Average Power Ratio of OFDM," IEEE Vehicular Technology Conference, Vol. 3, May 1998, pp. 2072-2076.
M. Pauli, and H.-P, Kuchenbecker, "Minimization of the Intermodulation Distortion of a Nonlinearly Amplified OFDM Signal," Wireless Personal Communications 4: pp. 90-101, 1996.
M. Birchler, S. Jasper and A. Tsiortzis, "Low-Splatter Peak-to-Average Signal Reduction with Interpolation," U. S. Patent 5.638.403, Motorola, Inc., June 10, 1997.
M. Birchler, "Low-Splatter Peak-to-Average Signal Reduction," U. S. Patent 5.287.387, Motorola, Inc., Feb. 15, 1994.

ERROR FEEDBACK

J. S. Chow, J. A. C. Bingham, and M. S. Flowers, "Mitigating clipping noise in multi-carrier systems," IEEE International Conference on Communications, Vol. 2, June 1997, pp. 715-719.
M. Hahm, "Device and Method for Limiting Peaks of a Signal," U. S. Patent 6.356.606, Lucent Technologies, Inc., Mar. 12, 2002.

H. Yang, "Method & Apparatus for Reducing the Peak Power Probability of a Spread Spectrum Signal," European Patent Application, EP1058400, Nortel Networks Limited, Dec. 6, 2000.

ADAPTIVE

S. Miller, and D. O'Flea, "Peak Power and Bandwidth Efficient Linear Modulation," IEEE Transactions on Communications, Vol. 46, No. 12, pp. 1639-1648, December 1998.
Y.-S. Park, and S. Miller, "Peak-to-Average Power Ratio Suppression Schemes in DFT based OFDM," IEEE Vehicular Technology Conference, Vol. 1, Sept. 2000, pp. 292-297.
R. Enright, and M. Darnell, "OFDM Modem with Peak-to-Mean Envelope Power Ratio Reduction Using Adaptive Clipping," International Conference on HF Radio Systems and Techniques, July 1997, pp. 44-49.
R. Berangi, and M. Faulkner, "Peak Power Reduction for Multi-code CDMA and Critically Sampled Complex Gaussian Signals," International Symposium on Telecommunications, Tehran, Iran, 2001.
J. McCoy, "Method and Apparatus for Peak Limiting in a Modulator," U. S. Patent 6.147.984, Motorola, Inc., Nov. 14, 2000.
S. Miller and D. O'Flea, "Radio with Peak Power and Bandwidth Efficient Modulation," U. S. Patent 5.621.762, Motorola, Inc., Apr. 15, 1997.

REDUCING CREST FACTOR BY ADDING UNUSED CHANNELIZATION CODES

K. Laird and J. Smith, "Method and Apparatus for Reducing Peak-to-Average Power Ratio of a Composite Carrier Signal," U. S. Patent 5.991.262, Motorola, Inc., Nov. 23, 1999.

REDUCTION OF PEAK-TO-AVERAGE POWER RATIO VIA ARTICIFAL SIGNALS

M. Lampe, and H. Rohling, "Reducing Out-of-Band Emissions due to Nonlinearities in OFDM Systems," Vehicular Technology Conference, Vol. 3, May 1999, pp. 2255-2259.
J. Yang, J. Yang, and J. Li, "Reduction of the Peak-to-Average Power Ratio of the Multicarrier Signal via Artificial Signals," International Conference on Communication Technology, Vol. 1, Aug. 2000, pp. 581-585.
G. Awater, R. van Nee, and A. de Wild, "Transmission System and Method Employing Peak Cancellation to Reduce the Peak-to-Average Power Ratio," U. S. Patent 6.175.551, Lucent Technologies, Inc., Jan. 16, 2001.

T. Beukema, "Method and Apparatus for Peak Suppression Using Complex Scaling Values," U. S. Patent 5.727.026, Motorola, Inc., Mar. 10, 1998.
G. Vannucci, "Methos and Apparatus for Tailored Distortion of a Signal Prior to Amplification," European Patent Application, EP0940911, Lucent Technologies, Inc., Sept. 8, 1999.
B. M. Popovic, "Synthesis of Power Efficient Multitone Signals with Flat Amplitude Spectrum," IEEE Transactions on Communications, Vol. 39, No. 7, pp. 1031-1033, July 1991.
J. Schoukens, Y. Rolain and P. Guillaume, "Design of Narrowband High-Resolution Multisines," IEEE Transactions on Instrumentation and Measurements, Vol. 45, No. 3, pp. 750-753, June 1996.

NONLINEAR SCALING

C.-S. Hwang, "A Peak Power Reduction Method for Multicarrier Transmission," IEEE International Conference on Communications, Vol. 5, June 2001, pp. 1496-1500.
D. L. Jones, "Peak Power Reduction in OFDM and DMT via Active Channel Modification," Conference Record of the Asilomar Conference on Signals, Systems and Computers, Vol. 2, Oct. 1999, pp. 1076-1079.
J. Harris, T. Giallorenzi, D. Matolak, and D. Griffin, "Data Transmission System with a Low Peak-to-Average Power Ratio Based on Distorting Frequently Occuring Signals," U. S. Patent 5.651.028, Unisys Corporation, Jul. 22, 1997.
T. Giallorenzi, D. Matolak, J. Harris, R. Steagall, and B. Williams, "Data Transmission System with a Low Peak-to-Average Power Ratio Based on Distorting Small Amplitude Signals," United States Patent, U. S. Patent 5.793.797, Unisys Corporation, Aug. 11, 1998.
Y. Arai, and T. Kanda, "Spread Spectrum Communication Apparatus," U. S. Patent 5.668.806, Canon Kabushiki Kaisha, Sept. 16, 1997.

CODE SELECTION

R. N. Braithwaite, "Using Walsh Code Selection to Reduce the Power Variance of Band-Limited Forward-Link CDMA Waveforms," IEEE Journal on Selected Areas in Communications, Vol. 18, No. 11, pp. 2260-2269, Nov. 2000.
V. K. N. Lau, "Peak-to-average ratio (PAR) Reduction by Walsh-Code Selection for IS-95 and CDMA2000 Systems," IEE Proceedings- Communications, Vol. 147, No. 6, pp. 361-364, Dec. 2000.

OTHERS

J. Armstrong, "New OFDM Peak-to-Average Power Reduction Scheme," IEEE Vehicular Technology Conference, 2001, Vol. 1, May 2001, pp. 756 - 760.

J. Tellado, and J. M. Cioffi, "Efficient Algorithms for Reducing PAR in Multicarrier Systems," IEEE International Symposium on Information Theory, Aug. 1998, pp. 191.

T. Wada, T. Yamazato, M. Katayama, and A. Ogawa, "A Constant Amplitude Coding for Orthogonal Multi-code CDMA Systems," IEICE Trans. fundamentals, Vol. E80-A. No. 12, pp. 2477-2484, Dec. 1997.

Chapter 19

19. APPENDIX: DERIVATION OF THE LAGRANGE INTERPOLATOR

The transfer function of a FIR filter is

$$H(z) = \sum_{n=0}^{N} h(n) z^{-n}, \tag{19.1}$$

where $h(n)$ is the impulse response of the filter. N is the degree of the filter, thus the number of taps in the filter are $L = N + 1$. The aim of the design procedure is to minimize the complex error function defined by

$$E(e^{i\omega}) = H(e^{i\omega}) - H_{id}(e^{i\omega}), \tag{19.2}$$

where $H(e^{i\omega})$ is the approximation and $H_{id}(e^{i\omega})$ is the ideal frequency response.

By setting the error function E and its N derivatives to zero at zero frequency, a maximally flat interpolation filter design at $\omega = 0$ is obtained. The FIR filter coefficients derived this way are the same as the weighting coefficient in the classical Lagrange interpolation. This can be written as

$$\left.\frac{d^j E(e^{i\omega})}{d\omega^j}\right|_{\omega=0} = 0, \text{ for } j = 0, 1, 2, \ldots, N. \tag{19.3}$$

This can be written

$$\frac{d^j}{d\omega^j}\left[\sum_{n=0}^{N} h(n) e^{-i\omega\omega} - e^{-i\omega\mu_k}\right]_{\omega=0} = 0, \quad \text{for } j = 0, 1, 2, \ldots, N. \tag{19.4}$$

By differentiating the following is obtained:

$$\sum_{n=0}^{N} n^j\, h(n) = \mu_k^j, \quad \text{for } j = 0, 1, 2, \ldots, N. \tag{19.5}$$

This set of $N + 1$ linear equations may be rewritten in the matrix form as

$$Vh = v, \tag{19.6}$$

where V is an $L \times L$ Vandermonde matrix

$$\mathbf{V} = \begin{bmatrix} 0^0 & 1^0 & 2^0 & \dots & N^0 \\ 0^1 & 1^1 & 2^1 & & N^1 \\ 0^2 & 1^2 & 2^2 & & N^2 \\ \vdots & & & \ddots & \vdots \\ 0^N & 1^N & 2^N & \dots & N^N \end{bmatrix} = \begin{bmatrix} 1 & 1 & 1 & \dots & 1 \\ 0 & 1 & 2 & & N \\ 0 & 1 & 4 & & N^2 \\ \vdots & & & \ddots & \vdots \\ 0 & 1 & 2^N & \dots & N^N \end{bmatrix} \tag{19.7}$$

h is the coefficient vector of the FIR filter

$$\mathbf{h} = [h(0) \quad h(1) \quad h(2) \quad \dots \quad h(N)]^T \tag{19.8}$$

and

$$\mathbf{v} = [1 \quad \mu_k \quad \mu_k^2 \quad \dots \quad \mu_k^N]^T \tag{19.9}$$

Because the Vandermonde matrix is known to be nonsingular, it has also an inverse matrix V^{-1}. Equation (19.6) can be expressed as

$$h = V^{-1} v \tag{19.10}$$

where V^{-1} can be evaluated using Cramer's rule. The solution is given in explicit form as

$$h(n) = \prod_{j=0,\, j \neq k}^{N} \frac{\mu_k - j}{n - j}, \text{ for } n = 0, 1, 2, \dots, N \tag{19.11}$$

where μ_k is the fractional delay and N is the order of the FIR filter.

There are also simpler approaches to derive the solution, but they are purely mathematical and will not bring out the signal processing aspects as clearly as this one.

INDEX

H

I

L

M

N

O

P

Q

R

S

T

V,W